Lecture Notes in Physics

Volume 826

For further volumes:
http://www.springer.com/series/5304

The Lecture Notes in Physics

The series Lecture Notes in Physics (LNP), founded in 1969, reports new developments in physics research and teaching—quickly and informally, but with a high quality and the explicit aim to summarize and communicate current knowledge in an accessible way. Books published in this series are conceived as bridging material between advanced graduate textbooks and the forefront of research and to serve three purposes:

- to be a compact and modern up-to-date source of reference on a well-defined topic
- to serve as an accessible introduction to the field to postgraduate students and nonspecialist researchers from related areas
- to be a source of advanced teaching material for specialized seminars, courses and schools

Both monographs and multi-author volumes will be considered for publication. Edited volumes should, however, consist of a very limited number of contributions only. Proceedings will not be considered for LNP.

Volumes published in LNP are disseminated both in print and in electronic formats, the electronic archive being available at springerlink.com. The series content is indexed, abstracted and referenced by many abstracting and information services, bibliographic networks, subscription agencies, library networks, and consortia.

Proposals should be sent to a member of the Editorial Board, or directly to the managing editor at Springer:

Christian Caron
Springer Heidelberg
Physics Editorial Department I
Tiergartenstrasse 17
69121 Heidelberg/Germany
christian.caron@springer.com

Bryce DeWitt

Bryce DeWitt's Lectures on Gravitation

Edited by Steven M. Christensen

Springer

Author
Bryce DeWitt
Vista Lane 2411
Austin, TX 78703
USA

Editor
Steven M. Christensen
Ellisfield Drive 4020
Durham, NC 27705
USA
e-mail: steve@smc.vnet.net

ISSN 0075-8450 e-ISSN 1616-6361

ISBN 978-3-540-36909-7 e-ISBN 978-3-540-36911-0

DOI 10.1007/978-3-540-36911-0

Springer Heidelberg Dordrecht London New York

Cover design: eStudio Calamar, Berlin/Figueres

Printed on acid-free paper

Springer is part of Springer Science+Business Media (www.springer.com)

Preface

In the summer of 1971, when I was traveling with Bryce DeWitt and his other new graduate students in Europe, I remember seeing him sitting at a desk for days on end writing lecture notes. Since we had met just a few weeks before, I was astonished at how focused and dedicated to this endeavor he was. We were staying at the International Centre for Theoretical Physics outside Trieste, Italy and we students had been given the task of reading a prerelease copy of Misner, Thorne, and Wheeler's new book Gravitation while Bryce wrote his notes. Bryce's lectures were to be given in the Fall of that year at Stanford where he would be visiting after leaving The University of North Carolina at Chapel Hill to take a new position at the Center for Relativity at The University of Texas at Austin.

We eventually all arrived in Austin in January of 1972. Later, when we got offices in the new physics, math, and astronomy building, I was situated in the little office outside his and was given the task of organizing his preprints, reprints, and other papers. This sounds like a menial job, appropriate perhaps for a new student, but it was one of the most memorable times of my life. Within these papers were his lecture notes, favorite papers written by others, and many unpublished calculations and manuscripts. Amongst these were more than three hundred pages of the Stanford course he had given and I had not seen. The fact that he had put them together in so short a time made me feel even more in awe of his abilities and anxious to take any class he might teach. In fact, some of what was in the Stanford notes eventually could be heard in his Theory of Everything course (long before this term became popular for other reasons).

Bryce was one of the most respected researchers in Quantum Gravity and related subjects of the last half of the twentieth century and into the twenty first and, as with many such people, his teaching was perhaps not well known. But like taking a class from Wheeler or Feynman, being a student of his was frankly historic in my mind. Each day after one of his lectures, I would go back to my tiny apartment and rewrite my notes line by line to try to absorb what he was saying in a much deeper way than in the classroom. I had the honor of having many wonderful teachers in my academic career, but few were as exciting to listen to as

Bryce. Somehow, I got more from his words and equations on the blackboard than ever before.

Bryce was not the kind of teacher who just taught in the classroom. He went to lunch with his students very often and challenged us all to think more deeply into any subject we were studying. I wish there had been recording technology like my iPhone at the time. Some of those lunches were fascinating in the extreme.

Bryce not only taught advanced graduate courses, but also basic physics to those large undergraduate classes so many universities have now. A few times, as his teaching assistant, I would take over a class when he was out of town. Once, one of the students came up to me after class to say something like, "We all love Professor DeWitt, but it is nice to sometimes have someone like you give the lecture. He is such a good lecturer and so awesome, we sometimes get caught up in that and not what we are trying to understand. Having someone at our level give the lecture makes us feel like if you can learn it to teach us, we have hope of getting it too." I smiled at her somewhat demeaning comment, but I did understand what she was getting at.

When Bryce's amazing and loving wife, Cecile, called me to see if I might edit his notes and put them into book form, I was both honored and flabbergasted. It took far too long to accomplish the task, I think mostly because I wanted perfection. Because Bryce was no longer around to make sure I did it right, there were months when I could not deal with the work. But, here are the results.

This book is not a textbook, though it contains lectures and problems. Like Bryce's other books and papers (all of which should be studied thoroughly by any serious person interested in quantum theory, relativity, and gravitation), these chapters and the calculations in the appendices will give you some insight into his thought processes and extraordinary talents as an equation manipulator.

Some of the chapters here are very complex. These are lecture notes after all and you will find extensive detail, including appendices containing various side calculations. It is certain that many of these lengthy derivations cannot be found in any other book or paper.

I found the chapters related to Special Relativity to be nostalgic in particular. The science fiction novel Tau Zero by Poul Anderson http://en.wikipedia.org/wiki/Tau_Zero was one that Bryce had read and used the ideas in the book to illustrate his lectures. He gave me his paperback copy of Tau Zero, which I still treasure.

These notes were written nearly forty years ago. Clearly much has happened since then and any subject that catches your interest should be followed up with reading the latest work. Cecile and I choose to keep the book in the form that Bryce wrote the notes and made as few changes or additions as possible. This is an historical document designed to preserve his thoughts.

The process of converting hundreds of pages of handwritten notes and equations into the manuscript was tedious in the extreme. My deepest appreciation goes to Steven Lyle and Christian Caron at Springer for their patience and major support with the TeX and publishing issues, to Stephen Fulling for TeX guidance, and to Jacob Bekenstein and Chris Eling for many comments on Chaps. 10–12.

Very special thanks go to Brandon DiNunno in Austin who spent a great many hours comparing the handwritten notes against the typeset text looking for issues and typos that I did not find. Any problems of any kind with the book are entirely mine. Errors and other comments can be sent to me at steve@smc.vnet.net. I will collect any and put them up on the web.

The manuscript was processed using the TeXShop software on a Macbook Pro computer. I drew illustrations using Adobe Illustrator.

My heartfelt affection and thanks to Cecile for her guidance and encouragement.

Several times I offered to give the work over to someone else and she would have none of it. I urge everyone to read the book, "The pursuit of Quantum Gravity–Memoirs of Bryce DeWitt from 1946 to 2004."

Finally, my gratitude and love to my wife, Sunny, for her constant support and encouragement in this and all my efforts.

July 2010 Steve Christensen

Contents

Chapter 1
Review of the Uses of Invariants in Special Relativity

1.1 Relative Velocity

The following standard summation and index notation is used with Greek indices running from 0 to 3:

$$A^2 = A \cdot A\,, \qquad A \cdot B = \eta_{\mu\nu}A^\mu B^\nu = \eta^{\mu\nu}A_\mu B_\nu = A_\mu B^\mu\,,$$

$$(\eta_{\mu\nu}) = (\eta^{\mu\nu}) = \mathrm{diag}(-1,1,1,1)\,, \qquad \eta_{\mu\sigma}\eta^{\sigma\nu} = \delta^\nu_\mu\,.$$

Let two observers be moving with constant 4-velocities u_1^μ and u_2^μ, respectively:

$$u_1^2 = u_2^2 = -1 \qquad (c = 1).$$

Each observer sees the other move with a velocity of magnitude $|v|$. This quantity must be a function of the only nontrivial invariant that can be constructed from u_1^μ and u_2^μ, namely $u_1 \cdot u_2$. This invariant may be computed in a rest frame of u_1:

$$(u_1^\mu) = (1,0,0,0)\,, \qquad (u_2^\mu) = (\gamma, \gamma\boldsymbol{v})\,,$$

where

$$\gamma = \frac{1}{\sqrt{1-v^2}} = -u_1 \cdot u_2\,.$$

Now

$$1 - v^2 = \frac{1}{\gamma^2} \quad \Longrightarrow \quad v^2 = 1 - \frac{1}{\gamma^2} = \frac{\gamma^2-1}{\gamma^2} = \frac{(u_1\cdot u_2)^2 - 1}{(u_1 \cdot u_2)^2}\,.$$

Problem 1 Let $\boldsymbol{v}_1$ and $\boldsymbol{v}_2$ be the 3-velocities of the two observers in an arbitrary inertial frame. Show that

B. DeWitt, *Bryce DeWitt's Lectures on Gravitation*,
Edited by Steven M. Christensen, Lecture Notes in Physics, 826,
DOI: 10.1007/978-3-540-36911-0_1, © Springer-Verlag Berlin Heidelberg 2011

$$|v| = \frac{\sqrt{(\boldsymbol{v}_1 - \boldsymbol{v}_2)^2 - (\boldsymbol{v}_1 \times \boldsymbol{v}_2)^2}}{1 - \boldsymbol{v}_1 \cdot \boldsymbol{v}_2} .$$

Proof Since $-u_1 \cdot u_2 = \gamma_1 \gamma_2 (1 - \boldsymbol{v}_1 \cdot \boldsymbol{v}_2)$,

$$\begin{aligned} v^2 &= \frac{\gamma_1^2 \gamma_2^2 (1 - \boldsymbol{v}_1 \cdot \boldsymbol{v}_2)^2 - 1}{\gamma_1^2 \gamma_2^2 (1 - \boldsymbol{v}_1 \cdot \boldsymbol{v}_2)^2} \\ &= \frac{1 - 2\boldsymbol{v}_1 \cdot \boldsymbol{v}_2 + (\boldsymbol{v}_1 \cdot \boldsymbol{v}_2)^2 - (1 - \boldsymbol{v}_1^2)(1 - \boldsymbol{v}_2^2)}{(1 - \boldsymbol{v}_1 \cdot \boldsymbol{v}_2)^2} \\ &= \frac{(\boldsymbol{v}_1 - \boldsymbol{v}_2)^2 - (\boldsymbol{v}_1 \times \boldsymbol{v}_2)^2}{(1 - \boldsymbol{v}_1 \cdot \boldsymbol{v}_2)^2} , \end{aligned}$$

where we have used the fact that

$$(\boldsymbol{v}_1 \times \boldsymbol{v}_2)^2 = \boldsymbol{v}_1^2 \boldsymbol{v}_2^2 - (\boldsymbol{v}_1 \cdot \boldsymbol{v}_2)^2 .$$

□

1.2 Doppler Shift

The amplitude of a plane monochromatic electromagnetic wave has a spacetime dependence of the form $Re(e^{ik \cdot x})$, where $(k^\mu) = (k^0, \boldsymbol{k})$, with $\boldsymbol{k}$ the propagation vector and $k^0(>0)$ the angular frequency, the two being equal in magnitude, i.e., $k^2 = 0$. Because hyperplanes of constant phase are physically determined, independently of the choice of coordinate system, $k \cdot x$ must be an invariant under Lorentz transformations, and hence k^μ must transform as a 4-vector.

Let an atom having 4-velocity u_0 emit a nearly monochromatic pulse of light (photon) at the event x_0 and let the pulse be detected at the event x by a detector moving with 4-velocity u. Suppose $|\boldsymbol{x} - \boldsymbol{x}_0| \gg$ wavelength. Then, the pulse will be nearly planar when it reaches x and will be characterized there by a propagation 4-vector k parallel to $x - x_0$. Evidently,

$$k^2 = 0 , \quad (x - x_0)^2 = 0 , \quad k \cdot (x - x_0) = 0 .$$

The angular frequency of the pulse in the atom's rest frame is

$$w_0 = -k \cdot u_0 .$$

This frequency is characteristic of the atom and is independent of the coordinate system. The angular frequency observed by the detector is

$$w = -k \cdot u .$$

Introduce a set of three orthonormal vectors n_i^μ $(i = 1, 2, 3)$ in the inertial frame carried by the detector. Together with the detector's 4-velocity u^μ, these vectors form what is called an *orthonormal tetrad* or *vierbein*:

$$n_i \cdot n_j = \delta_{ij} , \quad n_i \cdot u = 0 , \quad u^2 = -1 .$$

and such a tetrad is often referred to as defining a local rest frame, in this case, a local rest frame for the detector. The triad n_i^μ $(i = 1, 2, 3)$ is said to form a basis for the hyperplane of simultaneity of the detector and to generate a projection tensor

$$P^{\mu\nu} \equiv n_i^\mu n_i^\nu = \eta^{\mu\nu} + u^\mu u^\nu$$

on this hyperplane.

Let $\boldsymbol{v}$ be the 3-velocity of the atom relative to the local rest frame of the detector, and let $\boldsymbol{m}$ be the unit vector characterizing the direction from which the pulse appears to come in this frame. We have

$$v_i = \frac{u_0 \cdot n_i}{\gamma} = -\frac{u_0 \cdot n_i}{u_0 \cdot u} , \quad \gamma = \frac{1}{\sqrt{1 - v^2}} ,$$

$$m_i = \frac{k \cdot n_i}{k \cdot u} \quad (\text{remember } k \cdot u < 0) ,$$

$$m_i m_i = \frac{(k \cdot n_i)(n_i \cdot k)}{(k \cdot u)^2} = \frac{k^\mu P_{\mu\nu} k^\nu}{(k \cdot u)^2} = \frac{k^2 + (k \cdot u)^2}{(k \cdot u)^2} = 1 ,$$

since $k^2 = 0$. From these two 3-vectors, we may construct an important scalar, namely, the component of the 3-velocity of the atom along the line of sight as viewed from the local rest frame of the detector:

$$v_\mathrm{R} \equiv m_i v_i = -\frac{(k \cdot n_i)(n_i \cdot u_0)}{(k \cdot u)(u_0 \cdot u)} = -\frac{k \cdot u_0 + (k \cdot u)(u \cdot u_0)}{(k \cdot u)(u_0 \cdot u)} = \frac{\omega_0}{\omega \gamma} - 1 ,$$

whence

$$\frac{\omega_0}{\omega} = \gamma(1 + v_\mathrm{R}) ,$$

or, in terms of wavelength,

$$\frac{\lambda}{\lambda_0} = \gamma(1 + v_\mathrm{R}) ,$$

where γ is the time dilation factor and $1 + v_\mathrm{R}$ the 'true' Doppler shift. This is the special relativistic Doppler shift formula.

It has become conventional to express the Doppler shift in terms of the so-called *red shift parameter*:

$$z = \frac{\lambda - \lambda_0}{\lambda_0} = \gamma(1 + v_R) - 1 \, .$$

In the non-relativistic limit $v \to 0$, we have

$$z \underset{v \to 0}{\longrightarrow} v_R \, .$$

In the case in which the relative velocity is along the line of sight, so that $v_R = v$, the Doppler shift formula reduces to

$$\frac{\lambda}{\lambda_0} = \frac{1+v}{\sqrt{1-v^2}} = \sqrt{\frac{1+v}{1-v}} \, ,$$

where

$$v > 0 \quad \text{for recession (red shift)} \, ,$$
$$v < 0 \quad \text{for approach (blue shift)} \, .$$

1.3 Aberration

Suppose another detector, moving with 4-velocity u', observes, at (or near) the event x, a pulse emitted at (or near) the event x_0 by another atom, similar to the first and also moving with 4-velocity u_0. Introduce a local rest frame for the new detector, characterized by a triad $n_i'^{\mu}$:

$$n_i' \cdot n_j' = \delta_{ij} \, , \quad n_i' \cdot u' = 0 \, , \quad u'^2 = -1 \, ,$$

$$n_i'^{\mu} n_i'^{\nu} = P'^{\mu\nu} = \eta^{\mu\nu} + u'^{\mu} u'^{\nu} \, .$$

The components of the 3-velocity of the second detector as viewed in the local rest frame of the first are

$$\bar{v}_i = -\frac{u' \cdot n_i}{u' \cdot u} \, ,$$

whereas the components of the 3-velocity of the first detector as viewed in the local rest frame of the second are

$$\bar{v}_i' = -\frac{u \cdot n_i'}{u \cdot u'} \, .$$

Note that

$$\begin{aligned} \bar{v}_i \bar{v}_i &= \frac{(u' \cdot n_i)(n_i \cdot u')}{(u' \cdot u)^2} = \frac{u'^{\mu} P_{\mu\nu} u'^{\nu}}{(u' \cdot u)^2} = \frac{-1 + (u' \cdot u)^2}{(u' \cdot u)^2} = \bar{v}_i' \bar{v}_i' \\ &= 1 - \frac{1}{\bar{\gamma}^2} \, , \end{aligned}$$

where

$$\bar{\gamma} = \frac{1}{\sqrt{1-\bar{v}^2}} = -u \cdot u' ,$$

with $\bar{v}$ the magnitude of the 3-velocity of either detector as viewed in the local rest frame of the other. If desired, the local rest frames of the two detectors can be aligned so that

$$(\bar{v}_i) = (\bar{v}, 0, 0) , \quad (\bar{v}'_i) = (-\bar{v}, 0, 0) .$$

Now let θ be the angle between $\bar{v}_i$ and the unit vector m_i in the rest frame of the first detector. We have

$$\bar{v} \cos\theta = m_i \bar{v}_i = -\frac{(k \cdot n_i)(n_i \cdot u')}{(k \cdot u)(u \cdot u')} = -\frac{k \cdot u' + (k \cdot u)(u \cdot u')}{(k \cdot u)(u \cdot u')} = \frac{\omega'}{\omega \bar{\gamma}} - 1 ,$$

where ω' is the angular frequency of the pulse observed by the second detector. The corresponding angle θ' in the rest frame of the second detector satisfies the equation

$$-\bar{v} \cos\theta' = m'_i \bar{v}'_i = -\frac{(k \cdot n'_i)(n'_i \cdot u)}{(k \cdot u')(u' \cdot u)} = \frac{\omega}{\omega' \bar{\gamma}} - 1 .$$

We now have

$$\frac{\omega'}{\omega} = \bar{\gamma}(1 + \bar{v}\cos\theta) , \quad \frac{\omega}{\omega'} = \bar{\gamma}(1 - \bar{v}\cos\theta') ,$$

$$1 = \bar{\gamma}^2 (1 + \bar{v}\cos\theta)(1 - \bar{v}\cos\theta') ,$$

$$1 - \bar{v}^2 = 1 + \bar{v}(\cos\theta - \cos\theta') - \bar{v}^2 \cos\theta \cos\theta' ,$$

$$\cos\theta' - \cos\theta = \bar{v}(1 - \cos\theta\cos\theta') ,$$

and finally,

$$\cos\theta' = \frac{\bar{v} + \cos\theta}{1 + \bar{v}\cos\theta} .$$

1.3.1 Consistency Check

Since

$$\bar{v}^2 < 1 \quad \text{and} \quad (1 - \bar{v}^2)\cos^2\theta \leq 1 - \bar{v}^2 ,$$

it follows that

$$\begin{aligned}
\bar{v}^2 + \cos^2\theta &\le 1 + \bar{v}^2\cos^2\theta\,,\\
\bar{v}^2 + 2\bar{v}\cos\theta + \cos^2\theta &\le 1 + 2\bar{v}\cos\theta + \bar{v}^2\cos^2\theta\,,\\
|\bar{v} + \cos\theta| &\le |1 + \bar{v}\cos\theta|\,.
\end{aligned}$$

Note that as $\bar{v} \longrightarrow 1$, the apparent direction of the emitting atom as viewed by the second detector tends more and more toward the forward direction, i.e., the direction in which the second detector moves relative to the first.

A more elegant aberration formula may be obtained by writing

$$\theta = \frac{1}{2}(\theta+\theta') + \frac{1}{2}(\theta-\theta')\,, \quad \theta' = \frac{1}{2}(\theta+\theta') - \frac{1}{2}(\theta-\theta')\,,$$

$$\cos\theta' - \cos\theta = 2\sin\frac{1}{2}(\theta+\theta')\sin\frac{1}{2}(\theta-\theta')\,,$$

and also

$$\begin{aligned}
\bar{v} &= \frac{\cos\theta' - \cos\theta}{1 - \cos\theta\cos\theta'}\,,\\
1 - \bar{v}^2 &= \frac{1 - 2\cos\theta\cos\theta' + \cos^2\theta\cos^2\theta' - \cos^2\theta' + 2\cos\theta\cos\theta' - \cos^2\theta}{(1-\cos\theta\cos\theta')^2}\\
&= \frac{(1-\cos^2\theta)(1-\cos^2\theta')}{(1-\cos\theta\cos\theta')^2} = \frac{\sin^2\theta\sin^2\theta'}{(1-\cos\theta\cos\theta')^2}\,,
\end{aligned}$$

whence

$$\bar{\gamma} = \frac{1}{\sqrt{1-\bar{v}^2}} = \frac{1-\cos\theta\cos\theta'}{\sin\theta\sin\theta'}\,.$$

Remember here that θ and θ' lie between 0 and π so that both $\sin\theta$ and $\sin\theta'$ are positive. Finally,

$$2\sin\frac{1}{2}(\theta-\theta') = \bar{\gamma}\bar{v}\frac{\sin\theta\sin\theta'}{\sin\frac{1}{2}(\theta+\theta')}\,.$$

In the non-relativistic limit $\bar{v} \longrightarrow 0$, this reduces to the classical formula used by astronomers:

$$\theta - \theta' \underset{\bar{v}\to 0}{\longrightarrow} \bar{v}\sin\theta\,.$$

Problem 2 Show that

$$\sin(\theta-\theta') = \frac{\bar{v} - \frac{\bar{\gamma}-1}{\bar{\gamma}}\cos\theta'}{1-\bar{v}\cos\theta'}\sin\theta'\,.$$

Then, because of the invariance of the aberration formula under the changes $\theta \to \theta'$, $\theta' \to \theta$, $\bar{v} \to -\bar{v}$, it follows that

$$\sin(\theta - \theta') = \frac{\bar{v} + \frac{\bar{\gamma}-1}{\bar{\gamma}} \cos\theta}{1 + \bar{v}\cos\theta} \sin\theta \; .$$

Note that for *either* $\theta = \pi/2$ or $\theta' = \pi/2$, the aberration formula reduces to $\sin(\theta - \theta') = \bar{v}$.

Problem 3 Let the ecliptic latitude and longitude of a star (conventionally measured from the vernal equinox in the direction of the earth's orbital motion) be ψ and ϕ, respectively, as observed in the rest frame of the sun. The aberration effect produced by the motion (velocity $\bar{v}$) of the earth in its orbit will cause these angles to shift to new values, $\psi + \delta\psi$ and $\phi + \delta\phi$, as viewed in a local rest frame attached to the earth. Derive expressions for $\delta\psi$ and $\delta\phi$ in terms of $\psi, \theta, \bar{v}$ and the ecliptic longitude α of the sun (Figs. 1.1, 1.2). Assume that the orbit of the earth is a circle and that $\bar{v} \ll 1$. Show that, during a year, the apparent position of the star in the sky as viewed from earth executes a tiny ellipse whose semi-major axis has angular size $\bar{v}$ (radians) and is oriented parallel to the ecliptic plane and whose semi-minor axis has angular size $\bar{v}|\sin\psi|$. Compute $\bar{v}$ for the earth in seconds of arc.

With the notation established above,

$$(m_i) = (\cos\psi\cos\phi, \cos\psi\sin\phi, \sin\psi) \; , \quad (\bar{v}_i) = \bar{v}(\sin\alpha, -\cos\alpha, 0) \; ,$$

$$\cos\theta = \cos\psi(\cos\phi\sin\alpha - \sin\phi\cos\alpha) = \cos\psi\sin(\alpha - \phi) \; ,$$

$$m_i' = \frac{m_i + \xi\bar{v}_i}{\sqrt{(m_j + \xi\bar{v}_j)(m_j + \xi\bar{v}_j)}} = (1 - \xi\bar{v}\cos\theta)(m_i + \xi\bar{v}_i) \; ,$$

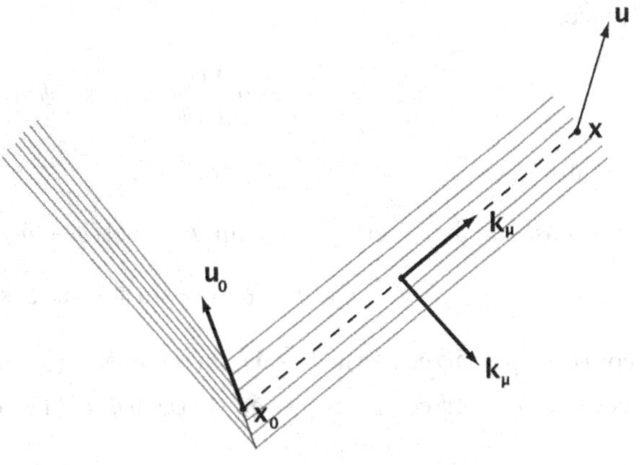

Fig. 1.1 Emitting atom and detector

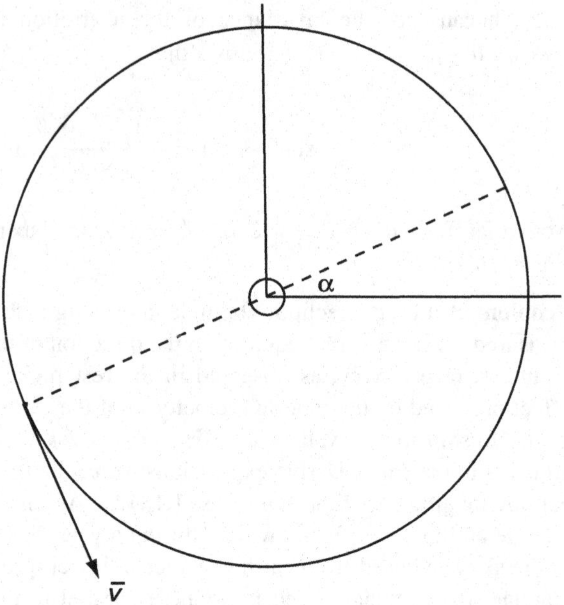

Fig. 1.2 Earth orbit

for some ξ, and

$$\cos\theta' = (1 - \xi\bar{v}\cos\theta)(\cos\theta + \xi\bar{v}) \ .$$

But

$$\cos\theta' = (1 - \bar{v}\cos\theta)(\bar{v} + \cos\theta) \ .$$

Therefore $\xi = 1$ and

$$(1 - \bar{v}\cos\theta)\sin\psi = \sin(\psi + \delta\psi) = \sin\psi + \delta\psi\cos\psi \ .$$

Hence,

$$\delta\psi = -\bar{v}\cos\theta\frac{\sin\psi}{\cos\psi} = -\bar{v}\sin\psi\sin(\alpha - \phi)$$

We also have

$$\cos(\psi + \delta\psi) = \cos\psi - \delta\psi\sin\psi \ , \quad \cos(\phi + \delta\phi) = \cos\phi - \delta\phi\sin\phi \ ,$$

$$\sin(\phi + \delta\phi) = \sin\phi + \delta\phi\cos\phi \ ,$$

$$\cos\psi\cos\phi - \delta\phi\cos\psi\sin\phi - \delta\psi\sin\psi\cos\phi = (1 - \bar{v}\cos\theta)\cos\psi\cos\phi + \bar{v}\sin\alpha \ ,$$
$$\cos\psi\sin\phi + \delta\phi\cos\psi\cos\phi - \delta\psi\sin\psi\sin\phi = (1 - \bar{v}\cos\theta)\cos\psi\sin\phi - \bar{v}\cos\alpha \ ,$$

which give

$$\delta\phi = \begin{cases} \frac{1}{\cos\psi \sin\phi}(-\delta\psi \sin\psi \cos\phi + \bar{v} \cos\theta \cos\psi \cos\phi - \bar{v} \sin\alpha) \\ \frac{1}{\cos\psi \cos\phi}(\delta\psi \sin\psi \sin\phi - \bar{v} \cos\theta \cos\psi \sin\phi - \bar{v} \cos\alpha) \end{cases}$$

$$= \begin{cases} \frac{\bar{v}}{\cos\psi \sin\phi}\left(\frac{\cos\theta\cos\phi}{\cos\psi} - \sin\alpha\right) \\ \frac{\bar{v}}{\cos\psi \cos\phi}\left(-\frac{\cos\theta\sin\phi}{\cos\psi} - \cos\alpha\right) \end{cases}$$

$$= \begin{cases} \frac{\bar{v}}{\cos\psi \sin\phi}[\sin(\alpha-\phi)\cos\phi - \sin\alpha] \\ \frac{\bar{v}}{\cos\psi \cos\phi}[-\sin(\alpha-\phi)\sin\phi - \cos\alpha] \end{cases}$$

$$= \begin{cases} \frac{\bar{v}}{\cos\psi \sin\phi}(-\sin\alpha \sin^2\phi - \cos\alpha \sin\phi \cos\phi) \\ \frac{\bar{v}}{\cos\psi \cos\phi}(-\sin\alpha \cos\phi \sin\phi - \cos\alpha \cos^2\phi) \end{cases}$$

Finally,

$$\delta\phi = -\frac{\bar{v}\cos(\alpha-\phi)}{\cos\psi} \qquad \bar{v} = 30\,\text{km/s} = 10^{-4}\ \text{rad} = 20.6''$$

$$\text{semi-major axis} = |\cos\psi\,\delta\phi|_{\max} = \bar{v}\ ,$$
$$\text{semi-minor axis} = |\delta\psi|_{\max} = \bar{v}|\sin\psi|\ .$$

1.4 Apparent Luminosity

Let a star having 4-velocity u_0 emit monochromatic photons at a steady rate uniformly in all directions. If ω_0 is the angular frequency of the photons in the star's rest frame and N_0 is the number emitted per unit time, then the power output or *absolute luminosity* of the star is

$$L_0 = N_0\omega_0 \quad (\hbar = 1)\ .$$

Let some of the photons that are emitted by the star at an event x_0 be detected by an observer at event x. If the 4-velocity of the observer is u, then

$$(x - x_0)^2 = 0\ , \quad -u_0 \cdot (x - x_0) = r_0\ , \quad -u \cdot (x - x_0) = r\ ,$$

where r_0 is the apparent distance of the detection event in the star's rest frame and r is the apparent distance of the emission event in the observer's rest frame. Since $x^\mu - x_0^\mu$ is parallel to the propagation vector k^μ of the photons arriving at the observer, it follows by arguments entirely similar to those used in deriving the Doppler shift formula that

$$\frac{r_0}{r} = \frac{\omega_0}{\omega} = \gamma(1 + v_R)\ ,$$

with the obvious notation.

Now introduce an orthonormal triad n^{μ}_{0i} to fix the orientation of the star's rest frame:

$$n_{0i}\cdot n_{0j}=\delta_{ij}\ ,\quad n_{0i}\cdot u_0=0\ ,\quad u_0^2=-1\ .$$

In this frame, the photons that reach the observer are propagated in the direction of the unit vector

$$\Omega_{0i}=r_0^{-1}n_{0i}\cdot(x-x_0)\ .$$

Suppose the spacetime point x suffers a displacement $\mathrm{d}x$. Then r_0 and Ω_{0i} suffer the changes

$$\mathrm{d}r_0=-u_0\cdot\mathrm{d}x\ ,\quad \mathrm{d}\Omega_{0i}=r_0^{-1}n_{0i}\cdot\mathrm{d}x+r_0^{-2}n_{0i}\cdot(x-x_0)u_0\cdot\mathrm{d}x\ .$$

Let $\mathrm{d}_1\Omega_{0i}$ and $\mathrm{d}_2\Omega_{0i}$ be the changes in Ω_{0i} corresponding to two such displacements d_1x and d_2x. Then

$$\mathrm{d}_1\Omega_{0i}\mathrm{d}_2\Omega_{0i}=r_0^{-2}\left[\mathrm{d}_1x+r_0^{-1}u_0\cdot\mathrm{d}_1x(x-x_0)\right]\cdot P_0\cdot\left[\mathrm{d}_2x+(x-x_0)r_0^{-1}u_0\cdot\mathrm{d}_2x\right]\ ,$$

where

$$P_0^{\mu\nu}=\eta^{\mu\nu}+u_0^{\mu}u_0^{\nu}\ .$$

But

$$P_0\cdot\mathrm{d}x=\mathrm{d}x+u_0u_0\cdot\mathrm{d}x\ ,\quad P_0\cdot(x-x_0)=x-x_0-r_0u_0\ ,$$

so

$$P_0\cdot\left[\mathrm{d}x+(x-x_0)r_0^{-1}u_0\cdot\mathrm{d}x\right]=\mathrm{d}x+(x-x_0)r_0^{-1}u_0\cdot\mathrm{d}x\ .$$

Furthermore,

$$P_0\cdot P_0=P_0\ ,$$

so

$$\mathrm{d}_1\Omega_{0i}\mathrm{d}_2\Omega_{0i}=r_0^{-2}\left[\mathrm{d}_1x+r_0^{-1}u_0\cdot\mathrm{d}_1x(x-x_0)\right]\cdot\left[\mathrm{d}_2x+(x-x_0)r_0^{-1}u_0\cdot\mathrm{d}_2x\right]\ .$$

Now suppose the displacements d_1x and d_2x lie in the observer's hyperplane of simultaneity and are at right angles to the line of sight to the star in the observer's rest frame. Then

$$u\cdot\mathrm{d}_{\alpha}x=0\ ,\quad (x-x_0)\cdot\mathrm{d}_{\alpha}x=0\ ,\quad \alpha=1,2\ ,$$

and, remembering that $(x-x_0)^2=0$, we have

$$\mathrm{d}_1\Omega_{0i}\mathrm{d}_2\Omega_{0i}=r_0^{-2}\mathrm{d}_1x\cdot\mathrm{d}_2x\ .$$

This means that the portion of the unit sphere containing Ω_{0i}, in the star's rest frame, is mapped via the photons themselves in a metric-preserving fashion with only an overall rescaling factor of r_0^{-2} onto the plane in the observer's rest frame that is at right angles to the line of sight to the star. Therefore the photons that pass through a surface element d^2S in this plane were emitted in a solid angle

$$d^2\Omega_0 = r_0^{-2} d^2 S$$

in the star's rest frame. The energy flux through this solid angle is

$$d^2\Phi_0 = L_0 \frac{d^2\Omega_0}{4\pi} = L_0 \frac{d^2 S}{4\pi r_0^2} .$$

To the observer, however, the energy flux through d^2S differs from this value in two respects. First, the apparent photon energy is changed from ω_0 to ω, i.e., by the Doppler shift factor $\omega/\omega_0 = \gamma^{-1}(1+v_R)^{-1}$. Secondly, the rate at which photons pass through d^2S is itself changed by the same factor. Therefore, the energy flux through d^2S as seen by the observer is

$$d^2\Phi = \left(\frac{\omega}{\omega_0}\right)^2 d^2\Phi_0 = \left(\frac{\omega}{\omega_0}\right)^2 L_0 \frac{d^2 S}{4\pi r_0^2} .$$

This flux may also be expressed in the form

$$d^2\Phi = L \frac{d^2 S}{4\pi r^2} ,$$

where L is the *apparent luminosity* of the star, i.e., the luminosity that a star at rest at a distance r relative to the observer would have to have in order to produce such a flux. Equating the two expressions, we find

$$L = \left(\frac{\omega}{\omega_0}\right)^2 \left(\frac{r}{r_0}\right)^2 L_0 = \frac{L_0}{\gamma^4 (1+v_R)^4} = \frac{L_0}{(1+z)^4} .$$

Because the final expression is independent of frequency, it holds for a star with an arbitrary spectrum. In the special case in which the relative 3-velocity of star and observer is along the line of sight, the above formula reduces to

$$L = \left(\frac{1-v}{1+v}\right)^2 L_0 , \quad \text{where} \quad \begin{cases} v > 0 & \text{for recession} , \\ v < 0 & \text{for approach} . \end{cases}$$

Using these formulas, it is important to distinguish between apparent luminosity and apparent brightness. The former depends only on the relative velocity of star and observer, whereas the latter varies inversely as the square of the distance r.

Chapter 2
Accelerated Motion in Special Relativity

Let a particle of mass m be acted upon by a force F^i in some inertial frame. Then

$$F^i = \frac{dp^i}{dt} = m\frac{\ddot{z}^i}{\dot{t}} = m\frac{\ddot{z}^i}{\gamma}, \qquad t = z^0,$$

where p^i is the three-vector portion of the energy–momentum four-vector of the particle in that frame:

$$p^\mu = m\dot{z}^\mu,$$

the world line of the particle being represented in the parametric form $z^\mu(\tau)$ and the dot denoting differentiation with respect to the proper time τ:

$$-1 = \dot{z}^2 = -(\dot{z}^0)^2 + \dot{z}^i\dot{z}^i = -\dot{t}^2\gamma^{-2},$$

and

$$\gamma = \left(1 - \frac{dz^i dz^i}{dt\ dt}\right)^{-1/2}.$$

The force that the particle actually 'feels' in its own instantaneous rest frame has magnitude given by

$$F_{\mathrm{R}} = ma,$$

where a is the *absolute acceleration* of the particle:

$$a^2 = \ddot{z}^2.$$

In general, $F_{\mathrm{R}} \neq F$, where $F^2 = F^i F^i$. However, when the three-acceleration is parallel (or antiparallel) to the three-velocity, the two magnitudes coincide, for we then have

B. DeWitt, *Bryce DeWitt's Lectures on Gravitation*,
Edited by Steven M. Christensen, Lecture Notes in Physics, 826,
DOI: 10.1007/978-3-540-36911-0_2, © Springer-Verlag Berlin Heidelberg 2011

$$0 = \dot{z} \cdot \ddot{z} = -\dot{z}^0 \ddot{z}^0 + \dot{z}^i \ddot{z}^i = -\dot{t}\left(\ddot{z}^0 \pm \left|\frac{\mathrm{d}z}{\mathrm{d}t}\right| |\ddot{z}|\right),$$

$$a^2 = -\left(\ddot{z}^0\right)^2 + \ddot{z}^i \ddot{z}^i = \left[1 - \left(\frac{\mathrm{d}z}{\mathrm{d}t}\right)^2\right] \ddot{z}^2 = \frac{\ddot{z}^2}{\gamma^2} = \frac{F^2}{m^2}.$$

Therefore, *a particle which starts from rest under the action of a constant force will experience a constant absolute acceleration.*

Let us determine the motion of such a particle under the initial conditions

$$z = 0, \qquad \dot{z} = 0, \qquad t = 0 \quad \text{at } \tau = 0.$$

Since the motion is in a straight line, we may retain only two coordinates, t and z. We have

$$-1 = -\dot{t}^2 + \dot{z}^2, \qquad \dot{t} = \sqrt{1 + \dot{z}^2},$$

$$0 = -\dot{t}\ddot{t} + \dot{z}\ddot{z}, \qquad \ddot{t} = \frac{\dot{z}\ddot{z}}{\dot{t}},$$

$$a^2 = -\ddot{t}^2 + \ddot{z}^2 = \left(1 - \frac{\dot{z}^2}{\dot{t}^2}\right)\ddot{z}^2 = \frac{\ddot{z}^2}{1 + \dot{z}^2}.$$

Let $u = \dot{z}$. Then $\ddot{z} = u\mathrm{d}u/\mathrm{d}z$ and

$$a\mathrm{d}z = \frac{u\mathrm{d}u}{\sqrt{1 + u^2}},$$

assuming motion in the positive z direction, whence

$$az = \sqrt{1 + u^2} - 1, \qquad u^2 = (1 + az)^2 - 1,$$

$$\mathrm{d}\tau = \frac{\mathrm{d}z}{\sqrt{(1 + az)^2 - 1}} = \frac{1}{a}\frac{\mathrm{d}(1 + az)}{\sqrt{(1 + az)^2 - 1}},$$

$$\tau = \frac{1}{a}\cosh^{-1}(1 + az),$$

$$z = \frac{1}{a}(\cosh a\tau - 1),$$

$$\dot{z} = \sinh a\tau,$$

$$\dot{t} = \cosh a\tau,$$

$$t = \frac{1}{a}\sinh a\tau,$$

$$z = \frac{1}{a}\left(\sqrt{1 + a^2t^2} - 1\right) \longrightarrow \begin{cases} \frac{1}{2}at^2 & \text{as } t \to 0, \\ t & \text{as } t \to \infty, \end{cases}$$

$$v \equiv \frac{\mathrm{d}z}{\mathrm{d}t} = \frac{at}{\sqrt{1 + a^2t^2}} \longrightarrow \begin{cases} at & \text{as } t \to 0, \\ 1 & \text{as } t \to \infty. \end{cases}$$

Table 2.1 Comparing proper time and inertial time at different distances during the voyage

τ (year)	z (ly)	t (year)
1	0.0628	1.01
5	1.777	6.41
10	10.26	24.2
15	40.5	85.0
20	146	297
30	1806	3 616
40	22 024	44 052
50	268 000	536 000
60	3 270 000	6 540 000

Problem 4 A cosmic spaceship departs from earth at a constant absolute acceleration of 950 cm/s^2 (slightly less than the acceleration due to gravity at the earth's surface). It maintains this acceleration for $\tau/4$ years of proper time, after which it decelerates at the same rate and in the same direction for another $\tau/4$ years of proper time. At the end of this time it is at rest with respect to the earth, but at a distance of z light years. Its crew at this point executes a certain assigned mission on a nearby planet, which takes a negligible amount of time compared to τ, and then returns to earth by an acceleration–deceleration procedure identical with that of the outward journey. The total voyage has required τ years of proper time. Let t be the number of years that have elapsed on earth since departure. Obtain expressions for z and t in terms of τ, and construct a table giving z and t for selected values of τ ranging from 1 to 60 years. (Hint: express the acceleration in light years per year and use symmetry arguments to simplify the problem Table 2.1)

Solution 4

$$1\ \text{ly/year}^2 = \frac{3 \times 10^{10}}{3.16 \times 10^7} = 950\ \text{cm/s}^2,$$

so $a = 1$. By symmetry, we have

$$z = 2\left(\cosh\frac{\tau}{4} - 1\right), \qquad t = 4\sinh\frac{\tau}{4}.$$

Problem 5 Suppose the spaceship of Problem 4 did not attempt to return to earth but merely executed a single acceleration–deceleration maneuver. How far would it have traveled in 50 years of proper flight time, and how much time would have elapsed back on earth?

Solution 5 We have

$$z = 2\left(\cosh\frac{\tau}{2} - 1\right), \qquad t = 2\sinh\frac{\tau}{2}.$$

For $\tau = 50$ year, we have

$$t \approx z \approx e^{\tau/2} = e^{25} = 72 \times 10^9 \begin{cases} \text{years,} \\ \text{light years.} \end{cases}$$

Problem 6 A cosmic spaceship makes use of the following propulsion mechanism. During an interval of proper time $d\tau$ the rest mass of the ship decreases by an amount $-dm$. This mass decrement is used in the following way. A fraction ξ ($0 < \xi \le 1$) is converted into kinetic energy (relative to the ship) of the remaining fraction. This remaining fraction is ejected from the ship in a constant (backward) direction, with the relative velocity v_e corresponding to the kinetic energy it has acquired. Express v_e as a function of ξ. What is the proper impulse dp imparted to the ship during the proper time $d\tau$ as a result of the ejection of the 'propellant'? (Express it as a function of v_e and dm.) What is the absolute acceleration a experienced by the ship as a result of this impulse?

Suppose the ship starts from rest (relative to some inertial frame) with an initial mass m_0, and suppose ξ (and hence v_e) remains constant in time. Obtain an expression for the velocity v of the ship at any instant as a function of v_e, m_0 and the mass m remaining at that instant. (Do not assume constant absolute acceleration.) For what value of v_e is the propulsion process most efficient, i.e., what physically allowable value of v_e yields the maximum value of v for a given m and m_0? To what value of ξ does this correspond? (To obtain the most efficient propulsion it will be necessary for the ship to carry antimatter as fuel.)

Solution 6 The kinetic energy of the fraction $1 - \xi$ is equated with the rest energy of the fraction ξ to give

$$(1-\xi)(-dm)(\gamma_e - 1) = \xi(-dm),$$

whence

$$(1-\xi)(\gamma_e - 1) = \xi, \qquad \gamma_e - 1 = \frac{\xi}{1-\xi}, \qquad \gamma_e = \frac{1}{1-\xi},$$

$$v_e^2 = 1 - \frac{1}{\gamma_e^2} = 1 - (1-\xi)^2 = \xi(2-\xi).$$

Therefore,

$$\boxed{v_e = \sqrt{\xi(2-\xi)}}$$

The proper impulse imparted to the ship during the proper time $d\tau$ is

$$\boxed{dp = (1-\xi)(-dm)\gamma_e v = (-dm)v_e}$$

The absolute acceleration experienced by the ship as a result of this impulse is then determined from

$$ma = \frac{dp}{d\tau} = -v_e \frac{dm}{d\tau},$$

so

$$\boxed{a = -v_e \frac{d}{d\tau} \log m}$$

We remember that

$$\begin{aligned} a &= \frac{\ddot{z}}{\dot{t}} = \frac{d}{dt}\dot{z} = \frac{d}{dt}(\dot{t}v) = \frac{d}{dt}(\gamma v) \\ &= \gamma^3 v^2 \frac{dv}{dt} + \gamma \frac{dv}{dt} = \gamma^3 \left(v^2 + \frac{1}{\gamma^2}\right) \frac{dv}{dt} \\ &= \gamma^3 \frac{dv}{dt} = \gamma^2 \frac{dv}{d\tau}, \end{aligned}$$

so that

$$\frac{dv}{1 - v^2} = a d\tau = -v_e d \log m,$$

and finally,

$$\tanh^{-1} v = v_e \log \frac{m_0}{m}.$$

The result is

$$\boxed{v = \tanh \log \left(\frac{m_0}{m}\right)^{v_e} = \frac{\left(\frac{m_0}{m}\right)^{v_e} - \left(\frac{m}{m_0}\right)^{v_e}}{\left(\frac{m_0}{m}\right)^{v_e} + \left(\frac{m}{m_0}\right)^{v_e}}}$$

For the most efficient propulsion, $\upsilon_e = 1$, $\xi = 1$, in which case we have

$$\boxed{v = \frac{m_0^2 - m^2}{m_0^2 + m^2}}$$

Problem 7 Suppose in Problem 6 that υ_e is chosen for most efficient propulsion, and suppose fuel is used at such a rate as to maintain constant absolute acceleration. Obtain an expression for the mass m remaining at the proper time τ ($\tau = 0$ when $m = m_0$). Obtain the corresponding expression for m as a function of the time t in the inertial frame in which the spaceship is at rest when $m = m_0$. (Choose the origin of time so that $t = 0$ at this instant.)

Suppose the spaceship executes a single acceleration–deceleration maneuver as in Problem 6, with $a = 950 \text{ cm/s}^2$. Let the total proper time elapsed from the beginning to the end of the maneuver be τ, and let the final mass of the ship (i.e., the *payload*) be m_1. Construct a table showing the values of z, t, and m_0/m for selected values of τ ranging from 1 to 50 years. Here z and t are respectively the total distance covered and the total time elapsed for the complete voyage in the frame in which the spaceship is at rest when $m = m_0$. Include also a column in your table giving the kinetic energy, in electron volts, of the interstellar hydrogen

Table 2.2 Parameters for Problem 7

τ (year)	z (ly)	t (year)	m_0/m_1	K_{max} (eV)
1	0.255	1.04	2.718	1.20×10^8
3	2.70	4.26	20.09	1.27×10^9
5	10.26	12.1	148.4	4.81×10^9
10	146	148	22 026	6.86×10^{10}
20	22 024	22 026	4.85×10^8	1.03×10^{13}
30	3 270 000	3 270 000	1.07×10^{13}	1.53×10^{15}
40	485 000 000	485 000 000	2.36×10^{17}	2.27×10^{17}
50	72 000 000 000	72 000 000 000	5.20×10^{21}	3.37×10^{19}

nuclei (protons) as seen from the ship at the midpoint of the journey when the relative velocity of ship and nuclei is a maximum. (Assume the hydrogen to be at rest in the original rest frame.) This is the bombardment energy against which the crew of the ship will have to be shielded (Table 2.2).

Solution 7 We have

$$v_e = 1, \qquad a = -\frac{d}{d\tau}\log m, \qquad a\tau = \log\frac{m_0}{m},$$

whence

$$\boxed{m = m_0 e^{-a\tau}}$$

Now

$$m = m_0(\cosh a\tau - \sinh a\tau),$$

so

$$\boxed{m = \begin{cases} m_0\left(\sqrt{1+a^2t^2} - at\right) & \underset{t\to 0}{\longrightarrow} \quad m_0(1-at), \\ m_0 at\left(\sqrt{1+\frac{1}{a^2t^2}} - 1\right) & \underset{t\to\infty}{\longrightarrow} \quad \frac{m_0}{2at}. \end{cases}}$$

For the acceleration–deceleration maneuver with $a = 1$ ly/year^2, we have (see solution to Problem 6)

$$z = 2\left(\cosh\frac{\tau}{2} - 1\right), \qquad t = 2\sinh\frac{\tau}{2}, \qquad \frac{m_0}{m_1} = e^{\tau},$$

$$\gamma_{max} = \dot{t}_{max} = \cosh\frac{\tau}{2},$$

$$K_{max} = m_p(\gamma_{max} - 1) = m_p\left(\cosh\frac{\tau}{2} - 1\right), \quad \text{where } m_p = 0.938 \times 10^9 \text{ eV}.$$

Problem 8 [Taken from *Tau Zero* by Paul Anderson, Doubleday, Garden City, New York (1970).] A spaceship is traveling between galaxies at a velocity v with respect to the intergalactic gas. (This gas is presumably mainly hydrogen, although it may consist of many other elements as well, including antihydrogen.) The ship is equipped with a scoop of cross-sectional area A with which it traps the gas in its path. The trapped gas is passed through a nuclear furnace which transmutes it (e.g., binding deuterium nuclei into helium, annihilating proton–antiproton pairs, etc.). The reaction products are then ejected, with no loss of total energy (relative to the ship), out the 'back' end of the ship. Let $\mathrm{d}m$ be the mass of gas trapped by the ship in a proper time interval $\mathrm{d}\tau$. This mass arrives with total energy $\gamma \mathrm{d}m$ ($\gamma = 1/\sqrt{1-v^2}$) relative to the ship, and the reaction products leave with total energy $(1-\xi)\gamma' \mathrm{d}m$ ($\gamma' = 1/\sqrt{1-v'^2}$), where ξ is the fractional decrease of rest mass under the transmutation and v' is the ejection velocity relative to the ship. By equating the two energies, obtain a relation between γ, γ' and ξ, and also an expression for the total proper impulse transmitted to the ship as a result of the transmutation of the mass $\mathrm{d}m$.

If the ship is traveling sufficiently fast, the above process may be used as a kind of ram-jet process whereby the ship is propelled without having to carry its own fuel supply, so that its mass M remains constant. Obtain an expression for the absolute acceleration a imparted to the ship by the ram-jet process as a function of v (or γ), A, ξ, M and the density ρ of the intergalactic gas (in its own rest frame). Obtain the limiting form of this expression as $v \to 1$, $\gamma \to \infty$. How big will a be in this limiting case if $A = 10^3\ \mathrm{m}^2$, $M = 10^4$ kg, $\xi = 0.01$, and $\rho = 10^{-26}\ \mathrm{kg/m^3}$? What implications does your answer have for the feasibility of such a ship? If the origin of proper time is chosen so that the ship's velocity is v_0 when $\tau = 0$, obtain an expression for v as a function of τ in the special case $\xi = 1$ (total conversion of matter). (Note: in solving this problem do not forget to take into account the compression of the gas, as seen from the ship's frame, resulting from the Lorentz contraction. This is crucial!)

Solution 8 We have

$$\boxed{(1-\xi)\gamma' = \gamma}$$

and

$$\gamma'^2 v'^2 = \gamma'^2 - 1 = \frac{\gamma^2}{(1-\xi)^2} - 1 = \frac{\gamma^2 - 1 + \xi(2-\xi)}{(1-\xi)^2}.$$

Hence,

$$\boxed{\begin{aligned} \mathrm{d}p &= [(1-\xi)\gamma' v' - \gamma v]\mathrm{d}m \\ &= \left[\sqrt{\gamma^2 - 1 + \xi(2-\xi)} - \sqrt{\gamma^2 - 1}\right]\mathrm{d}m \end{aligned}}$$

Since

$$\frac{\mathrm{d}m}{\mathrm{d}\tau} = A\rho\gamma v = A\rho\sqrt{\gamma^2 - 1},$$

we have

$$\boxed{\begin{aligned} a &= \frac{1}{M}\frac{\mathrm{d}p}{\mathrm{d}\tau} = \frac{A\rho}{M}(\gamma^2-1)\left[\sqrt{1+\frac{\xi(2-\xi)}{\gamma^2-1}}-1\right] \\ &\underset{\gamma\to\infty}{\longrightarrow}\frac{1}{2}\xi(2-\xi)\frac{A\rho}{M} \end{aligned}}$$

In MKS units,

$$a_{\mathrm{lim}} = \frac{1}{2}\xi(2-\xi)\frac{A\rho c^2}{M} = \frac{1}{2}\times 0.01\times 1.99\times\frac{10^3\times 10^{-26}\times 9\times 10^{16}}{10^4}\,\mathrm{m/s^2},$$

whence

$$\boxed{a_{\mathrm{lim}} = 9\times 10^{-13}\,\mathrm{m/s^2}\ !!}$$

Now

$$\gamma^2-1=\frac{1}{1-v^2}-1=\frac{v^2}{1-v^2},\qquad 1+\frac{1}{\gamma^2-1}=1+\frac{1-v^2}{v^2}=\frac{1}{v^2}.$$

When $\xi = 1$, we have (see also the solution to Problem 6)

$$\frac{\mathrm{d}v}{1-v^2} = a\mathrm{d}\tau = \frac{A\rho}{M}\frac{v^2}{1-v^2}\left(\frac{1}{v}-1\right)\mathrm{d}\tau,$$

$$\mathrm{d}v = \frac{A\rho}{M}v(1-v)\mathrm{d}\tau,$$

$$\left(\frac{1}{v}+\frac{1}{1-v}\right)\mathrm{d}v = \frac{A\rho}{M}\mathrm{d}\tau,$$

$$\log\frac{v}{v_0}-\log\frac{1-v}{1-v_0}=\frac{A\rho}{M}\tau,\qquad \frac{v}{1-v}=\frac{v_0}{1-v_0}\mathrm{e}^{A\rho\tau/M},$$

$$\left(1+\frac{v_0}{1-v_0}\mathrm{e}^{A\rho\tau/M}\right)v=\frac{v_0}{1-v_0}\mathrm{e}^{A\rho\tau/M},$$

$$\boxed{v=\frac{v_0\mathrm{e}^{A\rho\tau/M}}{1+v_0\left(\mathrm{e}^{A\rho\tau/M}-1\right)}=\frac{1}{1+\frac{1-v_0}{v_0}\mathrm{e}^{-A\rho\tau/M}}}$$

2.1 Accelerated Meter Stick

Let a meter stick be idealized as a line parallel to the x^1-axis in a certain Lorentz frame characterized by coordinates x^μ. The points of the meter stick may be labeled by a single parameter ξ. Let $x^i(\xi, t)$ be the coordinates of the point ξ at the time $t(= x^0)$. Suppose that

$$x^1(\xi, t) = \xi, \qquad 0 \leq \xi \leq 1 \quad \text{(range of meter stick)},$$
$$x^2(\xi, t) = f(t), \quad \text{for all } \xi, \text{ where } \quad |f'(t)| < 1 \text{ for all } t,$$
$$x^3(\xi, t) = 0, \quad \text{for all} \quad \xi \text{ and } t.$$

Under these conditions, the meter stick always appears to be straight and parallel to the x^1-axis and to move in the (x^1, x^2) plane in the x^2 direction according to a law of motion given by the arbitrary function $f(t)$. At least that is how it appears in the present Lorentz frame! Note that all points of the meter stick appear to move in unison in the x^2 direction in this frame. Because the concept of simultaneity is frame-dependent, we may expect it to behave in a different fashion in some other Lorentz frame. Let us see how it behaves in a Lorentz frame that moves with velocity $v(<1)$ in the x^1 direction relative to the present frame. The relevant Lorentz transformation is

$$t = \frac{\bar{t} + v\bar{x}^1}{\sqrt{1 - v^2}}, \qquad x^1 = \frac{v\bar{t} + \bar{x}^1}{\sqrt{1 - v^2}}, \qquad x^2 = \bar{x}^2, \qquad x^3 = \bar{x}^3,$$

which yields

$$\bar{x}^1(\xi, \bar{t}) = \sqrt{1 - v^2}x^1(\xi, t) - v\bar{t} = \sqrt{1 - v^2}\xi - v\bar{t},$$
$$\bar{x}^2(\xi, \bar{t}) = x^2(\xi, t) = f\left(\frac{\bar{t} + v\bar{x}^1(\xi, \bar{t})}{\sqrt{1 - v^2}}\right) = f\left(\sqrt{1 - v^2}\bar{t} + v\xi\right),$$
$$\bar{x}^3(\xi, \bar{t}) = x^3(\xi, t) = 0.$$

The first of these equations shows the meter stick moving in the $\bar{x}^1$ direction with velocity $-v$ and suffering a Lorentz contraction in that direction. The second equation shows the meter stick also moving in the $\bar{x}^2$ direction, at a rate reduced by the time dilation factor. This equation shows, moreover, that the motion is now not in unison. The points having the greater ξ values lead the others. Although the third equation shows that the motion continues to be in a plane, it is not possible to express the new appearance of the meter stick in terms of a simple tilt in this plane. This would be possible only if the function $f(t)$ were linear. More generally, *the meter stick now ceases to appear as a straight line.* But meter sticks do not bend just because we choose to look at them in a new reference frame! Or do they? In order to examine this question, we must study the general problem of rigidity.

2.2 Rigid Motions in Special Relativity

We shall study first the general motion of an arbitrary continuous medium in spacetime. We shall have occasion to consider continuous media several times in these lectures, and therefore the formalism developed here will have a utility extending beyond the present context.

Let the component particles of the medium be labeled by three parameters ξ^i, $i = 1, 2, 3$, and let the world line of particle ξ be given by four functions $x^\mu(\xi, \tau)$, $\mu = 0, 1, 2, 3$, where τ is its proper time. In the general theory of relativity the x^μ may be arbitrary coordinates in curved spacetime, but here we may assume them to be standard coordinates of some Lorentz frame.

Let $\xi^i + \delta\xi^i$ be the labels of a neighboring particle. Its world line is given by the functions

$$x^\mu(\xi + \delta\xi, \tau) = x^\mu(\xi, \tau) + x^\mu_{,i}(\xi, \tau)\delta\xi^i,$$

where the comma followed by a Latin index denotes partial differentiation with respect to the corresponding ξ. The four-vector $x^\mu_{,i}(\xi, \tau)\delta\xi^i$, representing the difference between the two sets of world-line functions, is not generally orthogonal to the world line of ξ. To get such a vector it is necessary to apply the projection tensor on the instantaneous hyperplane of simultaneity:

$$\delta x^\mu \equiv P^\mu_{\ \nu} x^\nu_{,i}\delta\xi^i, \qquad P^{\mu\nu} = \eta^{\mu\nu} + \dot{x}^\mu\dot{x}^\nu,$$

where the dot denotes partial differentiation with respect to τ, and we note that in general relativity the projection tensor will take the form $P^{\mu\nu} = g^{\mu\nu} + \dot{x}^\mu\dot{x}^\nu$, with $g^{\mu\nu}$ the metric tensor of spacetime. It is easy to verify that application of the projection tensor corresponds to a simple proper-time shift of amount

$$\delta\tau = \eta_{\mu\nu}\dot{x}^\mu x^\nu_{,i}\delta\xi^i,$$

so that

$$\delta x^\mu = x^\mu(\xi + \delta\xi, \tau + \delta\tau) - x^\mu(\xi, \tau).$$

The two particles ξ and $\xi + \delta\xi$ appear, in the instantaneous rest frame of either, to be separated by a distance δs given by

$$(\delta s)^2 = (\delta x)^2 = \gamma_{ij}\delta\xi^i\delta\xi^j,$$

where

$$\gamma_{ij} = P_{\mu\nu}x^\mu_{,i}x^\nu_{,j}.$$

The quantity γ_{ij} is called the *proper metric* of the medium. The medium undergoes *rigid motion* if and only if its proper metric is independent of τ. Under rigid motion the instantaneous separation distance between any pair of neighbouring particles is constant in time.

It is sometimes convenient to express the rigid motion condition $\dot{\gamma}_{ij} = 0$ in terms of derivatives with respect to the coordinates x^μ. Just as the x^μ are functions of the ξ^i and τ, so, inversely, may the ξ^i and τ be regarded as functions of the x^μ, at least in the domain of spacetime occupied by the medium. We shall write

$$u^\mu \equiv \dot{x}^\mu, \qquad u^2 = -1, \qquad P_{\mu\nu} = \eta_{\mu\nu} + u_\mu u_\nu.$$

If f is an arbitrary function over the domain occupied by the medium then

$$f_{,\mu} = f_{,i}\xi^i_{,\mu} + \dot{f}\tau_{,\mu},$$

where the comma followed by a Greek index denotes partial differentiation with respect to the corresponding x. We also have

$$\dot{x}\cdot\ddot{x} = 0 \quad \text{or} \quad u\cdot\dot{u} = 0,$$

$$u_\mu u^\mu_{,\nu} = 0, \qquad \dot{u}_\mu = u_{\mu,\nu}u^\nu, \qquad u_\mu u^\mu_{,i} = 0,$$

$$x^\mu_{,i}\xi^i_{,\nu} + \dot{x}^\mu\tau_{,\nu} = \delta^\mu_\nu,$$

$$\xi^i_{,\mu}x^\mu_{,j} = \delta^i_j, \qquad \xi^i_{,\mu}\dot{x}^\mu = 0,$$
$$\tau_{,\mu}x^\mu_{,i} = 0, \qquad \tau_{,\mu}\dot{x}^\mu = 1,$$

$$P_{\mu\nu}\dot{x}^\nu_{,i} = P_{\mu\nu}u^\nu_{,i} = u_{\mu,i}.$$

We now define the *rate-of-strain tensor* for the medium:

$$\begin{aligned}
r_{\mu\nu} &\equiv \dot{\gamma}_{ij}\xi^i_{,\mu}\xi^j_{,\nu} \\
&= \left(\dot{P}_{\sigma\tau}x^\sigma_{,i}x^\tau_{,j} + P_{\sigma\tau}\dot{x}^\sigma_{,i}x^\tau_{,j} + P_{\sigma\tau}x^\sigma_{,i}\dot{x}^\tau_{,j}\right)\xi^i_{,\mu}\xi^j_{,\nu} \\
&= (\dot{u}_\sigma u_\tau + u_\sigma\dot{u}_\tau)(\delta^\sigma_\mu - u^\sigma\tau_{,\mu})(\delta^\tau_\nu - u^\tau\tau_{,\nu}) \\
&\quad + u_{\tau,i}\xi^i_{,\mu}(\delta^\tau_\nu - u^\tau\tau_{,\nu}) + (\delta^\sigma_\mu - u^\sigma\tau_{,\mu})u_{\sigma,j}\xi^j_{,\nu} \\
&= \dot{u}_\mu u_\nu + u_\mu\dot{u}_\nu + \dot{u}_\mu\tau_{,\nu} + \tau_{,\mu}\dot{u}_\nu \\
&\quad + u_{\nu,\mu} - \dot{u}_\nu\tau_{,\mu} + u_{\mu,\nu} - \dot{u}_\mu\tau_{,\nu} \\
&= u_{\mu,\sigma}u^\sigma u_\nu + u_\mu u^\sigma u_{\nu,\sigma} + u_{\nu,\mu} + u_{\mu,\nu} \\
&= P^\sigma_\mu P^\tau_\nu(u_{\sigma,\tau} + u_{\tau,\sigma}).
\end{aligned} \tag{2.1}$$

The rate-of-strain tensor is seen to lie completely in the instantaneous hyperplane of simultaneity. It is the relativistic generalization of the nonrelativistic rate-of-strain tensor

$$r_{ij} = v_{i,j} + v_{j,i},$$

where v_i is a three-velocity field and the differentiation is with respect to ordinary Cartesian coordinates. Let us look for a moment at this tensor. The nonrelativistic condition for rigid motion is

$$r_{ij} = 0 \qquad \text{everywhere.}$$

This equation implies

$$0 = r_{ij,k} = v_{i,jk} + v_{j,ik}, \tag{2.2}$$

$$0 = r_{jk,i} = v_{j,ki} + v_{k,ji}. \tag{2.3}$$

Subtracting (2.3) from (2.2) and making use of the commutativity of partial differentiation, we find

$$v_{i,jk} - v_{k,ji} = 0, \tag{2.4}$$

which, upon permutation of the indices j and k, yields also

$$v_{i,kj} - v_{j,ki} = 0. \tag{2.5}$$

Adding (2.2) and (2.5) we finally get

$$v_{i,jk} = 0,$$

which has the general solution

$$v_i = -\omega_{ij} x_j + \beta_i, \tag{2.6}$$

where ω_{ij} and β_i are functions of time only. The condition $r_{ij} = 0$ constrains ω_{ij} to be antisymmetric, i.e.,

$$\omega_{ij} = -\omega_{ji},$$

and nonrelativistic rigid motion is seen to be, at each instant, a uniform rotation with angular velocity

$$\omega_i = \frac{1}{2}\varepsilon_{ijk}\omega_{jk}$$

about the coordinate origin, superimposed upon a uniform translation with velocity β_i. Because the coordinate origin may be located arbitrarily at each instant, rigid motion may alternatively be described as one in which an arbitrary particle in the medium moves in an arbitrary fashion while at the same time the medium as a whole rotates about this point in an arbitrary (but uniform) fashion. Such a motion has six degrees of freedom.

It turns out that relativistic rigid motion, which is characterized by the condition

$$r_{\mu\nu} = 0 \qquad \text{or} \qquad \dot{\gamma}_{ij} = 0,$$

has only *three* degrees of freedom! Pick an arbitrary particle in the medium and let it be the origin of the labels ξ^i. Let its world line $x^\mu(0,\tau)$ be arbitrary (but timelike). Introduce a local rest frame for the particle, characterized by an orthonormal triad $n_i^\mu(\tau)$:

$$n_i \cdot n_j = \delta_{ij}, \qquad n_i \cdot u_0 = 0, \qquad u_0^2 = -1, \qquad u_0^\mu \equiv \dot{x}^\mu(0,\tau).$$

Then let the world lines of all the other particles of the medium be given by

$$x^\mu(\xi,\tau) = x^\mu(0,\sigma) + \xi^i n_i^\mu(\sigma), \tag{2.7}$$

where σ is a certain function of the ξ^i and τ. To determine this function, write

$$u^\mu = \dot{x}^\mu(\xi,\tau) = (u_0^\mu + \xi^i \dot{n}_i^\mu)\dot{\sigma},$$

all arguments being suppressed in the final expression. Here and in what follows, it is to be understood that dots over u_0 and the n_i denote differentiation with respect to σ, while the dot over σ denotes differentiation with respect to τ. It will be convenient to expand $\dot{n}_i$ in terms of the orthonormal tetrad u_0, n_i:

$$\dot{n}_i^\mu = a_{0i} u_0^\mu + \Omega_{ij} n_j^\mu.$$

The coefficients a_{0i} are determined, from the identity

$$\dot{n}_i \cdot u_0 + n_i \cdot \dot{u}_0 = 0,$$

to be just the components of the absolute acceleration of the particle $\xi = 0$ in its local rest frame:

$$a_{0i} = n_i \cdot \dot{u}_0,$$

and the identity

$$\dot{n}_i \cdot n_j + n_i \cdot \dot{n}_j = 0$$

tells us that Ω_{ij} is antisymmetric:

$$\Omega_{ij} = -\Omega_{ji}.$$

We now have

$$u^\mu = \left[\left(1 + \xi^i a_{0i}\right) u_0^\mu + \xi^i \Omega_{ij} n_j^\mu \right] \dot{\sigma}.$$

But

$$-1 = u^2 = -\left[\left(1 + \xi^i a_{0i}\right)^2 - \xi^i \xi^j \Omega_{ik} \Omega_{jk} \right] \dot{\sigma}^2,$$

whence

$$\dot{\sigma} = \left[\left(1 + \xi^i a_{0i}\right)^2 - \xi^i \xi^j \Omega_{ik} \Omega_{jk} \right]^{-1/2}. \tag{2.8}$$

The right hand side of this equation is a function solely of σ and the ξ^i. Therefore the equation may be integrated along each world line $\xi =$ const., subject, say, to the boundary condition

$$\sigma(\xi, 0) = 0.$$

We shall, in particular, have the necessary condition

$$\sigma(0,\tau) = \tau.$$

We note that the medium must be confined to regions where

$$\left(1 + \xi^i a_{0i}\right)^2 > \xi^i \Omega_{ik} \xi^j \Omega_{jk} \quad (\geq 0).$$

Otherwise, some of its component particles will be moving faster than light.

Let us now compute the proper metric of the medium. We have

$$n_i \cdot u = -\Omega_{ij}\xi^j \dot{\sigma}, \tag{2.9}$$

$$\begin{aligned} x^\mu_{,i} &= n^\mu_i + (u^\mu_0 + \xi^j \dot{n}^\mu_j)\sigma_{,i} = n^\mu_i + u^\mu \dot{\sigma}^{-1}\sigma_{,i}, \\ u_\mu x^\mu_{,i} &= -\Omega_{ij}\xi^j \dot{\sigma} - \dot{\sigma}^{-1}\sigma_{,i}, \end{aligned}$$

$$\begin{aligned} \gamma_{ij} &= P_{\mu\nu} x^\mu_{,i} x^\nu_{,j} \\ &= \delta_{ij} - \Omega_{ik}\xi^k \sigma_{,j} - \Omega_{jk}\xi^k \sigma_{,i} - \dot{\sigma}^{-2}\sigma_{,i}\sigma_{,j} \\ &\quad + \left(\Omega_{ik}\xi^k \dot{\sigma} + \dot{\sigma}^{-1}\sigma_{,i}\right)\left(\Omega_{jl}\xi^l \dot{\sigma} + \dot{\sigma}^{-1}\sigma_{,j}\right) \\ &= \delta_{ij} + \dot{\sigma}^2 \Omega_{ik}\Omega_{jl}\xi^k \xi^l. \end{aligned} \tag{2.10}$$

From this expression and the expression (2.8) for $\dot{\sigma}$ on the preceding page, we see that the only way in which the motion of the medium can be rigid is either for all the Ω_{ij} to vanish or for all the Ω_{ij}, together with the a_{0i}, to be constants, independent of σ. In the latter case the motion is one of a six-parameter family (the Ω_{ij} and the a_{0i} are the parameters) of special motions known as *superhelical motions*, of which we shall study one simple example later (constant rotation about a fixed axis). For the present we concentrate on the case in which all the Ω_{ij} vanish.

2.3 Fermi–Walker Transport

When the Ω_{ij} vanish the triad n^μ_i is said to be *Fermi–Walker transported* along the world line of the particle $\xi = 0$. More generally, any tensor whose components relative to the tetrad u^μ_0, n^μ_i remain constant along the world line $\xi = 0$ is said to be Fermi–Walker transported along that world line. It is sometimes convenient to express the condition for Fermi–Walker transport without reference to the triad n^μ_i. Writing, for a vector A^μ along the world line,

$$A^\mu = A_u u^\mu_0 + A_i n^\mu_i,$$

where

$$A_u = -A \cdot u_0, \qquad A_i a_{0i} = A \cdot \dot{u}_0,$$

we have, if A^μ is Fermi–Walker transported,

$$\dot{A}^{\mu} \doteq A_u \dot{u}_0^{\mu} + A_i a_{0i} u_0^{\mu} = A \cdot \left(\dot{u}_0 u_0^{\mu} - u_0 \dot{u}_0^{\mu}\right),$$

an equation that admits of immediate generalization to tensors of arbitrary rank.

It is not possible to maintain the orientation (in spacetime) of the local-rest-frame triad n_i^{μ} constant along a world line unless that world line is straight. Under Fermi–Walker transport, however, the triad remains as constantly oriented, or as *rotationless*, as possible. The components of the $\dot{n}_i$ all vanish in the instantaneous hyperplane of simultaneity.

For a general non-Fermi–Walker transported triad, the Ω_{ij} are the components of the angular-velocity tensor that describes the instantaneous rate of rotation of the triad in the instantaneous hyperplane of simultaneity. The general motion of the medium introduced in (2.7) on p. 25 may be described formally as one in which the particle $\xi = 0$ moves in an arbitrary fashion and the medium as a whole executes an arbitrary rotation about this particle. But only if the rotation is absent is this motion truly rigid. Rigid motion in special relativity therefore possesses only the three degrees of freedom that the particle $\xi = 0$ itself possesses. Even these three degrees of freedom are not always attainable. In the case of superhelical motion, there are no degrees of freedom at all. Once the medium gets into superhelical motion, it must remain frozen into it if it wants to stay rigid.

Problem 9 A particle undergoes acceleration $\mathrm{d}\boldsymbol{v}/\mathrm{d}t$ in a certain inertial frame. 'Attached' to this particle is a four-vector S^{μ} that is orthogonal to the particle world line and Fermi–Walker transported along this line. The four-vector therefore satisfies the equations

$$S \cdot u = 0, \qquad \dot{S}^{\mu} = (S \cdot \dot{u})u^{\mu},$$

where u is the particle's four-velocity and the dot denotes differentiation with respect to the proper time. Instead of dealing with S^{μ}, it is often convenient to work with the three-vector part of

$$\overline{S}^{\mu} = L^{\mu}_{\ \nu} S^{\nu},$$

where $L^{\mu}_{\ \nu}$ is the Lorentz boost transformation to the local rest frame of the particle:

$$(L^{\mu}_{\ \nu}) = \begin{pmatrix} \gamma & -\gamma \boldsymbol{v} \\ -\gamma \boldsymbol{v} & \underline{1} + (\gamma - 1)\hat{\boldsymbol{v}}\hat{\boldsymbol{v}} \end{pmatrix},$$

where

$$\gamma = (1 - v^2)^{-1/2}, \qquad \hat{\boldsymbol{v}} = \boldsymbol{v}/|v|, \qquad \underline{1} = \text{unit dyadic.}$$

Show that the boost transformation is indeed a Lorentz transformation and that the inverse transformation $L^{-1\mu}_{\ \nu}$ back to the original frame is obtained from $L^{\mu}_{\ \nu}$ by making the replacement $\boldsymbol{v} \to -\boldsymbol{v}$. Show that in the boosted frame (rest frame), we have $\overline{S}^0 = 0$ and

$$\overline{u}^{\mu} = L^{\mu}_{\nu} u^{\nu} = (1, 0, 0, 0).$$

In the boosted frame, we may write

$$\dot{\overline{S}}^{\mu} = \dot{L}^{\mu}_{\nu} S^{\nu} + L^{\mu}_{\nu} \dot{S}^{\nu} = \dot{L}^{\mu}_{\nu} L^{-1\nu}_{\sigma} \overline{S}^{\sigma} + (S \cdot \dot{u}) \overline{u}^{\mu},$$

of which the three-vector part reduces to

$$\dot{\overline{S}}^{i} = \dot{L}^{i}_{\mu} L^{-1\mu}_{j} \overline{S}^{j}.$$

By straight forwardly computing $\dot{L}^{i}_{\mu} L^{-1\mu}_{j}$, show that this equation may be rewritten in the three-vector language

$$\frac{\mathrm{d}\overline{\boldsymbol{S}}}{\mathrm{d}t} = \boldsymbol{\Omega} \times \overline{\boldsymbol{S}},$$

and obtain an expression for $\boldsymbol{\Omega}$ in terms of γ, $\boldsymbol{v}$, and $\mathrm{d}\boldsymbol{v}/\mathrm{d}t$. Suppose that the particle moves with constant angular velocity ω around a circle of radius a. Obtain an expression for the precession frequency $|\boldsymbol{\Omega}|$ of the three-vector $\overline{\boldsymbol{S}}$ under these circumstances, and show that the precession is retrograde.

Solution 9 Note first that

$$\begin{pmatrix} \gamma & -\gamma \boldsymbol{v} \\ -\gamma \boldsymbol{v} & \underline{1} + (\gamma - 1)\hat{\boldsymbol{v}}\hat{\boldsymbol{v}} \end{pmatrix} \begin{pmatrix} \gamma & \gamma \boldsymbol{v} \\ \gamma \boldsymbol{v} & \underline{1} + (\gamma - 1)\hat{\boldsymbol{v}}\hat{\boldsymbol{v}} \end{pmatrix}$$

$$= \begin{pmatrix} \gamma^2 - \gamma^2 v^2 & \gamma^2 \boldsymbol{v} - \gamma \boldsymbol{v} - \gamma(\gamma - 1)\boldsymbol{v} \\ -\gamma^2 \boldsymbol{v} + \gamma \boldsymbol{v} + \gamma(\gamma - 1)\boldsymbol{v} & -\gamma^2 \boldsymbol{v}\boldsymbol{v} + \underline{1} + 2(\gamma - 1)\hat{\boldsymbol{v}}\hat{\boldsymbol{v}} + (\gamma - 1)^2 \hat{\boldsymbol{v}}\hat{\boldsymbol{v}} \end{pmatrix}$$

$$= \begin{pmatrix} 1 & 0 \\ 0 & \underline{1} \end{pmatrix},$$

because

$$-v^2\gamma^2 + 2(\gamma - 1) + (\gamma - 1)^2 = (1 - v^2)\gamma^2 - 1 = 0.$$

Therefore, $L^{-1\mu}_{\nu}$ is indeed obtained by making the replacement $\boldsymbol{v} \rightarrow -\boldsymbol{v}$. Now the condition that $\dot{L}^{\mu}_{\nu}$ be a Lorentz transformation may be expressed in the form

$$\eta_{\mu\nu} L^{\mu}_{\sigma} L^{\nu}_{\tau} = \eta_{\sigma\tau},$$

or, dropping indices,

$$L^{\mathrm{T}} \eta L = \eta, \qquad L^{\mathrm{T}} \eta = \eta L^{-1}, \qquad L = \eta L^{-1\mathrm{T}} \eta,$$

where the superscript T denotes transpose. But, in virtue of the form of η and the symmetry of L^{-1} in the present case, the last equation is obviously satisfied.

To show that $\overline{S}^0$ vanishes, we first show that

$$(\overline{u}^\mu) = (L^\mu_\nu u^\nu) = \begin{pmatrix} \gamma & -\gamma \boldsymbol{v} \\ -\gamma \boldsymbol{v} & \underline{1} + (\gamma - 1)\hat{\boldsymbol{v}}\hat{\boldsymbol{v}} \end{pmatrix} \begin{pmatrix} \gamma \\ \gamma \boldsymbol{v} \end{pmatrix}$$

$$= \begin{pmatrix} \gamma^2(1 - v^2) \\ [-\gamma^2 + \gamma + \gamma(\gamma - 1)]\boldsymbol{v} \end{pmatrix} = \begin{pmatrix} 1 \\ 0 \end{pmatrix}.$$

From this it follows immediately that

$$0 = S \cdot u = \overline{S} \cdot \overline{u} = -\overline{S}^0.$$

Finally,

$$\left(\dot{L}^i_\mu\right) = \left(-\gamma^3(\boldsymbol{v} \cdot \dot{\boldsymbol{v}})\boldsymbol{v} - \gamma\dot{\boldsymbol{v}},\ \gamma^3(\boldsymbol{v} \cdot \dot{\boldsymbol{v}})\hat{\boldsymbol{v}}\hat{\boldsymbol{v}} + (\gamma - 1)\left(\frac{\dot{\boldsymbol{v}}\boldsymbol{v}}{v^2} + \frac{\boldsymbol{v}\dot{\boldsymbol{v}}}{v^2} - 2\frac{\boldsymbol{v} \cdot \dot{\boldsymbol{v}}}{v^2}\hat{\boldsymbol{v}}\hat{\boldsymbol{v}}\right)\right),$$

$$\left(L_j^{-1\mu}\right)\begin{pmatrix} \gamma \boldsymbol{v} \\ \underline{1} + (\gamma - 1)\hat{\boldsymbol{v}}\hat{\boldsymbol{v}} \end{pmatrix},$$

$$\begin{aligned}
\left(\dot{L}^i_\mu L_j^{-1\mu}\right) &= -\gamma^4(\boldsymbol{v} \cdot \dot{\boldsymbol{v}})\boldsymbol{v}\boldsymbol{v} - \gamma^2\dot{\boldsymbol{v}}\boldsymbol{v} + \gamma^3(\boldsymbol{v} \cdot \dot{\boldsymbol{v}})\hat{\boldsymbol{v}}\hat{\boldsymbol{v}} \\
&\quad + (\gamma - 1)\left(\frac{\dot{\boldsymbol{v}}\boldsymbol{v}}{v^2} + \frac{\boldsymbol{v}\dot{\boldsymbol{v}}}{v^2} - 2\frac{\boldsymbol{v} \cdot \dot{\boldsymbol{v}}}{v^2}\hat{\boldsymbol{v}}\hat{\boldsymbol{v}}\right) + \gamma^3(\gamma - 1)(\boldsymbol{v} \cdot \dot{\boldsymbol{v}})\hat{\boldsymbol{v}}\hat{\boldsymbol{v}} \\
&\quad + (\gamma - 1)^2\left(\frac{\dot{\boldsymbol{v}}\boldsymbol{v}}{v^2} + \frac{\boldsymbol{v}\dot{\boldsymbol{v}}}{v^2}\hat{\boldsymbol{v}}\hat{\boldsymbol{v}} - 2\frac{\boldsymbol{v} \cdot \dot{\boldsymbol{v}}}{v^2}\hat{\boldsymbol{v}}\hat{\boldsymbol{v}}\right) \\
&= \left[-\gamma^4 v^2 + \gamma^3 - 2\frac{\gamma - 1}{v^2} + \gamma^3(\gamma - 1) - \frac{(\gamma - 1)^2}{v^2}\right](\boldsymbol{v} \cdot \dot{\boldsymbol{v}})\hat{\boldsymbol{v}}\hat{\boldsymbol{v}} \\
&\quad + \left[-\gamma^2 + \frac{\gamma - 1}{v^2} + \frac{(\gamma - 1)^2}{v^2}\right]\dot{\boldsymbol{v}}\boldsymbol{v} + \frac{\gamma - 1}{v^2}\boldsymbol{v}\dot{\boldsymbol{v}} \\
&= \left(\gamma^2 - \frac{\gamma^2 - 1}{v^2}\right)(\boldsymbol{v} \cdot \dot{\boldsymbol{v}})\hat{\boldsymbol{v}}\hat{\boldsymbol{v}} + (-\gamma^2 + 1 + \gamma^2 - \gamma)\frac{\dot{\boldsymbol{v}}\boldsymbol{v}}{v^2} + \frac{\gamma - 1}{v^2}\boldsymbol{v}\dot{\boldsymbol{v}} \\
&= \frac{\gamma - 1}{v^2}(\boldsymbol{v}\dot{\boldsymbol{v}} - \dot{\boldsymbol{v}}\boldsymbol{v}),
\end{aligned}$$

whence

$$\dot{\overline{\boldsymbol{S}}} = \frac{\gamma - 1}{v^2}(\boldsymbol{v}\dot{\boldsymbol{v}} - \dot{\boldsymbol{v}}\boldsymbol{v}) \cdot \overline{\boldsymbol{S}},$$

or

$$\frac{\mathrm{d}\overline{\boldsymbol{S}}}{\mathrm{d}t} = \boldsymbol{\Omega} \times \overline{\boldsymbol{S}},$$

where

$$\boxed{\Omega = -\frac{\gamma-1}{v^2}\boldsymbol{v} \times \frac{d\boldsymbol{v}}{dt} = -(\gamma - 1)\hat{\boldsymbol{v}} \times \frac{d\hat{\boldsymbol{v}}}{dt}}$$

In the case of the particle moving in a circle, we have

$$\frac{d\boldsymbol{v}}{dt} = \boldsymbol{\omega} \times \boldsymbol{v}, \qquad \boldsymbol{\omega} \cdot \boldsymbol{v} = 0,$$

$$\Omega = -\frac{\gamma - 1}{v^2}\boldsymbol{v} \times (\boldsymbol{\omega} \times \boldsymbol{v}) = -\frac{\gamma - 1}{v^2}\left[v^2\boldsymbol{\omega} - (\boldsymbol{\omega} \cdot \boldsymbol{v})\boldsymbol{v}\right],$$

so that

$$\boxed{\Omega = -(\gamma - 1)\boldsymbol{\omega}}$$

where

$$|\boldsymbol{v}| = a|\boldsymbol{\omega}|, \qquad \boxed{\gamma = (1 - a^2\omega^2)^{-1/2}}$$

2.4 Flat Proper Geometry

When the Ω_{ij} vanish the *proper geometry* of the medium is *flat* [see (2.10) on p. 26]:

$$\gamma_{ij} = \delta_{ij}.$$

Moreover, we have [see (2.9) on p. 26]

$$n_i \cdot u = 0,$$

so that the instantaneous hyperplane of simultaneity of the particle at $\xi = 0$ is an instantaneous hyperplane of simultaneity for all the other particles of the medium as well, and the triad n_i^μ serves to define a rotationless rest frame for the whole medium. In other words, the coordinate system defined by the parameters ξ^i may itself be regarded as being Fermi–Walker transported, and all the particles of the medium have a common designator of simultaneity in the parameter σ. Because σ is not generally equal to τ, however, it is not possible for the particles to have a common synchronization of standard clocks. The relation between σ and τ is given by (2.8) on p. 25 as

$$\dot{\sigma} = \left(1 + \xi^i a_{0i}\right)^{-1},$$

which permits us to compute the absolute acceleration a_i of an arbitrary particle in terms of a_{0i} and the ξ^i:

$$\begin{aligned} a_i &= n_i \cdot \dot{u} = n_i \cdot \frac{\partial u}{\partial \sigma}\dot{\sigma} = \dot{\sigma} n_i \cdot \frac{\partial}{\partial \sigma}\left[\left(1 + \xi^j a_{0j}\right) u_0 \dot{\sigma}\right] \\ &= \dot{\sigma}^2 (1 + \xi^j a_{0j}) n_i \cdot \dot{u}_0 \\ &= \frac{a_{0i}}{1 + \xi^j a_{0j}}. \end{aligned}$$

We see that, although the motion is rigid and 'rotationless', not all parts of the medium 'feel' the same acceleration.

When the Ω_{ij} vanish it is sometimes convenient to make use of σ and ξ^i as coordinates of spacetime. In these coordinates, the metric tensor takes the form

$$g_{00} = \left.\frac{\partial x^\mu}{\partial \sigma}\right|_\xi \left.\frac{\partial x^\nu}{\partial \sigma}\right|_\xi \eta_{\mu\nu} = u^2 \dot{\sigma}^{-2} = -(1 + \xi^i a_{0i})^2,$$

$$g_{i0} = g_{0i} = \left.\frac{\partial x^\mu}{\partial \xi^i}\right|_\sigma \left.\frac{\partial x^\nu}{\partial \sigma}\right|_\xi \eta_{\mu\nu} = (n_i \cdot u)\dot{\sigma}^{-1} = 0,$$

$$g_{ij} = \left.\frac{\partial x^\mu}{\partial \xi^i}\right|_\sigma \left.\frac{\partial x^\nu}{\partial \xi^j}\right|_\sigma \eta_{\mu\nu} = n_i \cdot n_j = \delta_{ij},$$

which has a simple diagonal structure. We note that this metric becomes *static*, i.e., time-independent, with the parameter σ now playing the role of 'time', in the special case in which the acceleration of each particle is constant.

Problem 10 The Rotationless Constantly Accelerating Medium

Suppose the particle at $\xi = 0$ undergoes constant absolute acceleration from rest in the x^1 direction in some inertial frame. One may choose initial conditions in such a way that this motion takes the form

$$x^0(0,\sigma) = \frac{1}{a}\sinh a\sigma, \qquad x^1(0,\sigma) = \frac{1}{a}\cosh a\sigma, \qquad x^2(0,\sigma) = 0 = x^3(0,\sigma).$$

Introduce a convenient Fermi–Walker transported triad with which to define the local rest frame of the particle, and let the spacetime coordinates of the remaining particles of the medium be defined, in terms of the σ and the ξ^i, as above. Obtain σ as a function of τ under the boundary condition $\sigma = 0$ when $\tau = 0$. Obtain also explicit forms for the functions $x^\mu(\xi, \tau)$ as well as the metric of spacetime in the coordinate system σ, ξ^i. Draw a flow diagram in the (x^0, x^1) plane, showing the world lines of the particles of the medium. Draw on this diagram some instantaneous hyperplanes of simultaneity and indicate the maximum region of spacetime accessible to the medium (Fig. 2.1).

Solution 10 We have

$$u_0^0 = \cosh a\sigma, \qquad u_0^1 = \sinh a\sigma, \qquad u_0^2 = 0 = u_0^3.$$

We may evidently choose

Fig. 2.1 Flow diagram of a rigid medium in inertial coordinates

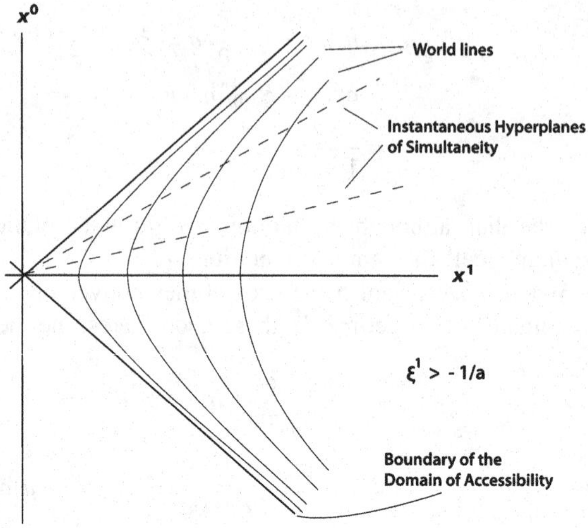

$$\boxed{\begin{array}{lll} n_1^0 = \sinh a\sigma & n_2^0 = 0 & n_3^0 = 0 \\ n_1^1 = \cosh a\sigma & n_2^1 = 0 & n_3^1 = 0 \\ n_1^2 = 0 & n_2^2 = 1 & n_3^2 = 0 \\ n_1^3 = 0 & n_2^3 = 0 & n_3^3 = 1 \end{array}}$$

$$a_{01} = n_1 \cdot \dot{u}_0 = a, \qquad a_{02} = n_2 \cdot \dot{u}_0 = 0, \qquad a_{03} = n_3 \cdot \dot{u}_0 = 0,$$

$$\dot{\sigma} = (1 + a\xi^1)^{-1}, \qquad \boxed{\sigma = \tfrac{\tau}{1+a\xi^1}}$$

$$\boxed{\begin{aligned} x^0(\xi,\tau) &= \frac{1}{a}\sinh a\sigma + \xi^1 \sinh a\sigma = \frac{1+a\xi^1}{a}\sinh\frac{a\tau}{1+a\xi^1} \\ x^1(\xi,\tau) &= \frac{1}{a}\sigma + \xi^1 \cosh a\sigma = \frac{1+a\xi^1}{a}\cosh\frac{a\tau}{1+a\xi^1} \\ x^2(\xi,\tau) &= \xi^2 \\ x^3(\xi,\tau) &= \xi^3 \end{aligned}}$$

$$\boxed{\left(g_{\mu\nu}\right)_{\sigma,\xi} = \operatorname{diag}\left(-(1+a\xi^1)^2, 1, 1, 1\right)}$$

2.5 Constant Rotation About a Fixed Axis

The simplest example of a medium undergoing rigid rotation is obtained by choosing

$$a_{0i} = 0, \qquad \Omega_{12} = \omega, \qquad \Omega_{23} = 0 = \Omega_{31}.$$

The world line of the particle at $\xi = 0$ is then straight, but the world lines of all the other particles are helices of constant pitch. We have

$$\dot{\sigma} = \left\{1 - \omega^2\left[(\xi^1)^2 + (\xi^2)^2\right]\right\}^{-1/2}$$

and the proper metric of the medium takes the form

$$(\gamma_{ij}) = \begin{pmatrix} 1 + (\dot{\sigma}\omega\xi^2)^2 & -(\dot{\sigma}\omega)^2\xi^1\xi^2 & 0 \\ -(\dot{\sigma}\omega)^2\xi^1\xi^2 & 1 + (\dot{\sigma}\omega\xi^1)^2 & 0 \\ 0 & 0 & 1 \end{pmatrix}.$$

It is convenient to relabel the particles by means of three new coordinates r, θ, z given by

$$\xi^1 = r\cos\theta, \qquad \xi^2 = r\sin\theta, \qquad \xi^3 = z.$$

In terms of these coordinates the proper distance δs between two particles separated by displacements δr, $\delta\theta$, and δz takes the form

$$\delta s^2 = (\delta r)^2 + \frac{r^2}{1 - \omega^2 r^2}(\delta\theta)^2 + (\delta z)^2.$$

The second term on the right of this equation may be understood as arising from the Lorentz contraction phenomenon. The problem of the rotating medium is sometimes posed as the so-called 'spinning disc paradox' and stated as follows. A disc of radius r is set spinning with angular frequency ω about its axis. Radial distances are unaffected, but distances in the direction of rotation become Lorentz contracted. In particular, the circumference of the disc gets reduced to the value $2\pi R\sqrt{1 - \omega^2 R^2}$. But this contradicts the Euclidean nature of the ordinary three-space that the disc inhabits!

What in fact happens is that, when set in rotation, the disc must suffer a strain that arises for kinematic reasons quite apart from any strains it suffers on account of centrifugal forces. In particular, it must undergo a stretching of amount $(1 - \omega^2 r^2)^{-1/2}$ in the direction of rotation, to compensate the Lorentz contraction factor $(1 - \omega^2 r^2)^{1/2}$ that appears when the disc is viewed in the inertial rest frame of its axis, thereby maintaining the Euclidean nature of three-space. It is this stretching factor that appears in the proper metric of the medium.

We note that the medium must be confined to regions where $r < \omega^{-1}$ and that its motion will not be rigid if ω varies with time. We note also that the proper geometry of the medium is *not flat*.

Problem 11 Verify that the proper metric of the rotating medium takes the form

$$\mathrm{diag}\left(1, \frac{r^2}{1-\omega^2 r^2}, 1\right)$$

with respect to the coordinates r, θ, z.

Solution 11 We have

$$\dot{\sigma}^2 = \frac{1}{1-\omega^2 r^2}.$$

Hence,

$$\begin{aligned}
\gamma_{rr} &= \frac{\partial \xi^i}{\partial r}\frac{\partial \xi^j}{\partial r}\gamma_{ij} \\
&= \cos^2\theta\left[1 + (\dot{\sigma})^2 \sin^2\theta\right] - 2(\dot{\sigma})^2 \sin^2\theta\cos^2\theta \\
&\quad + \sin^2\theta\left[1 + (\dot{\sigma}\omega r)^2 \cos^2\theta\right] \\
&= 1,
\end{aligned}$$

$$\begin{aligned}
\gamma_{r\theta} = \gamma_{\theta r} &= \frac{\partial \xi^i}{\partial r}\frac{\partial \xi^j}{\partial \theta}\gamma_{ij} \\
&= -r\sin\theta\cos\theta\left[1 + (\dot{\sigma})^2 \sin^2\theta\right] - r(\dot{\sigma})^2 \sin\theta\cos^3\theta \\
&\quad + r(\dot{\sigma})^2 \sin^3\theta\cos\theta + r\sin\theta\cos\theta\left[1 + (\dot{\sigma})^2\cos^2\theta\right] \\
&= 0,
\end{aligned}$$

$$\gamma_{rz} = \gamma_{zr} = \frac{\partial \xi^i}{\partial r}\frac{\partial \xi^j}{\partial z}\gamma_{ij} = 0, \langle \varphi_k \rangle_\phi = \langle \Im(\ln a_k)\rangle_\phi$$

$$\begin{aligned}
\gamma_{\theta\theta} &= \frac{\partial \xi^i}{\partial \theta}\frac{\partial \xi^j}{\partial \theta}\gamma_{ij} \\
&= r^2\sin^2\theta\left[1 + (\dot{\sigma}\omega r)^2 \sin^2\theta\right] + 2r^2(\dot{\sigma}\omega r)^2 \sin^2\theta\cos^2\theta \\
&\quad + r^2\cos^2\theta\left[1 + (\dot{\sigma})^2\cos^2\theta\right] \\
&= r^2\left[1 + (\dot{\sigma})^2\right] = r^2\left(1 + \frac{\omega^2 r^2}{1-\omega^2 r^2}\right) = \frac{r^2}{1-\omega^2 r^2}, \\
\gamma_{\theta z} = \gamma_{z\theta} &= \frac{\partial \xi^i}{\partial \theta}\frac{\partial \xi^j}{\partial z}\gamma_{ij} = 0, \\
\gamma_{zz} &= \frac{\partial \xi^i}{\partial z}\frac{\partial \xi^j}{\partial z}\gamma_{ij} = 1.
\end{aligned}$$

Problem 12 Show that the motion defined by the functions

$$x^0(\xi,\tau) = \frac{1+a\xi^3}{a}\sinh a\sigma,$$
$$x^1(\xi,\tau) = \xi^1 + v\sigma,$$
$$x^2(\xi,\tau) = \xi^2,$$
$$x^3(\xi,\tau) = \frac{1+a\xi^3}{a}\cosh a\sigma,$$

where a and v are constants, is rigid. Here σ is a function of τ and the ξ^i, having a form such that τ is the proper time along each world line $\xi = \text{const}$. Obtain the proper metric of the flow and state (with arguments) whether the proper geometry is flat.

Solution 12 We have

$$\dot{x}^0(\xi,\tau) = (1+a\xi^3)\dot{\sigma}\cosh a\sigma,$$
$$\dot{x}^1(\xi,\tau) = v\dot{\sigma},$$
$$\dot{x}^2(\xi,\tau) = 0,$$
$$\dot{x}^3(\xi,\tau) = (1+a\xi^3)\dot{\sigma}\sinh a\sigma,$$

whence

$$-1 = u^2 = -\left[(1+a\xi^3)^2 - v^2\right]\dot{\sigma}^2, \qquad \dot{\sigma} = \left[(1+a\xi^3)^2 - v^2\right]^{-1/2}.$$

The medium is confined to the region

$$\xi^3 > -\frac{1-v}{a}.$$

Now

$$\left(x^\mu_{,1}\right) = (0,1,0,0), \qquad \left(x^\mu_{,2}\right) = (0,0,1,0),$$

$$\left(x^\mu_{,3}\right) = \left(\sinh a\sigma + (1+a\xi^3)\sigma_{,3}\cosh a\sigma,\ v\sigma_{,3},\ 0,\ \cosh a\sigma + (1+a\xi^3)\sigma_{,3}\sinh a\sigma\right),$$

$$u\cdot x_{,1} = v\dot{\sigma}, \qquad u\cdot x_{,2} = 0,$$

$$\begin{aligned} u\cdot x_{,3} &= -(1+a\xi^3)\dot{\sigma}\sinh a\sigma\cosh a\sigma - (1+a\xi^3)^2\dot{\sigma}\sigma_{,3}\cosh^2 a\sigma \\ &\quad + v^2\dot{\sigma}\sigma_{,3} + (1+a\xi^3)\sinh a\sigma\cosh a\sigma + (1+a\xi^3)^2\dot{\sigma}\sigma_{,3}\sinh^2 a\sigma \\ &= -\left[(1+a\xi^3)^2 - v^2\right]\dot{\sigma}\sigma_{,3} = -\dot{\sigma}^{-1}\sigma_{,3}, \end{aligned}$$

$$\gamma_{11} = x_{,1} \cdot x_{,1} + (u \cdot x_{,1})^2 = 1 + v^2 \dot{\sigma}^2 = 1 + \frac{v^2}{(1 + a\xi^3)^2 - v^2}$$
$$= \frac{(1 + a\xi^3)^2}{(1 + a\xi^3)^2 - v^2},$$

whence

$$\boxed{\gamma_{11} = (1 - \overline{v}^2)^{-1}, \quad \overline{v} \equiv \tfrac{v}{1 + a\xi^3}}$$

$$\gamma_{12} = \gamma_{21} = x_{,1} \cdot x_{,2} + (u \cdot x_{,1})(u \cdot x_{,2}) \boxed{= 0},$$
$$\gamma_{13} = \gamma_{31} = x_{,1} \cdot x_{,3} + (u_{,1})(u \cdot x_{,3}) = v\sigma_{,3} - v\sigma_{,3} \boxed{= 0},$$
$$\gamma_{22} = x_{,2} \cdot x_{,2} + (u \cdot x_{,2})^2 \boxed{= 1},$$
$$\gamma_{23} = \gamma_{32} = x_{,2} \cdot x_{,3} + (u_{,2})(u \cdot x_{,3}) \boxed{= 0},$$
$$\begin{aligned}\gamma_{33} &= x_{,3} \cdot x_{,3} + (u_{,3})^2 \\ &= -\left[\sinh a\sigma + (1 + a\xi^3)\sigma_{,3} \cosh a\sigma\right]^2 + (v\sigma_{,3})^2 \\ &\quad + \left[\cosh a\sigma + (1 + a\xi^3)\sigma_{,3}\sigma\right]^2 + (\dot{\sigma}^{-1}\sigma_{,3})^2 \\ &= 1 - \left[(1 + a\xi^3)^2 - v^2 - \dot{\sigma}^{-2}\right](\sigma_{,3})^2 \boxed{= 1},\end{aligned}$$

or

$$\boxed{(\gamma_{ij}) = \mathrm{diag}\left(\tfrac{1}{1 - \overline{v}^2}, 1, 1\right)}$$

The stretching factor $1/(1 - \overline{v})^2$, although in only one direction, prevents the proper geometry from being flat.

2.6 Irrotational Flow

When the Ω_{ij} are nonvanishing, there exists no global hypersurface of simultaneity. This is a general property of rotational motion, whether rigid or not. The motion, or flow, of a fluid medium is said to be *irrotational* if and only if there exists a family of hypersurfaces that cut the world lines of all the particles of the medium orthogonally. In order that such a family exist, one must be able to write

$$u_\mu = \lambda \phi_{,\mu} \qquad (2.11)$$

for some scalar function ϕ. Here λ is a normalizing factor and the hypersurfaces ϕ = const. are global hypersurfaces of simultaneity. We have

$$\lambda = \left(-\phi_{,\mu}\phi^{\mu}_{,}\right)^{-1/2},$$

$$\lambda_{,\mu} = \left(-\phi_{,\mu}\phi^{\mu}_{,}\right)^{-3/2}\phi^{\nu}_{,}\phi_{,\nu\mu} = \lambda^2 u^{\nu}\phi_{,\nu\mu},$$

$$\phi_{,\mu\nu} = -\lambda^{-2}u_{\mu}\lambda_{,\nu} + \lambda^{-1}u_{\mu,\nu},$$

$$u_{\mu,\nu} = \lambda P^{\sigma}_{\mu}\phi_{,\sigma\nu}, \qquad P^{\sigma}_{\mu}P^{\tau}_{\nu}u_{\sigma,\tau} = \lambda P^{\sigma}_{\mu}P^{\tau}_{\nu}\phi_{,\sigma\tau}.$$

Because of the commutativity of partial differentiation, we have, as the necessary and sufficient condition for the integrability of (2.11),

$$P^{\sigma}_{\mu}P^{\tau}_{\nu}(u_{\sigma,\tau} - u_{\tau,\sigma}) = 0.$$

This is the condition for irrotational flow. Its nonrelativistic analog is (see p. 23)

$$v_{i,j} - v_{j,i} = 0,$$

which implies the existence of a scalar function ϕ such that

$$v_i = \phi_{,i}.$$

A nonrelativistic motion that is both rigid and irrotational reduces simply to [see (2.6) on p. 24]

$$v_i = \beta_i,$$

where β_i is a function of time alone.

Chapter 3
Realizations of Continuous Groups

General relativity replaces the flat spacetime of special relativity by a curved Riemannian manifold and extends the *invariance group* of the theory (i.e., the group of transformations that leave the forms of all dynamical equations invariant) from the Poincaré group to the group of general differentiable coordinate transformations, known to mathematicians as the *diffeomorphism group*. It will be helpful, in introducing the formal apparatus of general relativity, to develop first a little of the theory of groups of continuous transformations.

A continuous group, in the abstract, is a group whose elements may be regarded as points in a differentiable manifold. Moreover, if ξ_1 and ξ_2 are two group elements then their product $\xi_1\xi_2$, as a point in the manifold, must depend in a differentiable way on the points ξ_1 and ξ_2. In any sufficiently small region of the group (as in any differentiable manifold) a coordinate patch may be laid down which attaches labels ξ^a to the group elements ξ. The index a may come from a discrete finite set, in which case the group manifold is an ordinary finite dimensional manifold, or it may come from a continuous set, in which case the group is infinite dimensional. In the latter case the index a will typically stand for a collection of labels, some discrete and some denoting points in another (finite dimensional) manifold. When a comes from a continuous set the words 'derivative', 'differentiation' and 'differentiable' mean 'functional derivative', 'functional differentiation' and 'functionally differentiable', and the summation convention for repeated indices is extended to imply integration over the continuous set.

Continuous groups are generally encountered not in the abstract but as transformation groups in which each group element is envisaged as inducing a certain diffeomorphism in some other manifold. Let ϕ be a point of this other manifold. A coordinate patch may be laid down which attaches labels ϕ^i to this point. If ξ is sufficiently close to the identity it will induce a transformation $\phi \to \overline{\phi}$ such that $\overline{\phi}$ remains within the coordinate patch. In terms of coordinate labels, the transformation may be expressed in the explicit form

B. DeWitt, *Bryce DeWitt's Lectures on Gravitation*,
Edited by Steven M. Christensen, Lecture Notes in Physics, 826,
DOI: 10.1007/978-3-540-36911-0_3, © Springer-Verlag Berlin Heidelberg 2011

$$\overline{\phi}^i = \Phi^i(\xi, \phi),$$

where the Φ^i are differentiable functions of ξ and ϕ. We remark that the index i, like the index a, may come from either a discrete set or a continuous set. As an example from the first category, let the ϕ^i be polar angles on a sphere on which the rotation group acts. As an example from the second category, let the ϕ^i be the values of all components of some field (e.g., an electromagnetic field) at all points of spacetime, and let the group be either the Poincaré group or the full diffeomorphism group.

The functions Φ^i must satisfy the identities

$$\Phi^i(\xi_2, \Phi(\xi_1, \phi)) \equiv \Phi^i(\xi_2\xi_1, \phi), \quad \Phi^i(\text{identity}, \phi) = \phi^i.$$

They are then said to provide a realization of the group. In the special case in which the Φ^i are linear homogeneous in the ϕ^i, the realization is called a (matrix) *representation*. In these lectures, it will usually suffice to confine our attention to a single coordinate patch containing the identity in the group manifold. It is then convenient to take the origin of coordinates at the identity element and to expand the functions Φ^i as polynomials in the ξ, plus remainders:

$$\Phi^i(\xi, \phi) = \phi^i + \Phi^i_{a}(\phi)\xi^a + \frac{1}{2}\Phi^i_{ab}(\phi)\xi^a\xi^b + O(\xi^3).$$

Consider now the following successive transformations:

$$\begin{aligned}
\overline{\phi}^i &= \Phi^i(\xi_1, \phi),\\
\overline{\overline{\phi}}^i &= \Phi^i(\xi_2, \overline{\phi}) = \Phi^i(\xi_2\xi_1, \phi),\\
\overline{\overline{\overline{\phi}}}^i &= \Phi^i(\xi_1^{-1}, \overline{\overline{\phi}}) = \Phi^i(\xi_1^{-1}\xi_2\xi_1, \phi),\\
\overline{\overline{\overline{\overline{\phi}}}}^i &= \Phi^i(\xi_2^{-1}, \overline{\overline{\overline{\phi}}}) = \Phi^i(\xi_2^{-1}\xi_1^{-1}\xi_2\xi_1, \phi).
\end{aligned}$$

Expanding the first two equations to second order in the ξ, we find

$$\begin{aligned}
\overline{\overline{\phi}}^i &= \overline{\phi}^i + \Phi^i_{a}(\overline{\phi})\xi_2^a + \frac{1}{2}\Phi^i_{ab}(\overline{\phi})\xi_2^a\xi_2^b + O(\xi^3)\\
&= \phi^i + \Phi^i_{a}(\phi)\xi_1^a + \frac{1}{2}\Phi^i_{ab}(\phi)\xi_1^a\xi_1^b\\
&\quad + \Phi^i_{a}(\phi)\xi_2^a + \Phi^i_{a,j}(\phi)\Phi^j_{b}(\phi)\xi_1^b\xi_2^a + \frac{1}{2}\Phi^i_{ab}(\phi)\xi_2^a\xi_2^b + O(\xi^3),
\end{aligned}$$

the comma followed by a Latin index denoting differentiation with respect to a ϕ. In the special case $\xi_2 = \xi_1^{-1}$, we may infer from this

$$\Phi^i_{a}(\xi^a + \xi^{-1a}) + \frac{1}{2}\Phi^i_{ab}(\xi^a\xi^b + \xi^{-1a}\xi^{-1b}) + \Phi^i_{a,j}\Phi^j_{b}\xi^b\xi^{-1a} = O(\xi^3),$$

for all ϕ, suppressing the argument ϕ in the expression. If the realization is faithful, as we shall always assume it to be, we have

$$\Phi^i(\xi_1,\phi) = \Phi^i(\xi_2,\phi) \quad \text{for all } \phi \text{ if and only if } \xi_1 = \xi_2,$$

whence, as corollaries,

$$\boxed{\Phi^i_{,a}\xi^a = 0 \quad \text{forall } \phi \text{ if and only if } \xi^a = 0 \text{ for all } a}$$

$$\boxed{\xi^{-1a} = -\xi^a + O(\xi^2) \quad \text{forall } \xi \text{ near the identity}}$$

Continuing now, we have

$$\begin{aligned}
\overline{\overline{\overline{\phi}}}^i &= \overline{\overline{\phi}}^i + \Phi^i_{,a}\left(\overline{\overline{\phi}}\right)\left(\xi_1^{-1a} + \xi_2^{-1a}\right) + \frac{1}{2}\Phi^i_{,ab}\left(\overline{\overline{\phi}}\right)\left(\xi_1^{-1a}\xi_1^{-1b} + \xi_2^{-1a}\xi_2^{-1b}\right)\\
&\quad + \Phi^i_{,a,j}\left(\overline{\overline{\phi}}\right)\Phi^j_{,b}\left(\overline{\overline{\phi}}\right)\xi_1^{-1b}\xi_2^{-1a} + O(\xi^3)\\
&= \phi^i + \Phi^i_{,a}\left(\xi_1^a + \xi_2^a + \xi_1^{-1a} + \xi_2^{-1a}\right)\\
&\quad + \frac{1}{2}\Phi^i_{,ab}\left(\xi_1^a\xi_1^b + \xi_2^a\xi_2^b + \xi_1^{-1a}\xi_1^{-1b} + \xi_2^{-1a}\xi_2^{-1b}\right)\\
&\quad + \Phi^i_{,a,j}\Phi^j_{,b}\left[\xi_1^b\xi_2^a + \xi_1^{-1b}\xi_2^{-1a} + \left(\xi_1^b + \xi_2^b\right)\left(\xi_1^{-1a} + \xi_2^{-1a}\right)\right] + O(\xi^3)\\
&= \phi^i + \Phi^i_{,a,j}\Phi^j_{,b}\left(\xi_1^b\xi_2^a - \xi_2^b\xi_1^a\right) + O(\xi^3).
\end{aligned}$$

On the other hand, we have

$$\overline{\overline{\overline{\phi}}}^i = \phi^i + \Phi^i_{,a}\left(\xi_2^{-1}\xi_1^{-1}\xi_2\xi_1\right)^a + \cdots.$$

This tells us that $\xi_2^{-1}\xi_1^{-1}\xi_2\xi_1$, which is known as the *commutator* of the group elements ξ_1 and ξ_2, differs from the identity by a coordinate interval that is only of the second order in the ξ. We may in fact do the expanion

$$\left(\xi_2^{-1}\xi_1^{-1}\xi_2\xi_1\right)^a = C^a_{bc}\xi_2^b\xi_1^c + O(\xi^3),$$

where the necessary bilinearity of the first term in ξ_1^c and ξ_2^b is a consequence of the fact that it is antisymmetric in the labels 1 and 2, as follows from

$$\left(\xi_1^{-1}\xi_2^{-1}\xi_1\xi_2\right)^a = \left(\xi_2^{-1}\xi_1^{-1}\xi_2\xi_1\right)^{-1a} = -\left(\xi_2^{-1}\xi_1^{-1}\xi_2\xi_1\right)^a + O(\xi^4).$$

This implies that the coefficients C^a_{bc} are antisymmetric in the indices b and c.

Comparing the two expansions for $\overline{\overline{\overline{\phi}}}^i$, we finally get

$$\Phi^i_{,a,j}\Phi^j_{,b} - \Phi^i_{,b,j}\Phi^j_{,a} = \Phi^i_{,c}C^c_{ab}. \tag{3.1}$$

The C^a_{bc} are known as the *structure constants* of the group. They depend both on the group and on the choice of coordinates. For a change from one set of group coordinates to another, however, it is only what happens in the immediate vicinity of the identity that is relevant in determining how the C^a_{bc} change.

Problem 13 By differentiating the differential identity (3.1) satisfied by the $\Phi^i_{\,a}$, show that the structure constants satisfy the cyclic identity

$$C^d_{ae}C^e_{bc} + C^d_{be}C^e_{ca} + C^d_{ce}C^e_{ab} = 0.$$

Solution 13 We have

$$\begin{aligned}
&\Phi^i_{\,d}\left(C^d_{ae}C^e_{bc} + C^d_{be}C^e_{ca} + C^d_{ce}C^e_{ab}\right)\\
&= \left(\Phi^i_{\,a,j}\Phi^j_{\,e} - \Phi^i_{\,e,j}\Phi^j_{\,a}\right)C^e_{bc} + (bca) + (cab)\\
&= \Phi^i_{\,a,j}\left(\Phi^j_{\,b,k}\Phi^k_{\,c} - \Phi^j_{\,c,k}\Phi^k_{\,b}\right) - \left(\Phi^i_{\,b,k}\Phi^k_{\,c} - \Phi^i_{\,c,k}\Phi^k_{\,b}\right)_{,j}\Phi^j_{\,a}\\
&\quad + (bca) + (cab)\\
&= 0.
\end{aligned}$$

The identity itself then follows by faithfulness of the realization.

Problem 14 Obtain the structure constants of the rotation group $O(3)$ in the coordinate system in which ξ^a $(a = 1, 2, 3)$ represents a rotation through the angle $|\xi| = (\xi^a\xi^a)^{1/2}$ about the axis having direction cosines $\xi^a|\xi|$ relative to a Cartesian frame. Use the realization in which the ϕ^i $(i = 1, 2, 3)$ are the components of a 3-vector on which the rotation ξ acts. Note that in computing structure constants, it suffices to confine attention to infinitesimal group operations, i.e., group elements having infinitesimal coordinates $\delta\xi^a$. Such operations induce infinitesimal changes in the ϕ^i given by

$$\delta\phi^i = \Phi^i_{\,a}\delta\xi^a.$$

Solution 14 We have

$$\delta\phi^i = \varepsilon_{iaj}\delta\xi^a\phi^j.$$

Therefore,

$$\Phi^i_{\,a} = \varepsilon_{iaj}\phi^j,$$

$$\begin{aligned}
\Phi^i_{\,a,j}\Phi^j_{\,b} - \Phi^i_{\,b,j}\Phi^j_{\,a} &= \left(\varepsilon_{iaj}\varepsilon_{jbk} - \varepsilon_{ibj}\varepsilon_{jak}\right)\phi^k\\
&= \left(\delta_{ib}\delta_{ak} - \delta_{ik}\delta_{ab} - \delta_{ia}\delta_{bk} + \delta_{ik}\delta_{ba}\right)\phi^k\\
&= \varepsilon_{ick}\varepsilon_{cab}\phi^k = \Phi^i_{\,c}\varepsilon_{cab}.
\end{aligned}$$

Hence, finally,

$$C^c_{ab} = \varepsilon_{cab}.$$

3.1 Representations

When the realization is a representation, the functions (or functionals) Φ^i take the form

$$\Phi^i(\xi,\phi) = D^i_j(\xi)\phi^j,$$

where the matrices $D(\xi)$ satisfy (with suppression of indices)

$$D(\xi_1)D(\xi_2) = D(\xi_1\xi_2), \quad \text{for all } \xi_1,\ \xi_2.$$

Moreover,

$$\Phi^i_a(\phi) = G^i_{aj}\phi^j,$$

where the matrices G_a are given by

$$G_a = \left.\frac{\partial D(\xi)}{\partial \xi^a}\right|_{\xi^a=0}$$

and satisfy the commutation relation

$$[G_a, G_b] = G_c C^c_{ab}.$$

The G_a are known as the *generators* of the representation.

Problem 15 Obtain the commutation relation satisfied by the generators G^a_b of the full linear group GL(n) in n dimensions, in the coordinate system in which ξ^a_b represents the matrix $\delta^a_b + \xi^a_b$. (Note that we are now using a pair of indices on the generators and group coordinates, because it is inconvenient to attempt to map them into a single index.)

Solution 15 Use the representation provided by a contravariant vector ϕ^a in n dimensions. Then

$$\delta\phi^a = \delta\xi^a_b\phi^b = G^{ac}_{db}\phi^b\delta\xi^d_c,$$

$$G^{ac}_{db} = \delta^a_d\delta^c_b,$$

$$\begin{aligned}\left[G^a_b, G^c_d\right]^e_f &= G^{ea}_{bg}G^{gc}_{df} - G^{ec}_{dg}G^{ga}_{bf}\\ &= \delta^e_b\delta^a_g\delta^g_d\delta^c_f - \delta^e_d\delta^c_g\delta^g_b\delta^a_f\\ &= \left(\delta^a_d G^c_b - \delta^c_b G^a_d\right)^e_f,\end{aligned}$$

or

$$\left[G^a_b, G^c_d\right] = \delta^a_d G^c_b - \delta^c_b G^a_d.$$

3.2 Diffeomorphism Group

Suppose we have a differentiable manifold M. The diffeomorphism group on M, denoted Diff(M) by mathematicians, is the group of all one-to-one differentiable maps of M onto itself whose inverses are also differentiable. Mathematicians usually confine themselves to C^∞ maps, but we may assume merely differentiability up to the lowest order needed in any discussion. The maps themselves are known as diffeomorphisms.

Diffeomorphisms may be related to coordinate transformations as follows. Let a coordinate patch be laid down on some open set of M. (This open set must be homeomorphic to an open set of $\mathbb{R}^n$, where n is the dimensionality of M.) Denote the coordinates by x^μ. A diffeomorphism ξ that is sufficiently close to the identity may be regarded as a deformation of the coordinate mesh (involving a possible shift in the position of the coordinate mesh) in which every point is mapped into another point in such a way that its coordinates in the original coordinate system are identical with the coordinates of its image (under the mapping) in the deformed coordinate system. A natural coordinatization of the diffeomorphism group itself assigns to each diffeomorphism ξ a set of functions $\xi^\mu(x)$ that display the relation between the two coordinate systems in M:

$$\bar{x}^\mu = x^\mu + \xi^\mu(x).$$

Here the set of numbers μ, x^1, x^2, ..., x^n replace the index a in the general discussion about continuous groups in this Chapter. We see that the diffeomorphism group is infinite dimensional.

In the immediate vicinity of the identity, where the ξ^μ become infinitesimal, each point gets mapped into a neighbor that is reached by executing the displacement $-\xi^\mu$. The diffeomorphism itself thus generates a *flow* which, at the initial instant, is characterized by the contravariant vector having components $-\xi^\mu$.

Let us now compute the structure constants of the diffeomorphism group. As before, we consider the commutator of two group elements ξ_1 and ξ_2:

$$\begin{aligned}
\bar{x}^\mu &= x^\mu + \xi_1^\mu(x), \\
\bar{\bar{x}}^\mu &= \bar{x}^\mu + \xi_2^\mu(\bar{x}) \\
&= x^\mu + \xi_1^\mu(x) + \xi_2^\mu(x) + \xi^\mu_{2,\nu}(x)\xi_1^\nu(x) + O(\xi^3), \\
\bar{\bar{\bar{x}}}^\mu &= \bar{\bar{x}}^\mu + \xi_1^{-1\mu}(\bar{\bar{x}}),
\end{aligned}$$

$$\begin{aligned}
\bar{\bar{\bar{\bar{x}}}}^\mu &= \bar{\bar{\bar{x}}}^\mu + \xi_2^{-1\mu}\left(\bar{\bar{\bar{x}}}\right) \\
&= \bar{\bar{x}}^\mu + \xi_1^{-1\mu}(\bar{\bar{x}}) + \xi_2^{-1\mu}(\bar{\bar{x}}) + \xi_2^{-1\mu}{}_{,\nu}(\bar{\bar{x}})\xi_1^{-1\nu}(\bar{\bar{x}}) + O(\xi^3) \\
&= x^\mu + \xi_1^\mu + \xi_2^\mu + \xi_1^{-1\mu} + \xi_2^{-1\mu} + \left(\xi_1^{-1\mu}{}_{,\nu} + \xi_2^{-1\mu}{}_{,\nu}\right)\left(\xi_1^\nu + \xi_2^\nu\right) \\
&\quad + \xi^\mu_{2,\nu}\xi_1^\nu + \xi_2^{-1\mu}{}_{,\nu}\xi_1^{-1\nu} + O(\xi^3) \\
&= x^\mu + \xi^\mu_{2,\nu}\xi_1^\nu - \xi^\mu_{1,\nu}\xi_2^\nu + O(\xi^3),
\end{aligned}$$

where a comma followed by a Greek index denotes differentiation with respect to an x and, in passing to the final forms, we have suppressed the argument x and used the identity

$$\xi^{\mu} + \xi^{-1\mu} + \xi^{-1\mu}_{,\nu}\xi^{\nu} = O(\xi^3).$$

Writing

$$(\xi_2^{-1}\xi_1^{-1}\xi_2\xi_1)^{\mu} = \int \mathrm{d}x' \int \mathrm{d}x'' C^{\mu}_{\nu'\sigma''}\xi_2^{\nu'}\xi_1^{\sigma''} + O(\xi^3),$$

where one or more primes on an index indicates that the index is associated with a corresponding point x' or x'', etc., and $\int \mathrm{d}x$ denotes standard integration over the manifold M, we obtain

$$\int \mathrm{d}x' \int \mathrm{d}x'' C^{\mu}_{\nu'\sigma''}\xi_2^{\nu'}\xi_1^{\sigma''} = \xi^{\mu}_{2,\nu}\xi_1^{\nu} - \xi^{\mu}_{1,\nu}\xi_2^{\nu},$$

whence

$$C^{\mu}_{\nu'\sigma''} = \delta^{\mu}_{\nu}\delta(x,x'')\frac{\partial}{\partial x^{\sigma}}\delta(x,x') - \delta^{\mu}_{\sigma}\delta(x,x')\frac{\partial}{\partial x^{\nu}}\delta(x,x''),$$

where $\delta(x, x')$ is the delta function on M.

Instead of representing a contravariant vector, A say, by its components A^{μ} in a given coordinate system, mathematicians like to represent it as a differential operator

$$A = A^{\mu}\frac{\partial}{\partial x^{\mu}}$$

acting on the set of all (differentiable scalar) functions on M. This representation, in which a contravariant vector is determined by its action on functions over M, is coordinate independent.

It is sometimes convenient to use this representation for infinitesimal diffeomorphisms in which the ξ^{μ} become components of contravariant vectors. We then find that we may write

$$\int \mathrm{d}x' \int \mathrm{d}x'' C^{\mu}_{\nu'\sigma''}\xi_2^{\nu'}\xi_1^{\sigma''}\frac{\partial}{\partial x^{\mu}} = [\xi_1, \xi_2],$$

and the cyclic identity satisfied by the structure constants (see Problem 13) becomes a corollary of the ordinary Jacobi identity satisfied by commutators of linear operators.

3.3 Tensors and Tensor Densities

The irreducible matrix representations of the diffeomorphism group are given by the action of the group on the components of irreducible tensor fields (more correctly, *tensor density fields*) in a given coordinate system. The action is said to define the coordinate transformation law of the field in question. Thus let $\phi(x)$ stand for the set of components of some tensor density at the point x in a system of coordinates x^μ. The diffeomorphism ξ induces the coordinate transformation

$$\bar{x}^\mu = x^\mu + \xi^\mu(x),$$

and a corresponding transformation in ϕ:

$$\overline{\phi}(\bar{x}) = D(1 + \partial\xi/\partial x)\phi(x),$$

where $D(1 + \partial\xi/\partial x)$ is the representative of $\delta^\mu_\nu + \partial\xi^\mu/\partial x^\nu$ in the corresponding matrix representation of the full linear group.

In the infinitesimal case, the field transformation law takes the form

$$\overline{\phi}(x + \delta\xi) = \phi(x) + G^\mu_\nu\phi(x)\delta\xi^\nu{}_{,\mu}(x),$$

where the G^μ_ν are the generators of the representation D. Writing

$$\phi(x) + \delta\phi(x) \equiv \overline{\phi}(x),$$

we find

$$\delta\phi = -\mathcal{L}_{\delta\xi}\phi,$$

where, for any contravariant vector A, the operator $\mathcal{L}_A$ is defined by

$$\mathcal{L}_A = A^\mu\frac{\partial}{\partial x^\mu} - G^\mu_\nu A^\nu{}_{,\mu},$$

when acting on a tensor ϕ. The quantity $\mathcal{L}_{\delta\xi}\phi$ is called the *Lie derivative* of ϕ with respect to the vector $\delta\xi$. We note that $\mathcal{L}_A$, when acting on a scalar function, reduces to A itself.

Problem 16 For any two contravariant vectors A and B, show that

$$(\mathcal{L}_A B^\mu)\frac{\partial}{\partial x^\mu} = [A, B].$$

Solution 16 We have

$$\overline{B}^\mu(\bar{x}) = \frac{\partial\bar{x}^\mu}{\partial x^\nu}B^\nu(x), \quad \overline{B}^\mu(x + \delta\xi) = B^\mu(x) + B^\nu(x)\delta\xi^\mu_{,\nu}(x),$$

$$\delta B^\mu = -B^\mu_{,\nu}\delta\xi^\nu + B^\nu\delta\xi^\mu_{,\nu} = -\mathcal{L}_{\delta\xi}B^\mu,$$

$$\mathcal{L}_A B^\mu = B^\mu_{,\nu} A^\nu - A^\mu_{,\nu} B^\nu,$$

and hence,

$$(\mathcal{L}_A B^\mu)\frac{\partial}{\partial x^\mu} = [A, B].$$

Problem 17 Using the commutation relation obtained in Problem 15 for the generators of the linear group, show that

$$[\mathcal{L}_A, \mathcal{L}_B] = \mathcal{L}_{[A,B]}.$$

Solution 17 We have

$$\begin{aligned}
[\mathcal{L}_A, \mathcal{L}_B] &= \left[A^\mu \frac{\partial}{\partial x^\mu} - G^\mu_\nu A^\nu_{,\mu},\ B^\sigma \frac{\partial}{\partial x^\sigma} - G^\sigma_\tau B^\tau_{,\sigma}\right] \\
&= [A, B] - G^\sigma_\tau B^\tau_{,\sigma\mu} A^\mu + G^\mu_\nu A^\nu_{,\mu\sigma} B^\sigma + \left(\delta^\mu_\tau G^\sigma_\nu - \delta^\sigma_\nu G^\mu_\tau\right) A^\nu_{,\mu} B^\tau_{,\sigma} \\
&= [A, B] - G^\mu_\nu \left(B^\nu_{,\sigma\mu} A^\sigma + B^\nu_{,\sigma} A^\sigma_{,\mu} - A^\nu_{,\sigma\mu} B^\sigma - A^\nu_{,\sigma} B^\sigma_{,\mu}\right) \\
&= \left(B^\mu_{,\nu} A^\nu - A^\mu_{,\nu} B^\nu\right)\frac{\partial}{\partial x^\mu} - G^\mu_\nu \left(B^\nu_{,\sigma} A^\sigma - A^\nu_{,\sigma} B^\sigma\right)_{,\mu} \\
&= \mathcal{L}_{[A,B]}.
\end{aligned}$$

In practice, the transformation laws for the components of all types of tensor densities are most easily built up from the three prototypes:

$$\begin{aligned}
&\text{Contravariant vector} \quad \overline{\phi}^\mu(\overline{x}) = \frac{\partial \overline{x}^\mu}{\partial x^\nu}\phi^\nu(x), \\
&\text{Covariant vector} \quad \overline{\phi}_\mu(\overline{x}) = \frac{\partial x^\nu}{\partial \overline{x}^\mu}\phi_\nu(x), \\
&\text{Density of weight w} \quad \overline{\phi}(\overline{x}) = \left[\frac{\partial(x)}{\partial(\overline{x})}\right]^w \phi(x).
\end{aligned}$$

The corresponding infinitesimal laws are:

$$\begin{aligned}
&\text{Contravariant vector} \quad \delta\phi^\mu = -\phi^\mu_{,\nu}\delta\xi^\nu + \phi^\nu \delta\xi^\mu_{,\nu}, \\
&\text{Covariant vector} \quad \delta\phi_\mu = -\phi_{\mu,\nu}\delta\xi^\nu + \phi_\nu \delta\xi^\nu_{,\mu}, \\
&\text{Density of weight w} \quad \delta\phi = -\phi_{,\mu}\delta\xi^\mu - w\phi\delta\xi^\mu_{,\mu}.
\end{aligned}$$

Problem 18 Show that the n-dimensional permutation symbol may be regarded as representing, in every coordinate system, the components of either a completely antisymmetric contravariant tensor density of weight 1, in which case we it in the form ${}^1\varepsilon^{\mu_1 \cdots \mu_n}$, or a completely antisymmetric covariant tensor density of weight -1, in which case we write it as ${}^{-1}\varepsilon_{\mu_1 \dots \mu_n}$.

Solution 18 We have

$$^{1}\overline{\varepsilon}^{\mu_1\ldots\mu_n} = \frac{\partial(x)}{\partial(\overline{x})}\frac{\partial\overline{x}^{\mu_1}}{\partial x^{\nu_1}}\cdots\frac{\partial\overline{x}^{\mu_n}}{\partial x^{\nu_n}}{}^{1}\varepsilon^{\nu_1\ldots\nu_n} = {}^{1}\varepsilon^{\mu_1\ldots\mu_n},$$

$$^{-1}\overline{\varepsilon}_{\mu_1\ldots\mu_n} = \frac{\partial(\overline{x})}{\partial(x)}\frac{\partial x^{\nu_1}}{\partial\overline{x}^{\mu_1}}\cdots\frac{\partial x^{\nu_n}}{\partial\overline{x}^{\mu_n}}{}^{-1}\varepsilon_{\nu_1\ldots\nu_n} = {}^{-1}\varepsilon_{\mu_1\ldots\mu_n}.$$

Problem 19 Show that the Kronecker delta may be regarded as representing, in every coordinate system, the components of a mixed tensor having one covariant and one contravariant index.

Solution 19 This is proved by

$$\overline{\delta}^{\mu}_{\ \nu} = \frac{\partial\overline{x}^{\mu}}{\partial x^{\sigma}}\frac{\partial x^{\tau}}{\partial\overline{x}^{\nu}}\delta^{\sigma}_{\ \tau} = \frac{\partial\overline{x}^{\mu}}{\partial x^{\sigma}}\frac{\partial x^{\sigma}}{\partial\overline{x}^{\nu}} = \delta^{\mu}_{\ \nu}.$$

Problem 20 Show that an integral of the form $\int \rho \mathrm{d}x$ taken over the manifold is coordinate independent if and only if ρ is a density of weight 1.

Solution 20 We have

$$\int \rho \mathrm{d}x = \int \rho \frac{\partial(x)}{\partial(\overline{x})}\mathrm{d}\overline{x} = \int \overline{\rho}\mathrm{d}\overline{x}$$

if and only if

$$\overline{\rho}(\overline{x}) = \frac{\partial(x)}{\partial(\overline{x})}\rho(x).$$

Note that the integral may be carried out patch by patch so as to cover the whole manifold.

3.4 Bitensors, Tritensors, and *n*-Tensors

More complicated matrix representations of the diffeomorphism group may be obtained by forming direct products of tensor representations. Coordinate components of *bitensors* transform according to the law defined by the direct product of two tensor representations. Coordinate components of *tritensors* transform according to the law defined by the direct product of three tensor representations, and so on. In general, the coordinate components of n-tensors are functions of n independent points of the manifold M, i.e., they are functions over the n-fold Cartesian product of M with itself, viz.,

$$\underbrace{M \times M \times \cdots \times M}_{n \text{ times}}.$$

The simplest bitensor is the delta function, which is really a *bidensity* of total weight unity. The weight may be shared arbitrarily between the two points, but in a given context is usually well defined. The structure constants $C^{\mu}_{\nu'\sigma''}$ of the diffeomorphism group are components of a tritensor, transforming as a contravariant vector at the point x and as a covariant vector density of unit weight at x' and x''. Many other important examples of n-point tensors (especially bitensors) are encountered in the theory of geodesics and Green's functions.

Chapter 4
Riemannian Manifolds

A Riemannian manifold is a differentiable manifold in which a notion of length is introduced at the local level. If x^μ and $x^\mu + \mathrm{d}x^\mu$ are the coordinates in a given coordinate system of two infinitesimally close points, the infinitesimal distance $\mathrm{d}s$ between them is defined by

$$\mathrm{d}s^2 = g_{\mu\nu}\mathrm{d}x^\mu \mathrm{d}x^\nu \ ,$$

where $g_{\mu\nu}$ are the components of a special field associated with the manifold, the combination of field and manifold constituting the Riemannian manifold. If the notion of length is to be independent of the choice of coordinate system, the special field must be a covariant tensor. This tensor is known as the *metric tensor*. It may evidently be taken symmetric, because any antisymmetric part would make no contribution to the distance concept.

The metric tensor having been introduced, the notions of orthogonality and local parallelism may then be introduced by applying the classical Euclidean laws to infinitesimal triangles. For this purpose, the laws of similar triangles must be adopted ab initio as postulates, and the 'postulate of parallels' must be excluded, a procedure that is reasonable as far as physics is concerned both because the laws of similar triangles correspond immediately to the intuition of experience and because experience is always limited to finite regions. By defining the right angle as the angle of intersection of two lines that makes all four intersection angles equal, and by guaranteeing its uniqueness through further axiomatic refinements on the comparison of angles by means of the notions 'greater than' and 'less than' as well as 'equality', one may then derive the Pythagorean theorem in the well-known manner.

Conversely, the Pythagorean theorem may be invoked to *define* right angles. Thus, two 'displacements' d_1x^μ and d_2x^μ are said to be *orthogonal* if their lengths satisfy the relationship

$$\mathrm{d}_1s^2 + \mathrm{d}_2s^2 = \mathrm{d}s^2 \ ,$$

B. Dewitt, *Bryce DeWitt's Lectures on Gravitation*,
Edited by Steven M. Christensen, Lecture Notes in Physics, 826,
DOI: 10.1007/978-3-540-36911-0_4, © Springer-Verlag Berlin Heidelberg 2011

where

$$\mathrm{d}_1s^2 = g_{\mu\nu}\mathrm{d}_1x^\mu \mathrm{d}_1x^\nu\ , \qquad \mathrm{d}_2s^2 = g_{\mu\nu}\mathrm{d}_2x^\mu \mathrm{d}_2x^\nu\ ,$$
$$\mathrm{d}s^2 = g_{\mu\nu}(\mathrm{d}_1x^\mu - \mathrm{d}_2x^\mu)(\mathrm{d}_1x^\nu - \mathrm{d}_2x^\nu)\ .$$

This relationship is readily seen to reduce to

$$\mathrm{d}_1x \cdot \mathrm{d}_2x \equiv g_{\mu\nu}\mathrm{d}_1x^\mu \mathrm{d}_2x^\nu = 0\ .$$

In using the classical Euclidean laws, of course, one assumes that the quantities $\mathrm{d}_1\, s^2$, $\mathrm{d}_2\, s^2$, and $\mathrm{d}s^2$ are positive and hence that the components of the metric tensor at any point in any coordinate patch form a positive definite matrix. The formalism of tensors, however, allows one to abandon the Euclidean origins of the metric concept once it has outlived its usefulness as an initial guide. For an arbitrary Riemannian manifold,[1] we need only assume that $g_{\mu\nu}$ forms, in any coordinate patch, a (sufficiently) differentiable, nonsingular, but not necessarily positive definite, matrix, the inverse of which will be denoted by $g^{\mu\nu}$:

$$g_{\mu\sigma}g^{\sigma\nu} = \delta^\nu_\mu\ .$$

Since the Kronecker delta defines a mixed tensor (see Problem 19), it follows that the $g^{\mu\nu}$ are components of a symmetric contravariant tensor.

At any single point in a Riemannian manifold, a coordinate system may be introduced in which the components of the metric tensor take the *canonical form*

$$(g_{\mu\nu}) = \mathrm{diag}(-1,\ldots,-1,1,\ldots,1)$$

at that point. Since $g_{\mu\nu}$ must be nonsingular in every coordinate patch, it follows that the canonical form is an invariant of the manifold. Spacetime in relativity theory is assumed to be a four-dimensional Riemannian manifold in which the canonical form of the metric is that of Minkowski[2]:

$$(\eta_{\mu\nu}) = \mathrm{diag}(-1,1,1,1)\ .$$

4.1 Local Parallelism

A Riemannian manifold possesses not only the local distance concept but also a concept of local parallelism that arises naturally out of the classical Euclidean laws. These laws permit one first of all to define an infinitesimal parallelogram in a

[1] Mathematicians sometimes call the manifold *pseudo-Riemannian* when the metric is not positive definite.

[2] Mathematicians sometimes refer to Riemannian manifolds for which the metric tensor has one eigenvalue of one sign while all the others have opposite sign as *Lorentzian manifolds*.

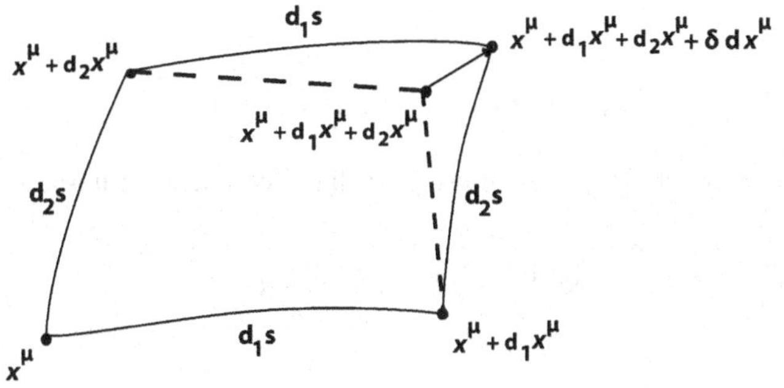

Fig. 4.1 Establishing local parallelism

Riemannian manifold as an infinitesimal plane quadrilateral having opposite sides of equal length. The relationship of such an object to an arbitrary coordinate mesh is indicated schematically in Fig. 4.1.

The quantity δdx^μ is the change, arising from the variability in the shape of the coordinate mesh from point to point as well as from changes in the intrinsic geometry of the manifold, in the numerical magnitude of the μ component of *either* of the infinitesimal intervals d_1x or d_2x as it is displaced in a parallel fashion along the other. It is evident from the figure that

$$\begin{aligned} d_1s^2 &= \left(g_{\mu\nu} + \frac{1}{2}g_{\mu\nu,\sigma}d_1x^\sigma\right)d_1x^\mu d_1x^\nu \\ &= \left[g_{\mu\nu} + g_{\mu\nu,\sigma}\left(d_2x^\sigma + \frac{1}{2}d_1x^\sigma\right)\right](d_1x^\mu + \delta dx^\mu)(d_1x^\nu + \delta dx^\nu) , \end{aligned}$$

correct to the third infinitesimal order. Keeping terms only up to this order, we find

$$2g_{\mu\nu}d_1x^\mu\delta dx^\nu + g_{\mu\nu,\sigma}d_1x^\mu d_1x^\nu d_2x^\sigma = 0 ,$$

and similarly,

$$2g_{\mu\nu}d_2x^\mu\delta dx^\nu + g_{\mu\nu,\sigma}d_2x^\mu d_2x^\nu d_1x^\sigma = 0 .$$

These are the parallelogram equations.

Now let λ_1 and λ_2 be two arbitrary parameters. Multiplying the first parallelogram equation by λ_1 and the second by λ_2, we find

$$\begin{aligned} g_{\mu\nu}(\lambda_1 d_1x^\mu + \lambda_2 d_2x^\mu)\delta dx^\nu &= -\frac{1}{2}(\lambda_1 d_1x^\mu g_{\mu\nu,\sigma} + \lambda_2 d_2x^\mu g_{\mu\sigma,\nu})d_1x^\nu d_2x^\sigma \\ &= -\frac{1}{2}(\lambda_1 d_1x^\mu + \lambda_2 d_2x^\mu)(g_{\mu\nu,\sigma} + g_{\mu\sigma,\nu} - g_{\nu\sigma,\mu})d_1x^\nu d_2x^\sigma \\ &= -\Gamma_{\mu\nu\sigma}(\lambda_1 d_1x^\mu + \lambda_2 d_2x^\mu)d_1x^\nu d_2x^\sigma , \end{aligned}$$

where

$$\Gamma_{\mu\nu\sigma} = \frac{1}{2}(g_{\mu\nu,\sigma} + g_{\mu\sigma,\nu} - g_{\nu\sigma,\mu}) = \Gamma_{\mu\sigma\nu} \,.$$

Since λ_1 and λ_2 are arbitrary, we must infer that the general solution of the above equation is

$$\delta \mathrm{d}x^\mu = -\Gamma^\mu_{\nu\sigma}\mathrm{d}_1x^\nu \mathrm{d}_2x^\sigma + Q^\mu \,,$$

where

$$\Gamma^\mu_{\nu\sigma} = g^{\mu\tau}\Gamma_{\tau\nu\sigma} \,,$$

and where Q^μ is a quantity of the second infinitesimal order, necessarily linear in d_1x^μ and d_2x^μ, and orthogonal to both d_1x^μ and d_2x^μ. The most general quantity of this kind has the form

$$Q^\mu = g^{\mu\nu}Q_\nu \,, \qquad Q_\mu = A_{\mu\nu\sigma}\mathrm{d}_1x^\nu \mathrm{d}_2x^\sigma \,,$$

where

$$A_{\mu\nu\sigma} = -A_{\nu\mu\sigma} = -A_{\sigma\nu\mu} \,.$$

However, this implies

$$A_{\mu\nu\sigma} = -A_{\nu\mu\sigma} = A_{\sigma\mu\nu} = -A_{\mu\sigma\nu} \,,$$

and therefore, Q^μ is antisymmetric under interchange of d_1x^μ and d_2x^μ. On the other hand, because the parallelogram is a plane figure, $\delta \mathrm{d}x^\mu$, and hence Q^μ, must be symmetric under interchange of d_1x^μ and d_2x^μ. Another way of saying this is that $\delta \mathrm{d}x^\mu$ can involve no preferred direction in the local subspace orthogonal to the plane defined by d_1x^μ and d_2x^μ. Therefore, we must have $Q^\mu = 0$ and

$$\delta \mathrm{d}x^\mu = -\Gamma^\mu_{\nu\sigma}\mathrm{d}_1x^\nu \mathrm{d}_2x^\sigma \,.$$

The objects $\Gamma^\mu_{\nu\sigma}$ and $\Gamma_{\mu\nu\sigma}$ are known as the *Christoffel symbols.*

4.2 Parallel Displacement of Tensors

The concept of parallel displacement can be extended to an arbitrary tensor density by considering an infinitesimal displacement $-\delta\xi^\mu$ of the entire coordinate mesh, which produces the new coordinate system $\bar{x}^\mu = x^\mu + \delta\xi^\mu$. Let us suppose that this displacement is locally parallel at the point x, as indicated in Fig. 4.2.

Comparing this figure with Fig. 4.1 shows us that we must have

$$[x'^\mu - \delta\xi^\mu(x')] - [x^\mu - \delta\xi^\mu(x)] = x'^\mu - x^\mu + \Gamma^\mu_{\nu\sigma}(x'^\nu - x^\nu)\delta\xi^\sigma + O(x' - x)^2 \,,$$

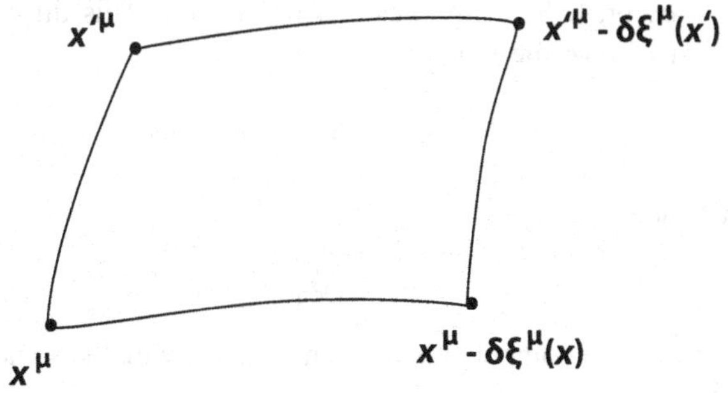

Fig. 4.2 Infinitesimal displacement of the coordinate mesh

for all neighboring points. Taking the limit $x' \to x$, we find

$$\delta\xi^\mu_{,\nu} = -\Gamma^\mu_{\nu\sigma}\delta\xi^\sigma\,, \quad \text{at } x\,.$$

Now let ϕ be the set of components of a tensor density field. We shall suppose that this field is displaced 'bodily' along with the coordinate mesh in the neighborhood of x, giving rise to a new field $\overrightarrow{\phi}$. In the new coordinate system $\bar{x}^\mu$, this field will take the form $\overrightarrow{\phi} + \delta\overrightarrow{\phi}$, where

$$\delta\overrightarrow{\phi} = -\mathcal{L}_{\delta\xi}\overrightarrow{\phi}\,.$$

However, the new field must have, at least in the neighborhood of x, the same form in the new coordinate system as the original field had in the old. That is,

$$\overrightarrow{\phi} + \delta\overrightarrow{\phi} = \phi\,, \quad \text{near } x\,,$$

or

$$\overrightarrow{\phi} = \phi - \delta\overrightarrow{\phi} = \phi + \mathcal{L}_{\delta\xi}\overrightarrow{\phi} = \phi + \mathcal{L}_{\delta\xi}\phi\,, \quad \text{near } x\,,$$

to first infinitesimal order.

This bodily displacement of the field is precisely the intuitive notion of parallel displacement. In order to get the change in the magnitude of the field components under parallel displacement, we must examine the displaced field $\overrightarrow{\phi}$ at the displaced point $x - \delta\xi$, and compare it with the original field ϕ at the original point x. The parallel displacement law for components is therefore

$$\begin{aligned}\delta\phi &= \overrightarrow{\phi}(x - \delta\xi) - \phi(x) = -\phi_{,\mu}\delta\xi^\mu + \mathcal{L}_{\delta\xi}\phi \\ &= -G^\nu_\mu\phi\delta\xi^\mu_{,\nu} = G^\nu_\mu\Gamma^\mu_{\nu\sigma}\phi\delta\xi^\sigma\,.\end{aligned} \tag{4.1}$$

4.3 Covariant Differentiation

Now the quantity $\phi + \delta\phi$ is a set of components, in the original coordinate system, of a tensor density at the point $x - \delta\xi$, and so is $\phi(x - \delta\xi)$. Therefore, the difference between the two sets is also a set of components of a tensor density at

$x - \delta\xi$ or, to first infinitesimal order, at x. This difference defines the *covariant derivative* of the field ϕ:

$$-\phi_{;\mu}\delta\xi^\mu = \phi(x - \delta\xi) - (\phi + \delta\phi) = -\left(\phi_{,\mu} + G^\nu_\sigma \Gamma^\sigma_{\nu\mu}\phi\right)\delta\xi^\mu ,$$

whence

$$\phi_{;\mu} = \phi_{,\mu} + G^\nu_\sigma \Gamma^\sigma_{\nu\mu}\phi .$$

The $\phi_{;\mu}$ are components of a tensor density that has one more covariant index than ϕ has.

In practice the covariant derivatives of all types of tensor densities are most easily built up from the three prototypes:

$$\begin{array}{rl} \text{Contravariant vector} & \phi^\mu_{;\nu} = \phi^\mu_{,\nu} + \Gamma^\mu_{\sigma\nu}\phi^\sigma , \\ \text{Covariant vector} & \phi_{\mu;\nu} = \phi_{\mu,\nu} - \Gamma^\sigma_{\mu\nu}\phi_\sigma , \\ \text{Density of weight } w & \phi_{;\mu} = \phi_{,\mu} - w\Gamma^\nu_{\nu\mu}\phi . \end{array}$$

Using the second of these, it is easy to show that the covariant derivatives of the metric tensor vanish:

$$\begin{aligned} g_{\mu\nu;\sigma} &= g_{\mu\nu,\sigma} - \Gamma^\tau_{\mu\sigma} g_{\tau\nu} - \Gamma^\tau_{\nu\sigma} g_{\mu\tau} \\ &= g_{\mu\nu,\sigma} - \Gamma_{\nu\mu\sigma} - \Gamma_{\mu\nu\sigma} \\ &= g_{\mu\nu,\sigma} - \frac{1}{2}(g_{\nu\mu,\sigma} + g_{\nu\sigma,\mu} - g_{\mu\sigma,\nu}) - \frac{1}{2}(g_{\mu\nu,\sigma} + g_{\mu\sigma,\nu} - g_{\nu\sigma,\mu}) \\ &= 0 . \end{aligned}$$

Problem 21 Show that the covariant derivatives of δ^μ_ν, ${}^1\varepsilon^{\mu_1\cdots\mu_n}$ and ${}^{-1}\varepsilon_{\mu_1\cdots\mu_n}$ all vanish.

Solution 21

$$\begin{aligned} \delta^\mu_{\nu;\sigma} &= \delta^\mu_{\nu,\sigma} + \Gamma^\mu_{\tau\sigma}\delta^\tau_\nu - \Gamma^\tau_{\nu\sigma}\delta^\mu_\tau \\ &= \Gamma^\mu_{\nu\sigma} - \Gamma^\mu_{\nu\sigma} = 0 , \end{aligned}$$

$$\begin{aligned} {}^1\varepsilon^{\mu_1\cdots\mu_n}_{;\nu} &= {}^1\varepsilon^{\mu_1\cdots\mu_n}_{,\nu} + \Gamma^{\mu_1}_{\sigma\nu}\,{}^1\varepsilon^{\sigma\mu_2\cdots\mu_n} \\ &\quad + \cdots + \Gamma^{\mu_n}_{\sigma\nu}\,{}^1\varepsilon^{\mu_1\cdots\mu_{n-1}\sigma} - \Gamma^\sigma_{\sigma\nu}\,{}^1\varepsilon^{\mu_1\cdots\mu_n} = 0 , \end{aligned}$$

$$\begin{aligned} {}^{-1}\varepsilon_{\mu_1\cdots\mu_n;\nu} &= {}^{-1}\varepsilon_{\mu_1\cdots\mu_n,\nu} - \Gamma^\sigma_{\mu_1\nu}\,{}^{-1}\varepsilon_{\sigma\mu_2\cdots\mu_n} \\ &\quad - \cdots - \Gamma^\sigma_{\mu_n\nu}\,{}^{-1}\varepsilon_{\mu_1\cdots\mu_{n-1}\sigma} + \Gamma^\sigma_{\sigma\nu}\,{}^{-1}\varepsilon_{\mu_1\cdots\mu_n} = 0 . \end{aligned}$$

The last two results follow from the fact that both sides of the equations are completely antisymmetric in the μ_i and hence the μ_i must be all different if one is to get something different from zero. In the terms involving $\Gamma^{\mu_m}_{\sigma\nu}$ or $\Gamma^\sigma_{\mu_m\nu}$, one then gets nonvanishing contributions only when σ takes on the value μ_m.

Covariant differentiation, like ordinary differentiation, obeys the Leibniz rule when applied to factors in a product. Since the covariant derivatives of δ^μ_ν vanish, covariant differentiation commutes with the process of *contraction* of an upper index with a lower index (setting the two equal and summing). By taking the covariant derivative of the identity $g_{\mu\sigma}g^{\sigma\nu} = \delta^\nu_\mu$, one infers that all the $g^{\mu\nu}_{;\sigma}$ vanish, and hence that covariant differentiation commutes with the operation of raising and lowering indices:

$$A^\mu = g^{\mu\nu}A_\nu\,, \qquad B_\mu = g_{\mu\nu}B^\nu\,,$$

and so on. Finally, because the covariant derivatives of the permutation symbols vanish it follows that the magnitude of the determinant of the metric tensor, which, when written in the form

$$g \equiv \left|\det(g_{\mu\nu})\right| = \frac{1}{n!}\left|{}^1\varepsilon^{\mu_1\ldots\mu_n}\,{}^1\varepsilon^{\nu_1\ldots\nu_n}\,g_{\mu_1\nu_1}\cdots g_{\mu_n\nu_n}\right|\,,$$

is seen to be a density of weight 2, also has vanishing covariant derivatives.[3]

4.4 Tensor Properties of the Lie Derivative

Owing to the symmetry of the Christoffel symbol $\Gamma^\mu_{\nu\sigma}$ in its lower indices, it is easy to verify that, in the expression for the Lie derivative $\mathcal{L}_A\phi$ of a tensor density ϕ with respect to a contravariant vector A, ordinary derivatives may be replaced by covariant derivatives:

$$\begin{aligned}\mathcal{L}_A\phi &= A^\mu\phi_{,\mu} - G^\mu_\nu A^\nu_{\ ,\mu}\phi\\ &= A^\mu(\phi_{,\mu} + G^\nu_\sigma\Gamma^\sigma_{\nu\mu}\phi) - G^\mu_\nu(A^\nu_{\ ,\mu} + \Gamma^\nu_{\sigma\mu}A^\sigma)\phi\\ &= A_\mu\phi_{;\mu} - G^\mu_\nu A^\nu_{\ ;\mu}\phi\,.\end{aligned}$$

This means that $\mathcal{L}_A\phi$ has the same coordinate transformation law as ϕ and that the tensor density whose components are the elements of $\mathcal{L}_A\phi$ is of the same type as the tensor density whose components are the elements of ϕ. This latter fact must obviously be independent of the introduction of any metric tensor into the manifold and could (with considerably more labor) have been verified directly prior to our discussion of Riemannian manifolds.

The above result permits us to express the infinitesimal coordinate transformation law for the metric tensor in a particularly simple and symmetric form:

[3] We confine our attention in these lectures to diffeomorphisms that may be connected continuously to the identity.

$$\begin{aligned}\delta g_{\mu\nu} &= -\mathcal{L}_{\delta\xi} g_{\mu\nu} \\ &= -g_{\mu\nu,\sigma}\delta\xi^\sigma - g_{\sigma\nu}\delta\xi^\sigma_{,\mu} - g_{\mu\sigma}\delta\xi^\sigma_{,\nu} \\ &= -\delta\xi_{\mu;\nu} - \delta\xi_{\nu;\mu} ,\end{aligned}$$

where

$$\delta\xi_\mu = g_{\mu\nu}\delta\xi^\nu .$$

4.5 The Curvature Tensor

Indices induced by covariant differentiation, like any other indices, may be raised and lowered by means of the metric tensor. Not all authors agree on the proper notation for this, some feeling that the semicolon should be raised and lowered along with the index. We shall keep the semicolon firmly fixed in the lower position thus:

$$g^{\mu\nu}\phi_{;\nu} = \phi^{\mu}_{;} .$$

There is disagreement also on the proper notation for repeated differentiation, some authors feeling that a proliferation of commas or semicolons is required along with the proliferation of indices. We believe that one comma, or semicolon as the case may be, should suffice and accordingly will use the abbreviations

$$\phi_{,\mu\nu\ldots} = (\phi_{,\mu})_{,\nu\ldots} , \qquad \phi_{;\mu\nu\ldots} = (\phi_{;\mu})_{;\nu\ldots} ,$$

for repeated ordinary and covariant differentiation with respect to the coordinates x^μ. One caution must be sounded, however, in using this notation. The order of the indices following a comma is obviously unimportant. On the other hand, covariant differentiation, unlike ordinary differentiation, is not generally commutative, and therefore, the order of the indices following a semicolon is usually very important.

We shall nevertheless often wish to change the order of covariant differentiations, and therefore, it will be extremely useful to know the commutation law for covariant derivatives. For an arbitrary tensor density ϕ, this may be computed in a straightforward manner as follows:

$$\begin{aligned}\phi_{;\mu\nu} - \phi_{;\nu\mu} &= (\phi_{;\mu})_{,\nu} + G^\sigma_\tau \Gamma^\tau_{\sigma\nu}\phi_{;\mu} - \Gamma^\sigma_{\mu\nu}\phi_{;\sigma} - (\mu \leftrightarrow \nu) \\ &= (\phi_{,\mu} + G^\sigma_\tau \Gamma^\tau_{\sigma\mu}\phi)_{,\nu} + G^\sigma_\tau \Gamma^\tau_{\sigma\nu}(\phi_{,\mu} + G^\rho_\lambda \Gamma^\lambda_{\rho\mu}\phi) - (\mu \leftrightarrow \nu) \\ &= G^\sigma_\tau \left(\Gamma^\tau_{\sigma\mu,\nu} - \Gamma^\tau_{\sigma\nu,\mu}\right)\phi + (\delta^\sigma_\lambda G^\rho_\tau - \delta^\rho_\tau G^\sigma_\lambda)\Gamma^\tau_{\sigma\nu}\Gamma^\lambda_{\rho\mu}\phi \\ &= G^\sigma_\tau \left(\Gamma^\tau_{\sigma\mu,\nu} - \Gamma^\tau_{\sigma\nu,\mu} + \Gamma^\tau_{\lambda\nu}\Gamma^\lambda_{\sigma\mu} - \Gamma^\tau_{\rho\mu}\Gamma^\rho_{\sigma\nu}\right)\phi \\ &= -G^\sigma_\tau R^\tau_{\sigma\mu\nu}\phi ,\end{aligned}$$

where

$$R^{\tau}_{\sigma\mu\nu} = \Gamma^{\tau}_{\sigma\nu,\mu} - \Gamma^{\tau}_{\sigma\mu,\nu} + \Gamma^{\tau}_{\rho\mu}\Gamma^{\rho}_{\sigma\nu} - \Gamma^{\tau}_{\rho\nu}\Gamma^{\rho}_{\sigma\mu} .$$

In practice, the commutation laws for covariant differentiation of specific tensor densities are built up from those of the three prototypes:

$$\begin{array}{ll} \text{Contravariant vector} & \phi^{\mu}_{;\nu\sigma} - \phi^{\mu}_{;\sigma\nu} = -R^{\mu}_{\tau\nu\sigma}\phi^{\tau} , \\ \text{Covariant vector} & \phi_{\mu;\nu\sigma} - \phi_{\mu;\sigma\nu} = R^{\tau}_{\mu\nu\sigma}\phi_{\tau} , \\ \text{Density of weight } w & \phi_{;\mu\nu} - \phi_{;\nu\mu} = 0 . \end{array}$$

The last of these follows from the identity

$$R^{\sigma}_{\sigma\mu\nu} = \Gamma^{\sigma}_{\sigma\nu,\mu} - \Gamma^{\sigma}_{\sigma\mu,\nu} = 0 ,$$

which in turn follows from

$$\begin{aligned} \Gamma^{\sigma}_{\sigma\mu} &= g^{\sigma\tau}\Gamma_{\sigma\tau\mu} = \frac{1}{2}g^{\sigma\tau}(g_{\sigma\tau,\mu} + g_{\sigma\mu,\tau} - g_{\tau\mu,\sigma}) \\ &= \frac{1}{2}g^{-1}g_{,\mu} = \frac{1}{2}(\log g)_{,\mu} . \end{aligned}$$

The $R^{\tau}{}_{\sigma\mu\nu}$ are components of a mixed tensor known as the *Riemann tensor* or *curvature tensor* of the manifold. Note that we are able to infer the tensor character without once computing the coordinate transformation law for the Christoffel symbols out of which the components of the curvature tensor are built.

The curvature tensor satisfies some important algebraic and differential identities. These are most easily derived by first introducing a special coordinate system. Let x_0 be an arbitrary point of the manifold and let $(\Gamma^{\mu}_{\nu\sigma})_0$ be the Christoffel symbol at that point in the coordinate system x^{μ}. Now introduce a new set of coordinates $\bar{x}^{\mu}$ related to the old ones by

$$x^{\mu} = \bar{x}^{\mu} - \frac{1}{2}(\Gamma^{\mu}_{\nu\sigma})_0\left(\bar{x}^{\nu} - \bar{x}^{\nu}_0\right)\left(\bar{x}^{\sigma} - \bar{x}^{\sigma}_0\right) .$$

We have

$$\bar{x}^{\mu}_0 = x^{\nu}_0 , \qquad \left(\frac{\partial x^{\mu}}{\partial \bar{x}^{\nu}}\right)_0 = \delta^{\mu}_{\nu} , \qquad \left(\frac{\partial^2 x^{\mu}}{\partial \bar{x}^{\nu}\partial \bar{x}^{\sigma}}\right)_0 = -(\Gamma^{\mu}_{\nu\sigma})_0 ,$$

$$(\bar{g}_{\mu\nu})_0 = \left(\frac{\partial x^{\sigma}}{\partial \bar{x}^{\mu}}\frac{\partial x^{\tau}}{\partial \bar{x}^{\nu}}g_{\sigma\tau}\right)_0 = (g_{\mu\nu})_0 ,$$

$$\begin{aligned} (\bar{g}_{\mu\nu,\sigma})_0 &= \left[\frac{\partial}{\partial \bar{x}^{\sigma}}\left(\frac{\partial x^{\tau}}{\partial \bar{x}^{\mu}}\frac{\partial x^{\rho}}{\partial \bar{x}^{\nu}}g_{\tau\rho}\right)\right]_0 \\ &= \left(\frac{\partial^2 x^{\tau}}{\partial \bar{x}^{\sigma}\partial \bar{x}^{\mu}}\frac{\partial x^{\rho}}{\partial \bar{x}^{\nu}}g_{\tau\rho} + \frac{\partial x^{\tau}}{\partial \bar{x}^{\mu}}\frac{\partial^2 x^{\rho}}{\partial \bar{x}^{\sigma}\partial \bar{x}^{\nu}}g_{\tau\rho} + \frac{\partial x^{\tau}}{\partial \bar{x}^{\mu}}\frac{\partial x^{\rho}}{\partial \bar{x}^{\nu}}\frac{\partial x^{\lambda}}{\partial \bar{x}^{\sigma}}g_{\tau\rho,\lambda}\right)_0 \end{aligned}$$

$$
\begin{aligned}
&= \left(-\Gamma^{\tau}_{\sigma\mu} g_{\tau\nu} - \Gamma^{\rho}_{\sigma\nu} g_{\mu\rho} + g_{\mu\nu,\sigma}\right)_0 \\
&= \left(-\Gamma_{\nu\sigma\mu} - \Gamma_{\mu\sigma\nu} + g_{\mu\nu,\sigma}\right)_0 \\
&= \left[-\frac{1}{2}(g_{\nu\sigma,\mu} + g_{\nu\mu,\sigma} - g_{\sigma\mu,\nu}) - \frac{1}{2}(g_{\mu\sigma,\nu} + g_{\mu\nu,\sigma} - g_{\sigma\nu,\mu}) + g_{\mu\nu,\sigma}\right]_0 \\
&= 0\,.
\end{aligned}
$$

In the new coordinate system, the first derivatives of the metric tensor, and hence also the Christoffel symbols, vanish at the point x_0. That is to say, we can always introduce a coordinate system in which the derivatives of $g_{\mu\nu}$ and the Christoffel symbols vanish at a given fixed point.

Now, dropping the subscript 0, let x be an arbitrary point of the manifold. Then in a coordinate system in which the Christoffel symbols vanish at that point, the curvature tensor takes the form

$$
\begin{aligned}
R_{\tau\sigma\mu\nu} &= \Gamma_{\tau\sigma\nu,\mu} - \Gamma_{\tau\sigma\mu,\nu} \\
&= \frac{1}{2}(g_{\tau\sigma,\nu\mu} + g_{\tau\nu,\sigma\mu} - g_{\sigma\nu,\tau\mu} - g_{\tau\sigma,\mu\nu} - g_{\tau\mu,\sigma\nu} + g_{\sigma\mu,\tau\nu}) \\
&= -\frac{1}{2}(g_{\tau\mu,\sigma\nu} + g_{\sigma\nu,\tau\mu} - g_{\tau\nu,\sigma\mu} - g_{\sigma\mu,\tau\nu})\,, \qquad (4.2)
\end{aligned}
$$

at that point, and its covariant derivative takes the form

$$
R_{\tau\sigma\mu\nu;\rho} = -\frac{1}{2}(g_{\tau\mu,\sigma\nu\rho} + g_{\sigma\nu,\tau\mu\rho} - g_{\tau\nu,\sigma\mu\rho} - g_{\sigma\mu,\tau\nu\rho})\,,
$$

at that point. From these forms, one may readily infer the identities:

$$
\begin{aligned}
&R_{\tau\sigma\mu\nu} = -R_{\tau\sigma\nu\mu} = R_{\mu\nu\tau\sigma}\,, \\
&R_{\tau\sigma\mu\nu} + R_{\tau\mu\nu\sigma} + R_{\tau\nu\sigma\mu} = 0\,, \\
&R_{\tau\sigma\mu\nu;\rho} + R_{\tau\sigma\nu\rho;\mu} + R_{\tau\sigma\rho\mu;\nu} = 0\,.
\end{aligned}
$$

Since the quantities in these identities are components of tensors, the identities hold not merely in the special coordinate system, but in any coordinate system at x. Moreover, because x is arbitrary, they actually hold everywhere. The differential identity is known as the *Bianchi identity*.

It is conventional to introduce special symbols for the contracted forms of the curvature tensor. There are only two nontrivial contractions:

$$
\begin{aligned}
R_{\mu\nu} &= R^{\sigma}_{\mu\sigma\nu} \\
&= \Gamma^{\sigma}_{\mu\nu,\sigma} - \Gamma^{\sigma}_{\sigma\mu,\nu} + \Gamma^{\sigma}_{\rho\sigma}\Gamma^{\rho}_{\mu\nu} - \Gamma^{\sigma}_{\rho\nu}\Gamma^{\rho}_{\mu\sigma} \\
&= R_{\nu\mu}\,,
\end{aligned}
$$

and

$$
R = R^{\mu}_{\mu}\,.
$$

The quantity $R_{\mu\nu}$ is known as the *Ricci tensor*, and R is known as the *Riemann scalar* or *curvature scalar*.

4.6 *n*-Beins, Tetrads and Flat Manifolds

The coordinate transformation which brings the metric tensor into canonical form at any point may be chosen linear. Denote the coefficients of such a transformation by e^{α}_{μ}. Then

$$e^{\alpha}_{\mu}e^{\beta}_{\nu}g^{\mu\nu} = \eta^{\alpha\beta} ,$$

where $\eta^{\alpha\beta}$ is the contravariant form of the canonical metric (and hence constant). The e^{α}_{μ} are known as the components of an n-bein or, in the case of spacetime, of a *tetrad*. The e^{α}_{μ} are not uniquely determined at any point but may be subjected to linear transformations of the form

$$\overline{e}^{\alpha}_{\mu} = L^{\alpha}_{\beta}e^{\beta}_{\mu}$$

that leave the canonical metric invariant:

$$L^{\alpha}_{\gamma}L^{\beta}_{\delta}\eta^{\gamma\delta} = \eta^{\alpha\beta} .$$

In the case of spacetime, these transformations are homogeneous Lorentz transformations.

The e^{α}_{μ} need not vary discontinuously from point to point. Since the $g^{\mu\nu}$ are differentiable, at least throughout a given coordinate patch, the e^{α}_{μ} may likewise be chosen differentiable throughout appropriate (overlapping) patches. The e^{α}_{μ} are then said to be components of an n-bein or tetrad field. For each α, e^{α}_{μ} ($\mu = 1, ..., n$ or $\mu = 0, 1, 2, 3$) are the components of a covariant vector. These vectors define a local canonical frame at each point of the manifold. We have already encountered examples of such local frames in the u^{μ}, n^{μ}_{i} (see Sect. 1.2), which together constitute a tetrad.

It will be a convenience to raise and lower indices α, β, etc., from the first part of the Greek alphabet by means of the covariant and contravariant forms of the canonical metric, viz., $\eta_{\alpha\beta}$ and $\eta^{\alpha\beta}$, just as we raise and lower indices μ, ν, etc., from the middle of the Greek alphabet by means of $g^{\mu\nu}$ and $g_{\mu\nu}$. We then have

$$e_{\alpha\mu}e^{\beta\mu} = \delta^{\beta}_{\alpha} ,$$

which implies

$$e^{\mu}_{\alpha}e^{\nu}_{\beta}g_{\mu\nu} = \eta_{\alpha\beta} , \qquad e_{\alpha\mu}e^{\alpha\nu} = \delta^{\nu}_{\mu} ,$$
$$e^{\alpha}_{\mu}e^{\beta}_{\nu}\eta_{\alpha\beta} = g_{\mu\nu} , \qquad e^{\mu}_{\alpha}e^{\nu}_{\beta}\eta^{\alpha\beta} = g^{\mu\nu} ,$$

and so on.

Suppose now the curvature tensor vanishes. It is then possible to introduce, throughout every coordinate patch, an n-bein field, i.e., n linearly independent vector fields, whose covariant derivatives all vanish:

$$e^{\alpha}_{\mu;\nu} = 0 \,.$$

This is so because the integrability condition for these equations is automatically satisfied:

$$0 = e^{\alpha}_{\mu;\nu\sigma} - e^{\alpha}_{\mu;\sigma\nu} = e^{\alpha}_{\tau} R^{\tau}_{\mu\nu\sigma} \,.$$

Such an n-bein field is obtained by taking a local canonical frame at any one point and displacing it in a parallel fashion throughout the coordinate patch. When the curvature tensor vanishes, parallel displacement becomes integrable, and the concept of local parallelism may be extended to a concept of distant parallelism throughout the patch. The resulting n-bein defines a canonical frame over the patch. The canonical coordinates $\bar{x}^{\alpha}$ may be obtained from the original coordinates x^{μ} by integrating the equations

$$\frac{\partial \bar{x}^{\alpha}}{\partial x^{\mu}} = e^{\alpha}_{\mu} \,,$$

the solubility of which is guaranteed by

$$\begin{aligned} \frac{\partial^2 \bar{x}^{\alpha}}{\partial x^{\mu} \partial x^{\nu}} - \frac{\partial^2 \bar{x}^{\alpha}}{\partial x^{\nu} \partial x^{\mu}} &= e^{\alpha}_{\nu,\mu} - e^{\alpha}_{\mu,\nu} \\ &= e^{\alpha}_{\nu;\mu} - e^{\alpha}_{\mu;\nu} + (\Gamma^{\sigma}_{\nu\mu} - \Gamma^{\sigma}_{\mu\nu}) e^{\alpha}_{\sigma} = 0 \,. \end{aligned}$$

The metric tensor in the canonical coordinates is obviously everywhere $\eta_{\alpha\beta}$. Any manifold that can be covered by canonical patches is said to be *flat*. The necessary and sufficient condition that a Riemannian manifold be flat in a given region is evidently that its curvature tensor vanish in that region.

Chapter 5
The Free Particle: Geodesics

We return now to physics by considering the simplest of all physical systems, the free particle. In special relativity, which is the theory of flat spacetime, the trajectory of a free particle is given, in a canonical (Minkowskian) coordinate system, by a set of linear equations. The world line of the particle is therefore *straight*. If the metric of special relativity were positive definite, the world line would be the shortest path between any two points located on it. The metric is actually Lorentzian and the world line of any real particle is time-like, i.e., $\eta_{\mu\nu}\mathrm{d}x^{\mu}\mathrm{d}x^{\nu} < 0$. Therefore, instead of working with an imaginary distance $\mathrm{d}s$ along the world line, one introduces the proper time $\mathrm{d}\tau$, where $\mathrm{d}\tau^2 = -\eta_{\mu\nu}\mathrm{d}x^{\mu}\mathrm{d}x^{\nu}$. By integrating $\mathrm{d}\tau$ along the world line, one then extends the distance concept from a local concept to a global idea of time along the world line. It then turns out that the straight world line is the path that *maximizes* the proper time between any two points on it.

Whether maximizing or minimizing something, the important point is that the world line of a particle is the solution of a variational problem in which a certain functional of the trajectory is extremized. In physics, it is conventional to call this functional the *action*.

Generalizing from flat spacetime to curved spacetime, general relativity chooses for the action functional of a free particle and the expression will be

$$S = -m \int \mathrm{d}\tau, \quad \mathrm{d}\tau^2 = -g_{\mu\nu}\mathrm{d}x^{\mu}\mathrm{d}x^{\nu}. \tag{5.1}$$

The factor m is the *rest mass* of the particle and is introduced in order to give S the dimensions of action. The minus sign is introduced so that the variational problem corresponds to a *least* action principle, as is conventional in physics.

It is convenient to express the world line of the particle in parametric form $x^{\mu} = z^{\mu}(\lambda)$, where λ is a parameter that increases monotonically as one moves along the line toward the future, but that is otherwise arbitrary. The action then takes the form

B. DeWitt, *Bryce DeWitt's Lectures on Gravitation*,
Edited by Steven M. Christensen, Lecture Notes in Physics, 826,
DOI: 10.1007/978-3-540-36911-0_5, © Springer-Verlag Berlin Heidelberg 2011

$$S = -m \int (-\dot{z}^2)^{1/2} \mathrm{d}\lambda,$$

where we use the abbreviated notation

$$A^2 = A \cdot A, \quad A \cdot B = g_{\mu\nu}(x) A^\mu(x) B^\nu(x), \quad \dot{z}^\mu(\lambda) = \frac{\mathrm{d}}{\mathrm{d}\lambda} z^\mu(\lambda).$$

In canonical coordinates in flat spacetime, if we set $\lambda = x^0 = t$, we obtain the familiar action of special relativity:

$$S = -m \int \sqrt{1 - \dot{z}^2} \mathrm{d}t.$$

In order to compute the variation equations in the general case, it will be advantageous first to introduce the concepts of *covariant variation* and covariant differentiation with respect to the parameter λ. Let ϕ be the set of components of an arbitrary tensor density defined along the world line $z^\mu(\lambda)$, which depends both on the world line and on the point selected along the world line, i.e., which is a functional of both $z^\mu(\lambda)$ (as a function of λ) and λ as a parameter. Let the world line be modified by a displacement $\delta z^\mu(\lambda)$ that vanishes outside a certain interval. The components of the tensor density will suffer a corresponding variation $\delta\phi$. Let this variation be compared with the variation $\delta_\| \phi$ that ϕ suffers under parallel displacement through the interval $\delta z^\mu(\lambda)$ [see Sect. 4.1 in Chap. 4]:

$$\delta_\| \phi = -G^\nu_\mu \Gamma^\mu_{\nu\sigma} \phi \delta z^\sigma.$$

The difference defines the covariant variation of ϕ:

$$\bar{\delta}\phi = \delta\phi - \delta_\| \phi = \delta\phi + G^\nu_\mu \Gamma^\mu_{\nu\sigma} \phi \delta z^\sigma.$$

If ϕ is a field defined throughout the manifold, then its dependence on the world line is a simple point dependence $\phi(z(\lambda))$, so that

$$\delta\phi = \phi_{,\mu} \delta z^\mu,$$

and its covariant variation is given by

$$\bar{\delta}\phi = \phi_{;\mu} \delta z^\mu.$$

We note, in particular, that the covariant variation of the metric tensor vanishes:

$$\bar{\delta} g_{\mu\nu} = 0.$$

By choosing the displacement δz^μ to be tangent to the world line, we may, in a similar manner, define the covariant derivative[1] of ϕ with respect to λ:

$$\dot{\phi} = \frac{\mathrm{D}}{\mathrm{D}\lambda}\phi = \frac{\mathrm{d}\phi}{\mathrm{d}\lambda} + G^\nu_\mu \Gamma^\mu_{\nu\sigma}\phi\dot{z}^\sigma.$$

If ϕ is a field we have

$$\dot{\phi} = \phi_{;\mu}\dot{z}^\mu,$$

and in particular,

$$\dot{g}_{\mu\nu} = 0.$$

When there is no chance of confusion, the dot will be used in preference to the symbol D/Dλ. For example, we shall write

$$\ddot{z}^\mu = \frac{\mathrm{D}}{\mathrm{D}\lambda}\dot{z}^\mu, \quad \dddot{z}^\mu = \frac{\mathrm{D}}{\mathrm{D}\lambda}\ddot{z}^\mu,$$

and so on.

Covariant variation and differentiation, like ordinary variation and differentiation, obey the Liebniz rule when applied to factors in a product. Moreover, when applied to scalars, they reduce to ordinary variation and differentiation. Using these facts, together with

$$\begin{aligned}\bar{\delta}\dot{z}^\mu &= \delta\dot{z}^\mu + \Gamma^\mu_{\nu\sigma}\dot{z}^\nu\delta z^\sigma \\ &= \frac{\mathrm{d}}{\mathrm{d}\lambda}\delta z^\mu + \Gamma^\mu_{\nu\sigma}\delta z^\nu\dot{z}^\sigma = \frac{\mathrm{D}}{\mathrm{D}\lambda}\delta z^\mu,\end{aligned}$$

we may now express the least action principle in the form

$$\begin{aligned}0 = \delta S &= m\int \left(-\dot{z}^2\right)^{-1/2} g_{\mu\nu}\dot{z}^\nu\bar{\delta}\dot{z}^\mu \mathrm{d}\lambda \\ &\underset{\lambda\to\tau}{\longrightarrow} m\int g_{\mu\nu}\dot{z}^\nu\frac{\mathrm{D}}{\mathrm{D}\tau}\delta z^\mu \mathrm{d}\lambda = -\int \dot{p}_\mu\delta z^\mu \mathrm{d}\lambda,\end{aligned}$$

where

$$p_\mu = m g_{\mu\nu}\dot{z}^\nu \qquad (\lambda = \tau),$$

[1] The covariant proper time derivative may be used to generalize the concept of Fermi–Walker transport (see Sect. 2.3 in Chap. 2) to arbitrary Riemannian manifolds. A tensor ϕ is said to be Fermi–Walker transported along a curve having unit tangent vector u^μ if it satisfies the equation $\dot{\phi} = (u^\mu\dot{u}_\nu - \dot{u}^\mu u_\nu)G^\nu_\mu\phi$.

and where, in passing to the last line, λ has been set equal to the proper time and an integration by parts has been carried out. We shall normally always set λ equal to the proper time immediately after performing all variations required by a given problem. We then have

$$u^2 = -1, \quad \text{where } u^\mu = \dot{z}^\mu$$

and

$$p^2 + m^2 = 0, \quad \text{with } p_\mu = m u_\mu.$$

The p_μ are the components of the four-momentum of the particle.

Because of the arbitrariness of δz^μ, the least action principle leads us to the dynamical equations

$$\dot{p}_\mu = 0.$$

Although in general relativity it is no longer possible to say that the four-momentum of a free particle is constant, we may say that its covariant proper time derivative vanishes.

By factoring out the rest mass and raising the index, one may rewrite the dynamical equations for the free particle in the form

$$0 = \ddot{z}^\mu = \frac{d^2 z^\mu}{d\tau^2} + \Gamma^\mu_{\nu\sigma} \frac{dz^\nu}{d\tau} \frac{dz^\sigma}{d\tau}. \tag{5.2}$$

An alternative version is obtained through multiplication by $d\tau$:

$$du^\mu = -\Gamma^\mu_{\nu\sigma} u^\nu dz^\sigma.$$

This version says that the world line of the particle may be generated by repeatedly displacing its tangent vector in a parallel fashion along itself. That is, not only does the world line maximize the proper time, but also it is "self-parallel." In a Riemannian manifold, a curve having these properties is called a *geodesic*. Equation 5.2 are known as the *geodesic equations*. Note that the geodesic equations involve no mass parameter. Hence, all bodies (particles) behave alike in free fall in a gravitational field. Only under collisions, or in the quantum theory, does mass make a difference in the absence of external (non-gravitational) forces.

Geodesics in spacetime may also be spacelike or null. They all satisfy the geodesic equations. The parameter τ, however, is no longer proper time but is known simply as an *affine parameter*. Spacelike geodesics neither maximize nor minimize the distance between any two points on them; they merely make it stationary. For spacelike geodesics, the affine parameter may be normalized to equal the arc length. There is no natural normalization of the affine parameter for null geodesics. It should be noted that the geodesic equations themselves guarantee the constancy of $\dot{z}^2$ along every geodesic.

5.1 Isometries and Conservation Laws

Although we have seen that the values of the components of the four-momentum of a particle are not generally conserved in general relativity, there are cases in which conserved quantities exist other than the trivial quantity m. Suppose there exists a contravariant vector ξ such that the Lie derivative of the metric tensor with respect to it vanishes everywhere:

$$0 = \mathcal{L}_\xi g_{\mu\nu} = \xi_{\mu;\nu} + \xi_{\nu;\mu}.$$

Such a field is known as a *Killing vector field* and the above equation is known as *Killing's equation*. The significance of a Killing vector field is the following. Suppose we carry out the coordinate transformation

$$\bar{x}^\mu = x^\mu + \varepsilon\xi^\mu,$$

where ε is infinitesimal. Then the functional form of the metric tensor will suffer the infinitesimal change

$$\delta g_{\mu\nu} = -\mathcal{L}_{\varepsilon\xi} g_{\mu\nu} = -\varepsilon\mathcal{L}_\xi g_{\mu\nu}.$$

However, in view of Killing's equations, this change vanishes. That is, the metric tensor looks the same in the new coordinate system as in the old. Speaking more physically, the geometry of spacetime looks the same from the point of view of the new coordinate system as it did from the point of view of the old. Now remember that coordinate transformations can be thought of as generated by diffeomorphisms, which in effect *drag* the coordinate mesh to a new location. In the case of a coordinate transformation generated by a Killing vector, we can go further. We can imagine the very manifold as moving or sliding on itself. The invariance of the geometry under this motion assures that the manifold remains congruent to itself in its new location. Such a motion is called an *isometry*.

The existence of a Killing vector field, and hence of an isometry, is evidently a property of the geometry of the manifold. Isometries may be performed successively on a given manifold, and the set of all isometries admitted by the manifold forms a group known as the *isometry group* or the *group of motions* of the manifold. An isometry group is always a Lie group, and in the case of spacetime, its dimensionality can never exceed 10. The maximum dimensionality is reached, for example, in the case of flat spacetime, whose isometry group is the Poincaré group.

If there exists a Killing vector field ξ, then any world line $z^\mu(\tau)$, whether or not it satisfies the geodesic equations, will encounter precisely the same physical (geometrical) environment after it has been displaced by an amount

$$\delta z^\mu(\tau) = \varepsilon\xi^\mu(z^\mu(\tau))$$

as it encountered before. This means that the action will remain invariant under such a displacement (see present chapter):

$$0 = \delta S = \int p_\mu \frac{\mathrm{D}}{\mathrm{D}\tau} \delta z^\mu \mathrm{d}\tau = \varepsilon \int p_\mu \dot{\xi}^\mu \mathrm{d}\tau.$$

If the geodesic equations *are* satisfied, this implies

$$\int \frac{\mathrm{d}}{\mathrm{d}\tau}(p \cdot \xi)\mathrm{d}\tau = 0,$$

for all integration intervals, which in turn implies that $p \cdot \xi$ is a *conserved quantity* or *constant of the motion*:

$$p \cdot \xi = \text{constant}.$$

The direct verification of this is immediate:

$$\begin{aligned} \frac{\mathrm{d}}{\mathrm{d}\tau}(p \cdot \xi) &= \dot{p} \cdot \xi + p \cdot \dot{\xi} = p_\mu \xi^\mu_{;\nu} \dot{z}^\nu \\ &= m u^\mu u^\nu \xi_{\mu;\nu} = \frac{1}{2} m u^\mu u^\nu (\xi_{\mu;\nu} + \xi_{\nu;\mu}) = 0. \end{aligned}$$

5.2 Geodesic Deviation

Covariant variation and covariant differentiation with respect to λ, unlike ordinary variation and differentiation, do not commute. Their commutation law is obtained by the following calculation:

$$\begin{aligned} \bar{\delta}\dot{\phi} - \frac{\mathrm{D}}{\mathrm{D}\lambda}\bar{\delta}\phi &= \delta\left(\frac{\mathrm{d}\phi}{\mathrm{d}\lambda} + G^\nu_\mu \Gamma^\mu_{\nu\sigma} \phi \dot{z}^\sigma\right) + G^\nu_\mu \Gamma^\mu_{\nu\sigma}\left(\frac{\mathrm{d}\phi}{\mathrm{d}\lambda} + G^\lambda_\rho \Gamma^\rho_{\lambda\tau} \phi \dot{z}^\tau\right)\delta z^\sigma \\ &\quad - \frac{\mathrm{d}}{\mathrm{d}\lambda}\left(\delta\phi + G^\nu_\mu \Gamma^\mu_{\nu\sigma} \phi \delta z^\sigma\right) - G^\lambda_\rho \Gamma^\rho_{\lambda\tau}\left(\delta\phi + G^\nu_\mu \Gamma^\mu_{\nu\sigma} \phi \delta z^\sigma\right)\dot{z}^\tau \\ &= G^\nu_\mu\left(\Gamma^\mu_{\nu\tau,\sigma} - \Gamma^\mu_{\nu\sigma,\tau}\right)\phi \dot{z}^\tau \delta z^\sigma + \left(\delta^\nu_\rho G^\lambda_\mu - \delta^\lambda_\nu G^\nu_\rho\right)\Gamma^\mu_{\nu\sigma}\Gamma^\rho_{\lambda\tau}\phi \dot{z}^\tau \delta z^\sigma \\ &= G^\nu_\mu\left(\Gamma^\mu_{\nu\tau,\sigma} - \Gamma^\mu_{\nu\sigma,\tau} - \Gamma^\mu_{\rho\sigma}\Gamma^\rho_{\nu\tau} - \Gamma^\mu_{\rho\tau}\Gamma^\rho_{\nu\sigma}\right)\phi \dot{z}^\tau \delta z^\sigma \\ &= G^\nu_\mu R^\mu_{\nu\sigma\tau} \phi \dot{z}^\tau \delta z^\sigma \\ &= -G^\nu_\mu R^\mu_{\nu\sigma\tau} \phi \dot{z}^\sigma \delta z^\tau. \end{aligned}$$

Problem 22 In the special case that the tensor ϕ is a field, the commutation law of covariant variation and differentiation of ϕ may be derived directly from the commutation law for covariant differentiation with respect to coordinates. Carry out this derivation.

Solution 22 We have

$$\bar{\delta}\dot{\phi} - \frac{D}{D\lambda}\bar{\delta}\phi = \bar{\delta}(\phi_{;\mu}\dot{z}^\mu) - \frac{D}{D\lambda}(\phi_{;\mu}\delta z^\mu)$$
$$= \phi_{;\mu\nu}\delta z^\nu \dot{z}^\mu + \phi_{;\mu}\bar{\delta}\dot{z}^\mu - \phi_{;\mu\nu}\dot{z}^\nu\delta z^\mu - \phi_{;\mu}\frac{D}{D\lambda}\delta z^\mu$$
$$= (\phi_{;\mu\nu} - \phi_{;\nu\mu})\dot{z}^\mu\delta z^\nu = -G^\sigma_\tau R^\tau_{\sigma\mu\nu}\phi\dot{z}^\mu\delta z^\nu.$$

A commutation law like the above may seem terribly abstract and of little practical importance. The appearance is misleading. In Riemannian geometry such commutation laws usually involve the curvature tensor, and in general relativity the curvature tensor gives a measure of the gravitational field. (Where it vanishes, the field is zero, etc.) Frequently, the derivation of some important physical result involving the curvature tensor can be most expeditiously carried out by performing a commutation operation that yields the curvature tensor. The derivation of the equation of geodesic deviation is a good illustration of this.

Suppose we have two free-particle world lines (geodesics) that are separated only by a very small (infinitesimal) distance. One may be regarded as a variation of the other. In order to use the variational calculus, we shall need (prior to the variation) to have the geodesic equations in the unconstrained form

$$\frac{D}{D\lambda}\left[\left(-\dot{z}^2\right)^{-1/2}\dot{z}^\mu\right] = 0,$$

in which the parameter λ has not yet been set equal to τ. The change from one world line to the other leaves these equations intact. The covariant mathematical statement of this fact is

$$\bar{\delta}\frac{D}{D\lambda}\left[\left(-\dot{z}^2\right)^{-1/2}\dot{z}^\mu\right] = 0,$$

with the basic variation $\delta z^\mu(\lambda)$ being understood to be the interval between points on the two world lines having the same value of λ. By interchanging the operations $\bar{\delta}$ and D/Dλ, we may convert this statement into the following:

$$0 = \frac{D}{D\lambda}\bar{\delta}\left[\left(-\dot{z}^2\right)^{-1/2}\dot{z}^\mu\right] - R^\mu_{\nu\sigma\tau}\left(-\dot{z}^2\right)^{-1/2}\dot{z}^\nu\dot{z}^\sigma\delta z^\tau$$
$$= \frac{D}{D\lambda}\left[\left(-\dot{z}^2\right)^{-3/2}\dot{z}^\mu(\dot{z}\cdot\bar{\delta}\dot{z}) + \left(-\dot{z}^2\right)^{-1/2}\bar{\delta}\dot{z}^\mu\right] - R^\mu_{\nu\sigma\tau}\left(-\dot{z}^2\right)^{-1/2}\dot{z}^\nu\dot{z}^\sigma\delta z^\tau$$
$$\underset{\lambda\to\tau}{\longrightarrow} \frac{D}{D\tau}\left(P^\mu_\nu\frac{D}{D\tau}\delta z^\nu\right) - R^\mu_{\nu\sigma\tau}u^\nu u^\sigma\delta z^\tau,$$

where

$$P^{\mu\nu} = g^{\mu\nu} + u^\mu u^\nu, \quad u^\mu = \dot{z}^\mu \quad (\lambda = \tau).$$

We now note that

$$\frac{\mathrm{D}}{\mathrm{D}\tau} P^{\mu}_{\ \nu} = 0,$$

because of the geodesic equations ($\dot{u}^{\mu} = 0$) themselves, and that

$$R^{\mu}_{\ \nu\sigma\tau} u^{\nu} u^{\sigma} P^{\tau}_{\ \rho} = R^{\mu}_{\ \nu\sigma\rho} u^{\nu} u^{\sigma},$$

because of the symmetries of the Riemann tensor. Therefore, the variational equation may be brought to its final form, known as the *equation of geodesic deviation*

$$\ddot{\eta}^{\mu} = R^{\mu}_{\ \nu\sigma\tau} u^{\nu} u^{\sigma} \eta^{\tau}, \tag{5.3}$$

where η^{μ} is the perpendicular displacement from one world line to the other:

$$\eta^{\mu} = P^{\mu}_{\ \nu} \delta z^{\nu}.$$

Chapter 6
Weak Field Approximation. Newton's Theory

In the region of spacetime containing the world lines of the sun and planets, spacetime is flat to a very high degree of approximation. This means that a coordinate system may be introduced for the whole solar system which is very nearly canonical, i.e., inertial. In such a coordinate system, the metric tensor may be written

$$g_{\mu\nu} = \eta_{\mu\nu} + h_{\mu\nu}, \quad \text{where } |h_{\mu\nu}| \lesssim \varepsilon \ll 1, \quad \forall \mu, \nu. \tag{6.1}$$

This coordinate system is not unique. In addition to rotations and weak (low velocity) Lorentz boosts, general coordinate transformations $\bar{x}^\mu = x^\mu + \xi^\mu$ may also be introduced all of which leave the quasi-canonical character of the coordinate system intact. In the latter case, we must only require

$$|\xi^\mu_{,\nu}| \lesssim \varepsilon \text{ and } |h_{\mu\nu,\sigma}\xi^\sigma| \lesssim \varepsilon^2, \quad \forall \mu, \nu. \tag{6.2}$$

To first order in small quantities, $h_{\mu\nu}$ suffers the transformation

$$\bar{h}_{\mu\nu} = h_{\mu\nu} - \xi_{\mu,\nu} - \xi_{\nu,\mu}, \quad \xi_\mu = \eta_{\mu\nu}\xi^\nu,$$

under such a change in coordinates.

It is sometimes convenient to fix the coordinate system partially by imposing the *supplementary condition*

$$l^{\mu\nu}_{,\nu} = 0,$$

where

$$l_{\mu\nu} = h_{\mu\nu} - \frac{1}{2}\eta_{\mu\nu}h, \quad h = h^\sigma_\sigma,$$

$$h_{\mu\nu} = l_{\mu\nu} - \frac{1}{2}\eta_{\mu\nu}l, \quad l = l^\sigma_\sigma = -h.$$

B. DeWitt, *Bryce DeWitt's Lectures on Gravitation*,
Edited by Steven M. Christensen, Lecture Notes in Physics, 826,
DOI: 10.1007/978-3-540-36911-0_6, © Springer-Verlag Berlin Heidelberg 2011

Here and elsewhere when working in the weak field approximation, we shall use the Minkowski metric to raise and lower indices. Because $l_{\mu\nu}$ obeys the coordinate transformation law

$$\bar{l}_{\mu\nu} = l_{\mu\nu} - \xi_{\mu,\nu} - \xi_{\nu,\mu} + \eta_{\mu\nu}\xi^{\sigma}_{,\sigma},$$

whence

$$\bar{l}^{\mu\nu}_{,\nu} = l^{\mu\nu}_{,\nu} - \xi^{\mu\nu}_{,\nu} - \xi^{\nu\mu}_{,\nu} - \xi^{\sigma\mu}_{,\sigma} = l^{\mu\nu}_{,\nu} + \xi^{\mu\nu}_{,\nu},$$

it is evident that $\bar{l}_{\mu\nu}$ can be made to satisfy the supplementary condition by choosing ξ^{μ} to be a solution of the wave equation with a source:

$$\Box^2\xi^{\mu} \equiv \xi^{\mu\nu}_{,\nu} = l^{\mu\nu}_{,\nu}.$$

The supplementary condition does not fix the coordinate system completely, for it is left intact by low velocity Lorentz transformations and by coordinate transformations that satisfy the homogeneous wave equation

$$\Box^2\xi^{\mu} = 0.$$

Such "coordinate waves" are sometimes eliminated by imposing boundary conditions, leaving only the Lorentz transformations.

In the case of the solar system, the spacetime geometry has additional properties not possessed by all weak-field geometries. These properties stem from the fact that the relative velocities of all the planets are very small when compared to the velocity of light. If (as we may) we assume that there is a negligible amount of gravitational radiation in the solar system, then the sun and planets themselves become the primary sources of the $h_{\mu\nu}$ term in the metric and, because of the slowness with which the planets move, it follows that there exist quasi-canonical coordinate systems, e.g., systems fixed with respect to the sun, in which all time derivatives $h_{\mu\nu,0}$ may be neglected when compared to spatial derivatives $h_{\mu\nu,i}$ ($i = 1, 2, 3$). Such coordinate systems may be called *quasi-stationary*, and the gravitational field itself may be called quasi-stationary. Any two quasi-stationary coordinate systems are connected by (a) a rotation, (b) a general transformation $\bar{x}^{\mu} = x^{\mu} + \xi^{\mu}$ in which the ξ, besides satisfying the previous conditions, have negligible time derivatives, or (c) some combination of the two. Under the restricted class of transformations (b), the components of $h_{\mu\nu}$ and $l_{\mu\nu}$ transform according to

$$\begin{aligned} \bar{h}_{00} &= h_{00}, & \bar{l}_{00} &= l_{00} - \xi_{i,i}, \\ \bar{h}_{0i} &= \bar{h}_{0i} - \xi_{0,i}, & \bar{l}_{0i} &= l_{0i} - \xi_{0,i}, \\ \bar{h}_{ij} &= h_{ij} - \xi_{i,j} - \xi_{j,i}, & \bar{l}_{ij} &= l_{ij} - \xi_{i,j} - \xi_{j,i} + \delta_{ij}\xi_{k,k}. \end{aligned} \tag{6.3}$$

The supplementary condition can be imposed upon $\bar{l}_{\mu\nu}$ by choosing ξ^{μ} to satisfy

$$\nabla^2 \xi^{\mu} \equiv \xi^{\mu}_{,ii} = l^{\mu i}_{,i}. \tag{6.4}$$

It will be noted that h_{00} is the same for all quasi-stationary coordinate systems, being unaffected either by rotations or by restricted transformations $\bar{x}^{\mu} = x^{\mu} + \xi^{\mu}$. This uniqueness makes it possible for us now to establish contact between formalism and observation – more precisely, between Einstein's theory of gravity (general relativity) and Newton's. Suppose we have a freely falling body (particle) moving slowly (compared to light) with respect to a quasi-stationary coordinate system in a quasi-stationary gravitational field. The world line $z^{\mu}(\tau)$ of this body satisfies the geodesic equation. Because of the slow motion and the fact that $|h_{\mu\nu}| \ll 1$, we have

$$\frac{\mathrm{d}^2 z^i}{\mathrm{d}\tau^2} = \frac{\mathrm{d}^2 z^i}{\mathrm{d}t^2}\ (t = z^0) \quad \text{and} \quad \left|\frac{\mathrm{d}z^i}{\mathrm{d}\tau}\right| \ll \left|\frac{\mathrm{d}z^0}{\mathrm{d}\tau}\right| = 1,$$

so that the geodesic equation becomes

$$\begin{aligned} 0 &= \frac{\mathrm{d}^2 z^i}{\mathrm{d}t^2} + \Gamma^i_{00} = \frac{\mathrm{d}^2 z^i}{\mathrm{d}t^2} + \frac{1}{2}(h_{i0,0} + h_{i0,0} - h_{00,i}) \\ &= \frac{\mathrm{d}^2 z^i}{\mathrm{d}t^2} - \frac{1}{2} h_{00,i}. \end{aligned}$$

In the case of quasi-stationary weak fields, we may evidently make the identification

$$h_{00} = -2\Phi, \tag{6.5}$$

where Φ is the *Newtonian gravitational potential.*

It is useful to examine the conditions under which we may expect the gravitational field to be quasi-stationary and weak. First, we must have

$$\Phi \ll 1.$$

But in Newton's theory,

$$|\Phi| \sim \frac{GM}{R},$$

where M is the mass of the object (or objects) producing the gravitational field, R is the distance from the object, and G is the gravity constant. Hence, it must not be possible to get closer to the object than a distance R satisfying

$$R \gg GM,$$

before the idealization of regarding the object as a mass point breaks down. Effectively this means that the mass of the object must be spread over a distance

R satisfying the above condition and hence that the density ρ of the object must satisfy

$$G\rho \sim \frac{GM}{R^3} \ll \frac{1}{(GM)^2}. \tag{6.6}$$

It will be noted that a test body falling from rest at infinity to the surface of this object then acquires a velocity of only

$$v \sim \left(\frac{GM}{R}\right)^{1/2} \ll 1,$$

and hence that the slow motion condition will generally be maintained even under near collisions.

Another point of contact between formalism and observation can be established by referring to the coordinate system attached to a rotationless constantly accelerated rigid medium, the metric of which is given in Chap. 2. This metric is static and, over a range of ξ^i satisfying

$$|\xi^i a_{0i}| \ll 1,$$

may be regarded as quasi-canonical. In this range we have

$$h_{00} = -2\xi^i a_{0i},$$

which corresponds to an effective Newtonian potential

$$\Phi = \xi^i a_{0i}.$$

This is a statement[1] of the *principle of equivalence*: over a small region, the effects of a constant acceleration cannot be distinguished from those of a uniform gravitational field. (Of course, no real gravitational field is everywhere uniform.)

Still another point of contact can be established by referring to the phenomenon of the tides. The description of the tides in Newton's theory is based on a comparison of the dynamical equations of two test bodies that are separated from one another by a small interval η:

[1] This is the original principle of equivalence. Nowadays it is often referred to as the *weak* principle of equivalence to distinguish it from the *strong* principle of equivalence, which says that every valid Lorentz invariant theory can be immediately generalized to a valid general relativistic theory by first presenting the formalism of the theory in standard Minkowski (canonical) coordinates and then converting all ordinary spacetime derivatives to covariant derivatives. Unfortunately, the strong principle of equivalence is not fool proof. It is sometimes possible to present a Lorentz invariant theory in two different forms that generalize to distinct general relativistic theories. The latter theories differ from one another by the presence or the absence of certain terms involving the curvature tensor.

$$\frac{d^2 z^i}{dt^2} = -\Phi_{,i}(z), \quad \frac{d^2}{dt^2}(z^i + \eta^i) = -\Phi_{,i}(z + \eta).$$

Subtracting one equation from the other, we obtain

$$\frac{d^2 \eta^i}{dt^2} = -\Phi_{,ij}\eta^j. \tag{6.7}$$

The analogous equation in general relativity is the equation of geodesic deviation. To obtain the form that this equation takes in quasi-canonical coordinates, we must first examine the Riemann tensor in these coordinates; in particular, we must examine the relative magnitudes of the various terms of which it is composed. Let L be defined by[2]

$$|h_{\mu\nu,\sigma\tau}| \lesssim \frac{\varepsilon}{L^2}, \quad \forall \mu, \nu, \sigma, \tau.$$

(L may be regarded as the minimum distance over which any of the $h_{\mu\nu}$ changes by an appreciable fraction of itself.) By integration, we may then infer that

$$|h_{\mu\nu,\sigma}| \lesssim \frac{\varepsilon}{L}, \quad \forall \mu, \nu, \sigma.$$

From this it follows that the terms involving first derivatives of the metric tensor are of order ε^2/L^2, while those involving second derivatives are of order ε/L^2. Therefore we keep only the latter, obtaining [see (4.2)]

$$R_{\mu\nu\sigma\tau} = -\frac{1}{2}\left(h_{\mu\sigma,\nu\tau} + h_{\nu\tau,\mu\sigma} - h_{\mu\tau,\nu\sigma} - h_{\nu\sigma,\mu\tau}\right). \tag{6.8}$$

We must also determine the form taken by the second covariant proper time derivative:

$$\begin{aligned}
\ddot{\eta}^\mu &= \frac{d}{d\tau}\dot{\eta}^\mu + \Gamma^\mu_{\nu\sigma}\dot{\eta}^\nu u^\sigma \\
&= \frac{d}{d\tau}\left(\frac{d\eta^\mu}{d\tau} + \Gamma^\mu_{\nu\sigma}\eta^\nu u^\sigma\right) + \Gamma^\mu_{\nu\sigma}\left(\frac{d\eta^\nu}{d\tau} + \Gamma^\nu_{\tau\rho}\eta^\tau u^\rho\right)u^\sigma \\
&= \frac{d^2\eta^\mu}{d\tau^2} + \Gamma^\mu_{\nu\sigma,\tau}\eta^\nu u^\sigma u^\tau + 2\Gamma^\mu_{\nu\sigma}\frac{d\eta^\nu}{d\tau}u^\sigma - \Gamma^\mu_{\nu\sigma}\Gamma^\sigma_{\tau\rho}(\eta^\nu u^\tau - \eta^\tau u^\nu)u^\rho.
\end{aligned}$$

In passing to the final form, we have used the geodesic equation satisfied by u^μ. By the arguments we applied to the Riemann tensor, the last term is now to be dropped in comparison with the term in $\Gamma^\mu_{\nu\sigma,\tau}$.

In the case of slow motion in quasi-canonical coordinates, we have $(u^\mu) = (1, 0, 0, 0)$, $\eta^0 = 0$, $\tau = t$, and the equation of geodesic deviation takes the form

[2] Because, as we shall see later, the $h_{\mu\nu}$ satisfy a set of differential equations, none of the $h_{\mu\nu,\sigma\tau}$ is ever infinite and therefore $L > 0$.

$$\frac{\mathrm{d}^2\eta^i}{\mathrm{d}t^2} + \Gamma^i_{j0,0}\eta^j + 2\Gamma^i_{j0}\frac{\mathrm{d}\eta^j}{\mathrm{d}t} = R^i_{00j}\eta^j .$$

If, in addition, the field and coordinates are quasi-stationary, we have

$$\Gamma^i_{j0,0} = 0,$$

$$\Gamma^i_{j0} = \Gamma_{ij0} = \frac{1}{2}(g_{ij,0} + g_{i0,j} - g_{j0,i}) = \frac{1}{2}(h_{i0,j} - h_{j0,i}),$$

$$R^i_{00j} = R_{i00j} = -\frac{1}{2}(h_{i0,0j} + h_{0j,i0} - h_{ij,00} - h_{00,ij}) = \frac{1}{2}h_{00,ij},$$

and hence,

$$\frac{\mathrm{d}^2\eta^i}{\mathrm{d}t^2} = (h_{0j,i} - h_{0i,j})\frac{\mathrm{d}\eta^j}{\mathrm{d}t} + \frac{1}{2}h_{00,ij}\eta^j .$$

We note that all the terms in this equation are invariant under the restricted class of coordinate transformations given by (6.3), which maintain the quasi-canonical and quasi-stationary character of the coordinate system. The second term on the right is just the Newtonian term found in (6.7), if we again make the identification $h_{00} = -2\Phi$. The first term on the right is a *new* term that is predicted by Einstein's theory but *not* by Newton's. The field $h_{0j,i} - h_{0i,j}$ appearing in this term is known as the *Lense–Thirring field*. We shall see later that it is normally extremely weak and becomes appreciable only in the vicinity of rapidly spinning matter.[3]

In the case of the quasi-canonical coordinate system attached to the constantly accelerating medium, we note that $h_{0i} = 0$ and $h_{00,ij} = 0$, and hence there is no tidal effect. This is not surprising, as we already know that spacetime is flat in this case, so that $R_{\mu\nu\sigma\tau} = 0$. But it emphasizes the fact that it is the presence of tidal forces that signals the presence of a real gravitational field as opposed to a uniform or acceleration field.

In the presence of a mass M located at the origin of coordinates, we have

$$\Phi = -\frac{GM}{|\boldsymbol{x}|}, \quad \Phi_{,i} = GM\frac{x^i}{|\boldsymbol{x}|^3},$$

$$-\Phi_{,ij} = \frac{GM}{|\boldsymbol{x}|^3}(3\hat{x}^i\hat{x}^j - \delta_{ij}), \quad \hat{x}^i = \frac{x^i}{|\boldsymbol{x}|}, \quad |\boldsymbol{x}| \neq 0.$$

From this one may easily see that the tidal forces exerted on any spherical body in the neighborhood of this mass tend to draw the body out into a prolate ellipsoid having its long axis in the direction of $\hat{\boldsymbol{x}}$.

---

[3] The gradient $(h_{0j,i} - h_{0i,j})_{,k}$ of the Lense–Thirring field is, in quasi-stationary coordinates, equal to $2R_{0kij}$. It can therefore be defined, in these coordinates, as a line integral involving the Riemann tensor.

Problem 23 Let a free falling fluid spheroid of mass m and radius r be subject to the gravitational action of a mass M at a distance $|\boldsymbol{x}| \gg r$ from the spheroid's center. Obtain an expression for the difference in height between high tide and low tide on the spheroid as a function of m, M, r, and $|\boldsymbol{x}|$. Hint: Let η^i be the components of the radius vector from the center of the spheroid to a fluid particle on its surface. Because Newtonian gravitational fields may be superposed, the dynamical equation of the particle under the combined action of the mass M and the rest of the spheroid is

$$\frac{\mathrm{d}^2\eta^i}{\mathrm{d}t^2} = \frac{GM}{|\boldsymbol{x}|^3}(3\hat{x}^i\hat{x}^j - \delta_{ij})\eta^j - \frac{GM}{|\boldsymbol{\eta}|^3}\eta^i + f^i = -\frac{\partial\Psi}{\partial\eta^i} + f^i,$$

where f^i is the supporting (pressure gradient) force per unit mass at the surface of the fluid and the "potential" Ψ is given by

$$\begin{aligned}\Psi &= -\frac{GM}{2|\boldsymbol{x}|^3}(3\hat{x}^i\hat{x}^j - \delta_{ij})\eta^i\eta^j - \frac{GM}{|\eta|} \\ &\approx -\frac{GM}{2|\boldsymbol{x}|^3}(3\hat{x}^i\hat{x}^j - \delta_{ij})\eta_0^i\eta_0^j - \frac{GM}{r} + \frac{GM}{r^3}\eta_0^i\delta\eta^i,\end{aligned}$$

where η_0^i is the value η^i would have if M were zero and $\delta\eta^i$ is the deviation from this value. At equilibrium, we must have $\mathrm{d}^2\eta^i/\mathrm{d}\,t^2 = 0$ and hence $f^i = \partial\Psi/\partial\eta^i$. Since f^i can act only perpendicularly to the surface, it follows that the surface must be equipotential, i.e., $\Psi =$ constant. Neglecting viscous forces (drag) and the effect of the continents, compute for the earth the difference in height between high and low spring tides, i.e., the tides when the sun and moon are either at opposition (full moon) or conjunction (new moon), using the following data:

$$\begin{aligned} m_\oplus &= 5.98 \times 10^{24}\,\mathrm{kg}, & r_\oplus &= 6.38 \times 10^6\,\mathrm{m}, \\ M_\odot &= 1.99 \times 10^{30}\,\mathrm{kg}, & |\boldsymbol{x}|_\odot &= 1.50 \times 10^{11}\,\mathrm{m}, \\ M_{\mathrm{moon}} &= 7.35 \times 10^{22}\,\mathrm{kg}, & |\boldsymbol{x}|_{\mathrm{moon}} &= 3.84 \times 10^8\,\mathrm{m}. \end{aligned}$$

Solution 23 High tide occurs when $\hat{\boldsymbol{\eta}}_0 = \hat{\boldsymbol{x}}$ and low tide occurs when $\hat{\boldsymbol{\eta}}_0 \cdot \hat{\boldsymbol{x}} = 0$ ($\hat{\eta}_0^i = \eta_0^i/|\boldsymbol{\eta}|$). Let the value of $\hat{\boldsymbol{\eta}}_0 \cdot \delta\boldsymbol{\eta}$ in these two cases be $\delta_{\mathrm{H}}\eta$ and $\delta_{\mathrm{L}}\eta$, respectively. Then

$$\begin{aligned}\Psi = \text{constant} &= -\frac{GMr^2}{|\boldsymbol{x}|^3} - \frac{GM}{r} + \frac{GM}{r^2}\delta_{\mathrm{H}}\eta \\ &= \frac{GMr^2}{2|\boldsymbol{x}|^3} - \frac{GM}{r} + \frac{GM}{r^2}\delta_{\mathrm{L}}\eta,\end{aligned}$$

whence

$$\frac{GM}{r^2}(\delta_{\mathrm{H}}\eta - \delta_{\mathrm{L}}\eta) = \frac{3GMr^2}{2|\boldsymbol{x}|^3},$$

and therefore

$$\delta_H \eta - \delta_L \eta = \frac{3}{2} \frac{M}{m} \frac{r^4}{|\boldsymbol{x}|^3}$$

For the earth we have

$$\begin{aligned}
\delta_H \eta - \delta_L \eta &= \frac{3}{2} \frac{r_\oplus^4}{m_\oplus} \left(\frac{M_\odot}{|\boldsymbol{x}|_\odot^3} + \frac{M_{\text{moon}}}{|\boldsymbol{x}|_{\text{moon}}^3} \right) \\
&= \frac{3}{2} \frac{(6.38)^4 \times 10^{24}}{5.98 \times 10^{24}} \left(\frac{1.99 \times 10^{30}}{(1.50)^3 \times 10^{33}} + \frac{7.35 \times 10^{22}}{(3.84)^3 \times 10^{24}} \right) \\
&= \frac{3}{2} \frac{(6.38)^4}{5.98} (0.59 \times 10^{-3} + 1.30 \times 10^{-3}) = \frac{3}{2} \frac{(6.38)^4}{5.98} \times 1.89 \times 10^{-3} \\
&= 0.785\,\text{m}.
\end{aligned}$$

Chapter 7
Ensembles of Particles

Suppose a region of spacetime is occupied by a large number N of identical particles. Suppose that they possess a wide variety of momenta and that there are so many of them that their spacetime and momentum distribution may be taken as effectively continuous. We may then introduce a continuous function $f(x, \boldsymbol{p})$ of x^μ and p_i such that $f(x,\boldsymbol{p})\mathrm{d}^3\boldsymbol{x}\mathrm{d}^3\boldsymbol{p}$ is the number of particles in the volume element $\mathrm{d}^3\boldsymbol{x}$ ($\equiv \mathrm{d}x^1\mathrm{d}x^2\mathrm{d}x^3$) at the point x^i at "time"[1] x^0 having momenta in the range $\mathrm{d}^3\boldsymbol{p}(\equiv \mathrm{d}p_1\mathrm{d}p_2\mathrm{d}p_3)$ around p_i. Let us ask the question: How does f transform under general coordinate transformations?

The particles specified by $x^\mu, p_i, \mathrm{d}^3\boldsymbol{x}$, and $\mathrm{d}^3\boldsymbol{p}$ are well defined and the number $f(x,\boldsymbol{p})\mathrm{d}^3\boldsymbol{x}\mathrm{d}^3\boldsymbol{p}$ therefore has significance independently of the choice of the coordinate system. That is to say, $f(x,\boldsymbol{p})\mathrm{d}^3\boldsymbol{x}\mathrm{d}^3\boldsymbol{p}$ is an invariant. Hence, if we find the coordinate transformation law for the product $\mathrm{d}^3\boldsymbol{x}\mathrm{d}^3\boldsymbol{p}$, then we shall have found it for f.

Let us look first at $\mathrm{d}^3\boldsymbol{p}$, but instead of considering the 3-vector p_i, let us consider the 4-vector p_μ. This is the 4-momentum or *energy–momentum* 4-vector of a particular particle at a particular spacetime point x^μ. It is therefore a local covariant vector at that point and transforms according to

$$\bar{p}_\mu = \frac{\partial x^\nu}{\partial \bar{x}^\mu} p_\nu .$$

For an actual particle p_μ is constrained by the conditions

$$p^0 > 0, \qquad p^2 + m^2 = 0,$$

[1] x^0 is not strictly a "time". However, we shall assume here that hypersurfaces of constant x^0 are spacelike [which means $\det(g_{ij}) > 0$ and $g^{00} < 0$] and that x^0 increases as one moves toward the future in any timelike direction.

B. DeWitt, *Bryce DeWitt's Lectures on Gravitation*,
Edited by Steven M. Christensen, Lecture Notes in Physics, 826,
DOI: 10.1007/978-3-540-36911-0_7, © Springer-Verlag Berlin Heidelberg 2011

and is said to lie on the *mass shell*. Let us, however, for the moment consider an unrestricted increment: dp_μ in p_μ. This increment satisfies the same transformation law as p_μ itself:

$$d\bar{p}_\mu = \frac{\partial x^\nu}{\partial \bar{x}^\mu} dp_\nu .$$

From this, we obtain the transformation law for the energy–momentum 4-volume element:

$$d^4\bar{p} = \frac{\partial(x)}{\partial(\bar{x})} d^4 p,$$

where $d^4 p \equiv dp_0 dp_1 dp_2 dp_3$. We now recall that the magnitude of the determinant of the metric tensor is a density of weight 2 and hence transforms according to

$$\bar{g} = \left[\frac{\partial(x)}{\partial(\bar{x})}\right]^2 g.$$

Therefore, the combination $g^{-1/2} d^4 p$ is coordinate invariant; it is the value of the volume element $d^4 p$ in a coordinate system in which the metric becomes locally canonical.

Now let $\phi(x, p)$ be an arbitrary scalar function of the x^μ and p_μ. We may restrict this function to the mass shell by multiplying it by

$$\theta(p^0)\delta(p^2 + m^2).$$

The latter quantity is invariant under coordinate transformations that maintain the orientation of x^0 and the spacelike character of the surfaces x^0 = constant. Hence the integral

$$\Phi(x) = g^{-1/2}(x) \int \phi(x,p)\theta(p^0)\delta(p^2 + m^2) d^4 p$$

is a scalar. Now under a change dp_μ in the 4-momentum, the quantity $p^2 + m^2$ suffers the change $2p^\mu dp_\mu$. Hence, if we hold the p_i fixed and integrate over p_0, then we may write

$$d(p^2 + m^2) = 2p^0 dp_0, \qquad \text{or } dp_0 = \frac{d(p^2 + m^2)}{2p^0},$$

and

$$\Phi(x) = g^{-1/2} \int \phi \frac{d^3\boldsymbol{p}}{2p^0},$$

where ϕ and p^0 are now restricted to the mass shell. However, ϕ, thus restricted, is still a scalar. Therefore, it follows that

$$g^{-1/2}\frac{\mathrm{d}^3\boldsymbol{p}}{p^0} \quad \text{is coordinate invariant.}$$

To build an analogous invariant out of $\mathrm{d}^3\boldsymbol{x}$, we begin by assuming that all the particles in the phase space volume element at (x^μ, p_μ) suddenly stop interacting with one another (if they interacted before) and start moving like free particles. Apart from some slight fuzzing due to the distribution of momenta in $\mathrm{d}^3\boldsymbol{p}$, the world lines of these particles will then fill a tube as shown in Fig. 1.

A natural invariant associated with this tube is the 3-volume of its orthogonal cross section, i.e., its size as viewed in its own rest frame. To determine this cross section, we introduce a timelike unit vector n_μ orthogonal to the hypersurface $x^0 = \text{constant}$:

$$(n_\mu) = \left((-g^{00})^{-1/2}, 0, 0, 0\right).$$

From the laws of minors and inverse matrices, we remember that

$$g^{00} = (-g)^{-1}\det(g_{ij}).$$

We also remember that the magnitude of the 3-volume of the intersection of the tube with the hypersurface $x^0 = \text{constant}$ is

$$\left[\det(g_{ij})\right]^{1/2}\mathrm{d}^3\boldsymbol{x} = g^{1/2}(-g^{00})^{1/2}\mathrm{d}^3\boldsymbol{x}.$$

Now this hypersurface does not generally intersect the tube orthogonally. In order to get the 3-volume of the orthogonal section, we must multiply the above expression by

$$n \cdot u = (-g^{00})^{-1/2}u^0.$$

Multiplying by an additional factor m, we find therefore that

$$g^{1/2}p^0\mathrm{d}^3\boldsymbol{x} \quad \text{is coordinate invariant.}$$

Finally, multiplying this with the invariant 3-momentum element previously obtained, we see that the factors $g^{1/2}p^0$ cancel and that the ordinary phase space volume element

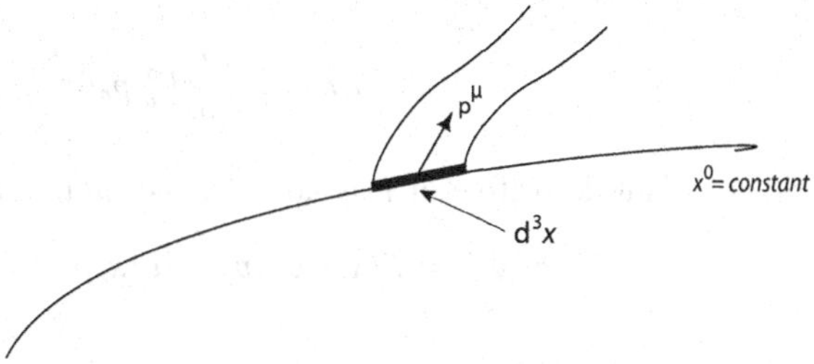

Fig. 7.1 World tube of non-interacting particles

$$\mathrm{d}^3\boldsymbol{p}\mathrm{d}^3\boldsymbol{x} \quad \text{is coordinate invariant.}$$

This in turn implies that the distribution function $f(x, \boldsymbol{p})$ is a scalar.

In order to display the scalar property of f explicitly, it is often convenient to replace f by a scalar function F of the x^μ and all four of the p_μ, viz.,

$$f(x,\boldsymbol{p}) = F(x,p)\big|_{p^0 > 0, p^2+m^2=0}.$$

As there is an infinite ways of extending f off the mass shell, there is no unique way of choosing F. However, any two choices, F_1 and F_2, will be related by

$$F_1(x,p) = F_2(x,p) + Q(x,p,p^2+m^2), \quad \text{where } Q(x,p,0) = 0, \quad \forall x^\mu, p_\mu,$$

and the scalar function Q has no effect on any physical computations involving F_1 or F_2 on the mass shell. From now on we assume some particular choice has been made for F.

The function F, although a scalar, is a generalization of the usual idea of a scalar function because of its dependence on the p_μ. This has the consequence that its covariant derivative with respect to a coordinate is not just a simple gradient as it is for an ordinary scalar function. It is easy enough to see that its derivative with respect to p_μ transforms as a contravariant vector. Its ordinary derivative with respect to x^μ, however, generally has no simple transformation character.

To obtain the appropriate definition for the covariant derivative we return, as always, to the idea of parallel displacement. As F is a scalar, it must suffer no change in magnitude under parallel displacement through an interval $\mathrm{d}x^\mu$:

$$\delta F = 0.$$

However, we may break δF up into a part that arises from its dependence on the x^μ and a part that arises from its dependence on the p_μ:

$$\delta F = \delta_x F + \frac{\partial F}{\partial p_\mu}\delta p_\mu.$$

As the p_μ are the components of a covariant vector, we have

$$\delta p_\mu = \Gamma^\sigma_{\mu\nu} p_\sigma \mathrm{d}x^\nu.$$

Therefore,

$$\delta_x F = -\frac{\partial F}{\partial p_\mu}\Gamma^\sigma_{\mu\nu} p_\sigma \mathrm{d}x^\nu.$$

The covariant derivative of F is now defined, in the usual manner, by

$$F_{;\mu}\mathrm{d}x^\mu = F(x+\mathrm{d}x,p) - [F(x,p) + \delta_x F(x,p)],$$

which yields

$$F_{;\mu} = \frac{\partial F}{\partial x^\mu} + \Gamma^\sigma_{\nu\mu} p_\sigma \frac{\partial F}{\partial p_\nu}.$$

Problem 24 Show that $(p^2)_{;\mu} = 0$ and use this result to demonstrate that if

$$F_1(x,p) = F_2(x,p) + Q(x,p,p^2+m^2), \quad \text{where } Q(x,p,0) = 0, \quad \forall x^\mu, p_\mu,$$

then $F_{1;\mu}$ and $F_{2;\mu}$ coincide on the mass shell.

Solution 24 We have

$$\begin{aligned}(p^2)_{;\mu} &= \frac{\partial p^2}{\partial x^\mu} + \Gamma^\sigma_{\nu\mu} p_\sigma \frac{\partial (p^2)}{\partial p_\nu} \\ &= g^{\nu\sigma}_{,\mu} p_\nu p_\sigma + 2\Gamma^\sigma_{\nu\mu} p_\sigma p^\nu \\ &= (-g_{\nu\sigma,\mu} + 2\Gamma_{\sigma\nu\mu}) p^\nu p^\sigma \\ &= (-g_{\nu\sigma,\mu} + g_{\sigma\nu,\mu} + g_{\sigma\mu,\nu} - g_{\nu\mu,\sigma}) p^\nu p^\sigma = 0.\end{aligned}$$

For a quicker derivation, one may simply introduce a coordinate system in which the $g_{\mu\nu,\sigma}$ vanish at x. As $Q(x, p, 0) = 0$, it follows that derivatives of Q with respect to its first two arguments vanish on the mass shell. Therefore, covariant differentiation, like ordinary differentiation, obeys the chain rule, we have

$$\begin{aligned}F_{1;\mu}|_{p^2+m^2=0} &= F_{2;\mu}|_{p^2+m^2=0} + Q_3(p^2+m^2)_{;\mu}|_{p^2+m^2=0} \\ &= F_{2;\mu}|_{p^2+m^2=0},\end{aligned}$$

where Q_3 denotes the derivative of Q with respect to its third argument.

Suppose none of the particles in the ensemble is either created or annihilated in the course of time, so that the total number of particles remains constant. This number is given by

$$\begin{aligned}N &= \int \mathrm{d}^3\boldsymbol{x} \int \mathrm{d}^3\boldsymbol{p} f(x,\boldsymbol{p}) \\ &= \int j^0 \mathrm{d}^3\boldsymbol{x} = \int_\Sigma j^\mu \mathrm{d}\Sigma_\mu,\end{aligned}$$

where

$$\begin{aligned}j^\mu &= \int u^\mu f(x,\boldsymbol{p}) \frac{\mathrm{d}^3\boldsymbol{p}}{u^0} \\ &= 2\int p^\mu F(x,p)\theta(p^0)\delta(p^2+m^2)\mathrm{d}^4p\end{aligned}$$

and Σ is the hypersurface $x^0 =$ constant. The quantity j^μ is evidently a contravariant vector density of unit weight. From this fact, it follows that the integral

$\int_\Sigma j^\mu \mathrm{d}\Sigma_\mu$, with Σ fixed in spacetime, is invariant under coordinate transformations. In order to show this, we introduce another contravariant vector density k^μ which is identical with j^μ in a finite neighborhood of Σ but which vanishes at a finite distance to the past of Σ. We may then write

$$\int_\Sigma j^\mu \mathrm{d}\Sigma_\mu = \int_\Sigma k^\mu \mathrm{d}\Sigma_\mu = \int_{-\infty}^{\Sigma} k^\mu_{,\mu} \mathrm{d}^4x,$$

and the question reduces to whether or not $k^\mu_{,\mu}$ transforms as a density of unit weight. That it does follows from the fact that the covariant divergence of a contravariant vector density of unit weight reduces to the ordinary divergence:

$$k^\mu_{;\mu} = k^\mu_{,\mu} + \Gamma^\mu_{\nu\mu} k^\nu - \Gamma^\nu_{\nu\mu} k^\mu = k^\mu_{,\mu}.$$

Therefore, $\int_\Sigma j^\mu \mathrm{d}\Sigma_\mu$ is independent of the choice of coordinates.

Now not only is $\int_\Sigma j^\mu \mathrm{d}\Sigma_\mu$ independent of the coordinate system, but it is also independent of Σ. This follows from the fact that, for any spacelike hypersurface, a coordinate system can be found in which that hypersurface is given by $x^0 =$ constant and in which the integral reduces to $\int j^0 \mathrm{d}^3\boldsymbol{x} = N$. Therefore,

$$0 = \int_{\Sigma_1} j^\mu \mathrm{d}\Sigma_\mu - \int_{\Sigma_2} j^\mu \mathrm{d}\Sigma_\mu = \int_{\Sigma_2}^{\Sigma_1} j^\mu_{;\mu} \mathrm{d}^4x,$$

for all spacelike Σ_1 and Σ_2. As Σ_1 and Σ_2 are arbitrary, the spacetime region between them can be reduced to an arbitrarily small neighborhood of any spacetime point. In order that the above relation hold for all Σ_1 and Σ_2, therefore, we must have[2]

$$j^\mu_{;\mu} = 0.$$

Since j^μ is a sum over timelike future-pointing 4-vectors, it is itself timelike. Therefore, we may introduce the quantities

$$\rho = (-j^2)^{1/2}, \qquad \rho_0 = g^{-1/2}\rho, \qquad \overline{u}^\mu = \rho^{-1} j^\mu,$$

whence

$$\overline{u}^2 = -1, \qquad j^\mu = \rho \overline{u}^\mu, \qquad (\rho \overline{u}^\mu)_{;\mu} = 0.$$

Here, $\overline{u}^\mu$ is the mean 4-velocity of the particles at x and may be regarded as the *flow vector* at x of the fluid medium formed by the particle ensemble. ρ_0 is the

[2] After we have this, we may let Σ be any connected hypersurface that intersects completely the support of the vector density j^μ, cutting all its flow lines an odd number of times. Then, Σ need not be spacelike, even in part.

proper number density of the particles at x, i.e., the particle density as viewed in a local Minkowskian rest frame of the medium at x.

In the special case that the particles of the ensemble are free particles that do not interact with each other, i.e., the world line of each is a geodesic, the conservation law $j^{\mu}_{;\mu} = 0$ reduces to a simple condition on the distribution function $F(x, p)$. As geodesic motion is derivable from a variational principle, a Hamiltonian formulation of the basic equations can be set up, although we shall not do this here. As of the canonical invariance[3] of the phase space volume element $\mathrm{d}^3\boldsymbol{x}\mathrm{d}^3\boldsymbol{p}$ and because the motion in phase space of any particle may, through the Hamilton equations, be regarded as the unfolding-in-time of a canonical transformation, the phase space volume occupied by the particles originally in $\mathrm{d}^3\boldsymbol{x}\mathrm{d}^3\boldsymbol{p}$ remains constant in time. However, because the number of particles in this volume element is $f\mathrm{d}^3\boldsymbol{x}\mathrm{d}^3\boldsymbol{p}$ and because this number is constant, one may conclude in turn that f has vanishing total time derivative. This is Liouville's theorem.

As it makes no difference whether the total derivative is taken with respect to time or with respect to proper time, we may recast Liouville's theorem in the covariant form:

$$\begin{aligned} 0 = \frac{\mathrm{d}F}{\mathrm{d}\tau} &= \frac{\partial F}{\partial x^{\mu}}\dot{x}^{\mu} + \frac{\partial F}{\partial p_{\mu}}\frac{\mathrm{d}p_{\mu}}{\mathrm{d}\tau} \\ &= \frac{\partial F}{\partial x^{\mu}}\dot{x}^{\mu} + \frac{\partial F}{\partial p_{\mu}}\left(\dot{p}_{\mu} + \Gamma^{\sigma}_{\mu\nu}p_{\sigma}\dot{x}^{\nu}\right) \\ &= F_{;\mu}\dot{x}^{\mu} + \frac{\partial F}{\partial p_{\mu}}\dot{p}_{\mu} = m^{-1}F_{;\mu}p^{\mu}, \end{aligned}$$

using $\dot{p}_{\mu} = 0$.

Problem 25 Let $G(x, p)$ be an arbitrary scalar function of the x^{μ} and p_{μ} having effectively compact support in p-space. By making use of the definition of the covariant derivative of such a function, together with integration by parts, show that

$$\frac{\partial}{\partial x^{\mu}}\int p^{\mu}G(x,p)\mathrm{d}^4p = \int p^{\mu}G_{;\mu}(x,p)\mathrm{d}^4p.$$

(Be sure to write $p^{\mu} = g^{\mu\nu}p_{\nu}$.) Use this result together with the first result of Problem 24 to show that the condition $F_{;\mu}p^{\mu} = 0$ guarantees that $j^{\mu}_{;\mu} = 0$.

Solution 25 We have

$$\begin{aligned} \frac{\partial}{\partial x^{\mu}}\int p^{\mu}G\mathrm{d}^4p &= \int\left(g^{\mu\nu}_{,\mu}p_{\nu}G + p^{\mu}\frac{\partial G}{\partial x^{\mu}}\right)\mathrm{d}^4p \\ &= \int\left(-g^{\mu\sigma}g_{\sigma\nu,\mu}p^{\nu}G + p^{\mu}G_{;\mu} - p^{\mu}\Gamma^{\sigma}_{\nu\mu}p_{\sigma}\frac{\partial G}{\partial p_{\nu}}\right)\mathrm{d}^4p \end{aligned}$$

---

[3] Coordinate transformations can be shown to be canonical transformations, and hence the coordinate invariance of $\mathrm{d}^3\boldsymbol{x}\mathrm{d}^3\boldsymbol{p}$ is a special case of this.

$$= \int \left(p^\mu G_{;\mu} - p^\mu g^{\nu\sigma} g_{\mu\nu,\sigma} G + g^{\mu\nu} \Gamma^\sigma_{\nu\mu} p_\sigma G + p^\mu \Gamma^\nu_{\nu\mu} G\right) \mathrm{d}^4 p$$
$$= \int p^\mu \left[G_{;\mu} + g^{\nu\sigma}(-g_{\mu\nu,\sigma} + \Gamma_{\mu\nu\sigma} + \Gamma_{\nu\sigma\mu}) G\right] \mathrm{d}^4 p$$
$$= \int p^\mu G_{;\mu} \mathrm{d}^4 p.$$

Therefore, using $(p^2)_{;\mu} = 0$ and $\delta(p^0)\delta(p^2 + m^2) = 0$, we have

$$j^\mu_{;\mu} = j^\mu_{,\mu} = 2 \int p^\mu F_{;\mu}(x, p) \theta(p^0) \delta(p^2 + m^2) \mathrm{d}^4 p = 0.$$

For a quicker derivation, one may introduce a coordinate system in which the $g_{\mu\nu,\sigma}$ vanish at x.

Problem 26 Milne's Cosmology. Show that the distribution function

$$F(x, p) = Z \int_{-\infty}^{\infty} \delta(x - \lambda p) \mathrm{d}\lambda$$

satisfies the condition $F_{,\mu} p^\mu = 0$ for a free particle gas. Here spacetime is assumed to be flat, the coordinates are assumed to be canonical (Minkowskian), Z is a normalizing constant, and the integrand is the 4-dimensional delta function of $x^\mu - \lambda p^\mu$. Obtain the function $f(x, \boldsymbol{p})$, and show that the momentum distribution is isotropic, with all momenta represented. Show that, at time $x^0 = 0$, all the particles are concentrated at $\boldsymbol{x} = 0$, while at time $x^0 = t$, those particles having velocity $\boldsymbol{v}$ are located at $\boldsymbol{x} = \boldsymbol{v}t$. The distribution is evidently one that corresponds to an explosion that takes place at the origin at time $x^0 = 0$. For any particle, the proper time lapse since the explosion is therefore

$$\tau = \left[\theta(x^0) - \theta(-x^0)\right](-x^2)^{1/2}.$$

Show that

$$j^\mu = Zm^2 \left[\theta(x^0) - \theta(-x^0)\right] \theta(-x^2) x^\mu (-x^2)^{-2},$$
$$\rho_0 = (-j^2)^{1/2} = Zm^2 \theta(-x^2)(-x^2)^{-3/2} = Zm^2 \theta(-x^2)|\tau|^{-3},$$

and hence that the world lines of the particles fill the entire interior of the light cone through the origin ($-x^2 > 0$).

By taking the particles to be galaxies having roughly identical masses, we obtain a model for a Big Bang cosmology in which gravity is neglected. As the 4-dimensional delta function is a scalar under Lorentz transformations, the distribution function F has the same form in all Lorentz frames having the same origin. This means that the "universe" has the same appearance as viewed from every galaxy. To determine this appearance, we may choose an "observer" galaxy having zero velocity. As optical information comes to this galaxy along its past light cone, one may conveniently take this cone as the hypersurface Σ. Let (x^μ) be a point on Σ and let $\boldsymbol{r}$ be the apparent position of the galaxy whose world line passes through that point, i.e., the actual position of the galaxy at the time it emitted the radiation received by the observer galaxy at rest at the origin. If the vertex of the cone Σ is at $(t, 0, 0, 0)$ with $t > 0$, then $(x^\mu) = (t - r, \boldsymbol{r})$. The surface element of Σ may be taken in the form: $(\mathrm{d}\Sigma_\mu) = (\mathrm{d}^3\boldsymbol{r}, \hat{\boldsymbol{r}}\mathrm{d}^3\boldsymbol{r})$, where $\hat{\boldsymbol{r}} = \boldsymbol{r}/r$. Use this to express the integral $\int_\Sigma j^\mu \mathrm{d}\Sigma_\mu$ as an integral over the volume elements $\mathrm{d}^3\boldsymbol{r}$ of apparent position. Although the latter integral diverges, showing that the total number of galaxies is infinite, the integrand has an immediate interpretation as the apparent or optical density of galaxies as seen by the observer galaxy at the origin. Obtain this optical density as a function of r and t.

Solution 26 We have

$$F_{,\mu}p^\mu = Z\int_{-\infty}^{\infty} p^\mu \delta_{,\mu}(x - \lambda p)\mathrm{d}\lambda = -Z\int_{-\infty}^{\infty} \frac{\mathrm{d}}{\mathrm{d}\lambda}\delta(x - \lambda p)\mathrm{d}\lambda$$
$$= -Z[\delta(x - \lambda p)]_{\lambda=-\infty}^{\lambda=\infty} = 0,$$

for all finite x^μ, and

$$f(x,\boldsymbol{p}) = Z\int_{-\infty}^{\infty} \delta(\boldsymbol{x} - \lambda\boldsymbol{p})\delta(x^0 - \lambda p^0)\mathrm{d}\lambda$$
$$= \frac{Z}{p^0}\delta\left(\boldsymbol{x} - \frac{\boldsymbol{p}}{p^0}x^0\right),$$

whence

$$\boxed{f(x,\boldsymbol{p}) = \frac{Z}{p^0}\delta(\boldsymbol{x} - \boldsymbol{v}x^0)}$$

Then

$$j^\mu = 2\int p^\mu F(x,\boldsymbol{p})\theta(p^0)\delta(p^2 + m^2)\mathrm{d}^4p$$
$$= 2Z\int \mathrm{d}^4p \int_{-\infty}^{\infty} \mathrm{d}\lambda p^\mu \delta(x - \lambda p)\theta(p^0)\delta(p^2 + m^2)$$
$$= 2Z\int_{-\infty}^{\infty} \lambda^{-5}x^\mu \theta(\lambda^{-1}x^0)\delta(\lambda^{-2}x^2 + m^2)\mathrm{d}\lambda.$$

To evaluate this last integral, let $\xi = \lambda^{-2}$. Then

$$\lambda = \pm\xi^{-1/2}, \qquad \lambda^{-5} = \pm\xi^{5/2}, \qquad \mathrm{d}\lambda = \mp\frac{1}{2}\xi^{-3/2}\mathrm{d}\xi, \qquad \lambda^{-5}\mathrm{d}\lambda = -\frac{1}{2}\xi\mathrm{d}\xi,$$

and

$$j^\mu = Z\left[\theta(x^0) - \theta(-x^0)\right]x^\mu \int_0^\infty \delta(\xi x^2 + m^2)\xi \mathrm{d}\xi$$
$$= Zm^2\left[\theta(x^0) - \theta(-x^0)\right]\theta(-x^2)x^\mu(-x^2)^{-2},$$
$$\rho_0 = (-j^2)^{1/2} = Zm^2\theta(-x^2)(-x^2)^{-3/2} = Zm^2\theta(-x^2)|\tau|^{-3}.$$

On Σ, we have

$$-x^2 = (t-r)^2 - r^2 = t^2 - 2tr = t(t-2r),$$
$$\int_\Sigma j^\mu \mathrm{d}\Sigma_\mu = Zm^2 \int_{r<t/2} \frac{t - r + \boldsymbol{r}\cdot\hat{\boldsymbol{r}}}{t^2(t-2r)^2}\mathrm{d}^3r = \int \rho_{\mathrm{op}}\mathrm{d}^3r,$$

where

$$\boxed{\rho_{\mathrm{op}} = \frac{Zm^2}{t(t-2r)^2}\theta(t-2r)}$$

Problem 27 As the total number of galaxies in the preceding problem is infinite, one might argue that, if all the galaxies are assumed to have the same absolute luminosity L_0, then the universe will appear to be infinitely bright. Let L be the apparent luminosity of a galaxy at an apparent distance r. If the time of observation (made by the observer galaxy at the origin) is t, then the recession velocity of this galaxy is $r/(t-r)$. Express L as a function of L_0, t and $\boldsymbol{r}$. Now let $\mathrm{d}^2\Phi$ be the radiation energy flux from the galaxy through a surface element d^2S orthogonal to the line of sight and located at the origin. The apparent brightness of the galaxy is then

$$B \equiv \frac{\mathrm{d}^2\Phi}{\mathrm{d}^2S} = \frac{L}{4\pi r^2}.$$

By summing this quantity over all the galaxies lying in a solid angle $\mathrm{d}^2\Omega$ from the origin, obtain the total brightness of the sky per unit solid angle, i.e., $\mathrm{d}^2B_{\mathrm{tot}}/\mathrm{d}^2\Omega$, and show that it is finite for $t > 0$. (Note that the calculation is somewhat unrealistic in that it assumes L_0 to be the same for all galaxies and hence fails to allow for aging effects.)

The immediate data received by an observer at the origin who views the above "universe" are red shift z and brightness B of individual galaxies, and the galactic distribution function D defined so that $D\mathrm{d}^2\Omega\mathrm{d}z$ is the number of galaxies in the solid angle $\mathrm{d}^2\Omega$ (from the observer) that are observed to have red shifts lying between z and $z + \mathrm{d}z$. Obtain B and D as functions of Z, m, L_0, t, and z. Check your answers by obtaining $\mathrm{d}^2B_{\mathrm{tot}}/\mathrm{d}^2\Omega$ directly from B and D.

Solution 27 We have

$$v = \frac{r}{t-r}, \qquad 1 - v = \frac{t-2r}{t-r}, \qquad 1 + v = \frac{t}{t-r},$$

$$\boxed{L = \left(\frac{1-v}{1+v}\right)^2 L_0 = \frac{(t-2r)^2}{t^2} L_0}$$

Hence,

$$\begin{aligned}
\frac{\mathrm{d}^2 \mathrm{B}}{\mathrm{d}^2 \Omega} &= \int_0^\infty \rho B r^2 \mathrm{d}r = \frac{1}{4\pi} \int_0^\infty \rho L \mathrm{d}r \\
&= \frac{Zm^2 L_0}{4\pi} \int_0^\infty \theta(t-2r) \frac{1}{t(t-2r)^2} \frac{(t-2r)^2}{t^2} \mathrm{d}r \\
&= \frac{Zm^2 L_0}{4\pi t^3} \int_0^{t/2} \mathrm{d}r,
\end{aligned}$$

and finally

$$\boxed{\frac{\mathrm{d}^2 \mathrm{B}_{\mathrm{tot}}}{\mathrm{d}^2 \Omega} = \frac{Zm^2 L_0}{8\pi t^2}}$$

Now

$$(1+z)^2 = \frac{1+v}{1-v} = \frac{t}{t-2r}, \qquad t - 2r = \frac{t}{(1+z)^2}.$$

$$\left[(1+z)^2 - 1\right] t = 2(1+z)^2 r, \qquad r = \frac{z(2+z)}{2(1+z)^2} t, \qquad L = \frac{L_0}{(1+z)^4},$$

so that

$$\boxed{B = \frac{L}{4\pi r^2} = \frac{L_0}{\pi t^2} \frac{1}{z^2(2+z)^2}}$$

Further,

$$\rho_{\mathrm{op}} = \frac{Zm^2}{t^3}(1+z)^4,$$

and

$$D\mathrm{d}z = \rho_{\mathrm{op}} r^2 \mathrm{d}r = \frac{Zm^2}{t^3}(1+z)^4 \frac{z^2(2+z)^2}{4(1+z)^4} t^3 \left[\frac{1}{1+z} - \frac{z(2+z)}{(1+z)^3}\right] \mathrm{d}z,$$

so

$$\boxed{D = \frac{1}{4} Z m^2 \frac{z^2 (2+z)^2}{(1+z)^3}}$$

Finally,

$$\frac{\mathrm{d}^2 \mathrm{B}_{\mathrm{tot}}}{\mathrm{d}^2 \Omega} = \int_0^\infty D B \mathrm{d}z = \frac{Z m^2 L_0}{4\pi t^2} \int_0^\infty \frac{\mathrm{d}z}{(1+z)^3} = \frac{Z m^2 L_0}{8\pi t^2},$$

as above.

7.1 Gases at Equilibrium

An ensemble of interacting particles that has attained a state of quasi-equilibrium is called a *gas*. True equilibrium, in which the parameters of the gas remain constant in time, cannot be reached unless spacetime possesses a timelike Killing vector field ξ^μ. If ξ^μ is the only Killing vector possessed by spacetime, then, at equilibrium, the flow lines of the gas must be parallel to ξ^μ. Moreover, if the gas is sufficiently dilute or if the interparticle forces are sufficiently weak that the time average of the interaction energy is negligible compared to the kinetic energy of the particles (Boltzmann's point-collision approximation), then the standard arguments of statistical mechanics lead to the conclusion that the equilibrium distribution function of the gas must be simply a function of the only constant of motion possessed by the free particles, namely $\xi \cdot p$:

$$F(x,p) = \Phi(\xi \cdot p).$$

The form of the function Φ will depend on the temperature and total number of the particles and on their nature: classical, quantum, boson, fermion, massive, massless, conserved, not conserved, etc. If spacetime possesses more than one independent Killing vector field, then the vector ξ above may, but need not, be replaced by an arbitrary linear combination of these. The only requirement is that the combination be everywhere timelike in the region occupied by the gas.

It is easy to verify that the above distribution function leads to a mean 4-velocity or flow vector for the gas that is parallel to ξ^μ. Let

$$\Psi(y) = \int^y \Phi(y) \mathrm{d}y.$$

Then

$$\begin{aligned} j^\mu &= 2 \int p^\mu \Phi(\xi \cdot p) \theta(p^0) \delta(p^2 + m^2) \mathrm{d}^4 p \\ &= 2 \frac{\partial}{\partial \xi_\mu} \int \Psi(\xi \cdot p) \theta(p^0) \delta(p^2 + m^2) \mathrm{d}^4 p. \end{aligned}$$

The last integral is a scalar density at x which depends only on $g_{\mu\nu}$ and ξ^μ. It must necessarily have the form:

$$\int \Psi(\xi \cdot p)\theta(p^0)\delta(p^2 + m^2)\mathrm{d}^4p = g^{1/2}X_m(\xi^2),$$

for some function X_m. Therefore,

$$j^\mu = 4g^{1/2}X_m'(\xi^2)\xi^\mu,$$

and hence j^μ and $\overline{u}^\mu$ are parallel to ξ^μ. We have, in fact,

$$\overline{u}^\mu = (-\xi^2)^{-1/2}\xi^\mu.$$

Problem 28 Suppose the particles of a gas at equilibrium, having a distribution function of the form $F(x,p) = \Phi(\xi \cdot p)$ where ξ^μ is a timelike Killing vector, suddenly ceased to interact with one another. Show that the distribution function would remain unchanged. (Hint: Show that F is an allowable distribution function for an ensemble of non-interacting particles.)

Solution 28 We have

$$(\xi \cdot p)_{;\mu} = \xi^\nu_{,\mu}p_\nu + \Gamma^\sigma_{\nu\mu}p_\sigma\xi^\nu = \xi^\nu_{;\mu}p_\nu.$$

Therefore,

$$\begin{aligned} F_{;\mu}p^\mu &= \Phi'(\xi \cdot p)_{;\mu}p^\mu = \Phi'\xi^\nu_{;\mu}p_\nu p^\mu \\ &= \frac{1}{2}\Phi' p^\mu p^\nu(\xi_{\mu;\nu} + \xi_{\nu;\mu}) = 0, \end{aligned}$$

as required.

In equilibrium statistical mechanics, the temperature enters through a Lagrange multiplier that is introduced, in the computation of the most probable distribution, to account for the conservation of the sum $\sum \xi{\cdot}p$ over all the gas particles. We may introduce it here through a simple rescaling of ξ^μ and Φ so that the distribution function takes the form:

$$F(x,p) = A\Phi(\beta_0\xi \cdot p),$$

where Φ is a universal function depending only on the nature of the gas and A is a constant proportional to the total particle number when this number is conserved under a temperature change, and unity when it is not. The Lagrange multiplier β_0 has the physical interpretation

$$\beta_0 = \frac{1}{kT_0},$$

where T_0 is the local temperature of the gas (measured by a thermometer!) at a point where ξ^μ is chosen to have the normalization $\xi^2 = -1$. The local temperature at an arbitrary point is given by

$$\beta = \frac{1}{kT},$$

where

$$\beta = (-\xi^2)^{1/2}\beta_0,$$

so that

$$F(x,p) = A\Phi(\beta\overline{u}\cdot p).$$

Evidently,

$$T = (-\xi^2)^{-1/2}T_0,$$

so that the local temperature is not generally a constant, even at equilibrium, but varies from point to point. Although this result was obtained by analyzing an idealized gas it must also hold true for real gases, since they may be placed in contact with idealized gases, at least in thought experiments. The local temperature at equilibrium satisfies a differential equation that is an immediate consequence of the Killing property of ξ^μ. To obtain it, we first compute

$$\overline{u}_{\mu;\nu} = (-\xi^2)^{-1/2}\xi_{\mu;\nu} + (-\xi^2)^{-3/2}\xi_\mu\xi^\sigma\xi_{\sigma;\nu},$$

$$\begin{aligned}\overline{u}_{\mu;\nu}\overline{u}^\nu &= (-\xi^2)^{-1}\xi_{\mu;\nu}\xi^\nu + (-\xi^2)^{-2}\xi_\mu\xi^\sigma\xi^\nu\xi_{\sigma;\nu}\\ &= (-\xi^2)^{-1}\xi_{\mu;\nu}\xi^\nu.\end{aligned}$$

The differential equation immediately follows:

$$\begin{aligned}T_{;\mu} + \overline{u}_{\mu;\nu}\overline{u}^\nu T &= (-\xi^2)^{-3/2}\xi^\nu\xi_{\nu;\mu}T_0 + (-\xi^2)^{-1}\xi_{\mu;\nu}\xi^\nu T\\ &= (-\xi^2)^{-1}\xi^\nu(\xi_{\mu;\nu} + \xi_{\nu;\mu})T\\ &= 0.\end{aligned}$$

Note that the temperature *is* constant along any flow line:

$$T_{;\mu}\overline{u}^\mu = 0.$$

Problem 29 Suppose the constantly accelerated medium of Problem 10 is a gas. By finding a Killing vector that generates the flow lines show that this gas can be at thermal equilibrium, and obtain an expression for the variation of T with the Lagrangian coordinate ξ^i. (Hint: Use the metric for the system of coordinates σ, ξ^i.)

Solution 29 As the Killing vector in the coordinate system σ, ξ^i try

$$(\xi^\mu) = (1,0,0,0).$$

Check:

$$\xi_\mu = g_{\mu 0},$$

$$\begin{aligned}\xi_{\mu;\nu} + \xi_{\nu;\mu} &= \xi_{\mu,\nu} + \xi_{\nu,\mu} - 2\Gamma^{\sigma}_{\mu\nu}\xi_\sigma \\ &= g_{\mu 0,\nu} + g_{\nu 0,\mu} - 2\Gamma_{0\mu\nu} \\ &= g_{\mu\nu,0} = 0,\end{aligned}$$

because the metric is static. Now

$$\xi^2 = g_{00} = -(1 + a\xi')^2,$$

whence

$$\boxed{T = (-\xi^2)^{-1/2} T_0 = \frac{T_0}{1 + a\xi'}}$$

Note that the temperature decreases with increasing ξ' at the same rate as the absolute acceleration.

Chapter 8
Production of Gravitational Fields by Matter

We have examined, in an introductory way, the effect of gravity on matter, or at least on particles. In order to study the complete interaction between gravity and matter, in particular the production of gravitational fields by matter,[1] we must endow the gravitational field with dynamical properties. For many reasons, e.g., conservation laws, quantization, etc., it is desirable to do this by means of a variational or least action principle. To the action functional S_{M} for the matter, suitably generalized (by the strong equivalence principle) to curved spacetime, one adds an action functional S_{G} for the gravitational field. S_{G} must be coordinate invariant and a functional of the metric field $g_{\mu\nu}$ alone. The most general such functional that leads to differential equations of order no higher than the second has the form

$$S_{\mathrm{G}} = \kappa \int g^{1/2} R \mathrm{d}^4 x + \lambda \int g^{1/2} \mathrm{d}^4 x,$$

where κ and λ are certain constants.[2] The total action is then

$$S = S_{\mathrm{G}} + S_{\mathrm{M}}, \tag{8.1}$$

and the dynamical (field) equations are

$$0 = \frac{\delta S}{\delta g_{\mu\nu}} = \frac{\delta S_{\mathrm{G}}}{\delta g_{\mu\nu}} + \frac{\delta S_{\mathrm{M}}}{\delta g_{\mu\nu}},$$

$$0 = \frac{\delta S}{\delta \Phi^A} = \frac{\delta S_{\mathrm{M}}}{\delta \Phi^A},$$

[1] The term 'matter' will here include electromagnetic fields.

[2] The minus sign attached to the λ is conventional. λ can, in fact, have either sign.

B. Dewitt, *Bryce DeWitt's Lectures on Gravitation*,
Edited by Steven M. Christensen, Lecture Notes in Physics, 826,
DOI: 10.1007/978-3-540-36911-0_8, © Springer-Verlag Berlin Heidelberg 2011

where $\delta/\delta g_{\mu\nu}$ and $\delta/\delta\Phi^A$ are the operators of functional differentiation and the Φ^A are the dynamical variables that describe the matter.

In order to compute the functional derivative $\delta S_{\mathrm{G}}/\delta g_{\mu\nu}$, we first compute the variations

$$
\begin{aligned}
\delta\Gamma^{\mu}_{\nu\sigma} &= \delta g^{\mu\tau}\Gamma_{\tau\nu\sigma} + \frac{1}{2}g^{\mu\tau}(\delta g_{\tau\nu,\sigma} + \delta g_{\tau\sigma,\nu} - \delta g_{\nu\sigma,\tau}) \\
&= -g^{\mu\tau}\Gamma^{\rho}_{\nu\sigma}\delta g_{\tau\rho} + \frac{1}{2}g^{\mu\tau}\big(\delta g_{\tau\nu;\sigma} + \delta g_{\tau\sigma;\nu} - \delta g_{\nu\sigma;\tau} \\
&\quad + \Gamma^{\rho}_{\tau\sigma}\delta g_{\rho\nu} + \Gamma^{\rho}_{\nu\sigma}\delta g_{\tau\rho} \\
&\quad + \Gamma^{\rho}_{\rho\tau\nu}\delta g_{\rho\sigma} + \Gamma^{\rho}_{\sigma\nu}\delta g_{\tau\rho} \\
&\quad - \Gamma^{\rho}_{\nu\tau}\delta g_{\rho\sigma} - \Gamma^{\rho}_{\sigma\tau}\delta g_{\nu\rho}\big) \\
&= \frac{1}{2}g^{\mu\tau}\big(\delta g_{\tau\nu,\sigma} + g_{\tau\sigma,\nu} - \delta g_{\nu\sigma,\tau}\big),
\end{aligned}
$$

$$
\begin{aligned}
\delta R^{\tau}_{\sigma\mu\nu} &= \delta\Gamma^{\tau}_{\sigma\nu,\mu} - \delta\Gamma^{\tau}_{\sigma\mu,\nu} \\
&\quad + \delta\Gamma^{\tau}_{\rho\mu}\Gamma^{\rho}_{\sigma\nu} + \Gamma^{\tau}_{\rho\mu}\delta\Gamma^{\rho}_{\sigma\nu} - \delta\Gamma^{\tau}_{\rho\nu}\Gamma^{\rho}_{\sigma\mu} - \Gamma^{\tau}_{\rho\nu}\delta\Gamma^{\rho}_{\rho\sigma\mu} \\
&= \delta\Gamma^{\tau}_{\sigma\nu;\mu} - \delta\Gamma^{\tau}_{\sigma\mu;\nu} \\
&\quad - \Gamma^{\tau}_{\rho\mu}\delta\Gamma^{\rho}_{\sigma\nu} + \Gamma^{\rho}_{\sigma\mu}\delta\Gamma^{\tau}_{\rho\nu} + \Gamma^{\rho}_{\nu\mu}\delta\Gamma^{\tau}_{\sigma\rho} \\
&\quad + \Gamma^{\tau}_{\rho\nu}\delta\Gamma^{\rho}_{\sigma\mu} - \Gamma^{\rho}_{\sigma\nu}\delta\Gamma^{\tau}_{\rho\mu} - \Gamma^{\rho}_{\mu\nu}\delta\Gamma^{\tau}_{\sigma\rho} \\
&\quad + \Gamma^{\rho}_{\sigma\nu}\delta\Gamma^{\tau}_{\rho\mu} + \Gamma^{\tau}_{\rho\mu}\delta\Gamma^{\rho}_{\sigma\nu} - \Gamma^{\rho}_{\sigma\mu}\delta\Gamma^{\tau}_{\tau\rho\nu} - \Gamma^{\tau}_{\rho\nu}\delta\Gamma^{\rho}_{\rho\sigma\mu} \\
&= \delta\Gamma^{\tau}_{\sigma\nu;\mu} - \delta\Gamma^{\tau}_{\sigma\mu;\nu} \\
&= \frac{1}{2}g^{\tau\rho}(\delta g_{\rho\sigma;\nu\mu} + \delta g_{\rho\nu;\sigma\mu} - \delta g_{\sigma\nu;\rho\mu} - \delta g_{\rho\sigma;\mu\nu} - \delta g_{\rho\mu;\sigma\nu} + \delta g_{\sigma\mu;\rho\nu}) \\
&= -\frac{1}{2}g^{\tau\rho}(\delta g_{\rho\mu;\sigma\nu} + \delta g_{\sigma\nu;\rho\mu} - \delta g_{\rho\nu;\sigma\mu} - \delta g_{\sigma\mu;\rho\nu} \\
&\quad + R^{\lambda}_{\rho\mu\nu}\delta g_{\lambda\sigma} + R^{\lambda}_{\sigma\mu\nu}\delta g_{\rho\lambda}),
\end{aligned} \tag{8.2}
$$

$$
\begin{aligned}
\delta R_{\tau\sigma\mu\nu} &= R^{\lambda}_{\lambda\sigma\mu\nu}\delta g_{\tau\lambda} + g_{\tau\lambda}\delta R^{\lambda}_{\sigma\mu\nu} \\
&= -\frac{1}{2}(\delta g_{\tau\mu;\sigma\nu} + \delta g_{\sigma\nu;\tau\mu} - \delta g_{\tau\nu;\sigma\mu} - \delta g_{\sigma\mu;\tau\nu} \\
&\quad + R^{\lambda}_{\tau\mu\nu}\delta g_{\lambda\sigma} - R^{\lambda}_{\sigma\mu\nu}\delta g_{\tau\lambda}) \\
&= -\frac{1}{4}\Big(\delta g_{\tau\mu;\sigma\nu} + \delta g_{\tau\mu;\nu\sigma} + \delta g_{\sigma\nu;\tau\mu} + \delta g_{\sigma\nu;\mu\tau}\cdot \\
&\quad - \delta g_{\tau\nu;\sigma\mu} - \delta g_{\tau\nu;\mu\sigma} - \delta g_{\sigma\mu;\tau\nu} - \delta g_{\sigma\mu;\nu\tau} \\
&\quad + R^{\lambda}_{\tau\sigma\nu}\delta g_{\lambda\mu} + R^{\lambda}_{\mu\sigma\nu}\delta g_{\tau\lambda} + R^{\lambda}_{\sigma\tau\mu}\delta g_{\lambda\nu} + R^{\lambda}_{\nu\tau\mu}\delta g_{\sigma\lambda} \\
&\quad - R^{\lambda}_{\tau\sigma\mu}\delta g_{\lambda\nu} - R^{\lambda}_{\nu\sigma\mu}\delta g_{\tau\lambda} - R^{\lambda}_{\sigma\tau\nu}\delta g_{\lambda\mu} - R^{\lambda}_{\mu\tau\nu}\delta g_{\sigma\lambda} \\
&\quad + 2R^{\lambda}_{\tau\mu\nu}\delta g_{\lambda\sigma} - 2R^{\lambda}_{\sigma\mu\nu}\delta g_{\tau\lambda}\Big)
\end{aligned}
$$

$$
\begin{aligned}
&= -\frac{1}{4}\Big(\delta g_{\tau\mu;\sigma\nu} + \delta g_{\tau\mu;\nu\sigma} + \delta g_{\sigma\nu;\tau\mu} + \delta g_{\sigma\nu;\mu\tau}\cdot \\
&\quad - \delta g_{\tau\nu;\sigma\mu} - \delta g_{\tau\nu;\mu\sigma} - \delta g_{\sigma\mu;\tau\nu} - \delta g_{\sigma\mu;\nu\tau} \\
&\quad - R^{\lambda}_{\nu\tau\sigma}\delta g_{\lambda\mu} + R^{\lambda}_{\mu\tau\sigma}\delta g_{\lambda\nu} - R^{\lambda}_{\sigma\mu\nu}\delta g_{\lambda\tau} + R^{\lambda}_{\tau\mu\nu}\delta g_{\lambda\sigma}\Big),
\end{aligned}
$$

$$
\begin{aligned}
\delta R_{\mu\nu} &= \delta R^{\sigma}_{\mu\sigma\nu} = \delta\Gamma^{\sigma}_{\mu\nu;\sigma} - \delta\Gamma^{\sigma}_{\mu\sigma;\nu} \\
&= \frac{1}{2}g^{\sigma\tau}\left(\delta g_{\tau\mu;\nu\sigma} + \delta g_{\tau\nu;\mu\sigma} - \delta g_{\mu\nu;\tau\sigma} - \delta g_{\tau\mu;\sigma\nu} - \delta g_{\tau\sigma;\mu\nu} + \delta g_{\mu\sigma;\tau\nu}\right) \\
&= \frac{1}{2}g^{\sigma\tau}\left(\delta g_{\mu\sigma;\nu\tau} + \delta g_{\nu\sigma;\mu\tau} - \delta g_{\mu\nu;\sigma\tau} - \delta g_{\sigma\tau;\mu\nu}\right),
\end{aligned}
$$

$$
\begin{aligned}
\delta R &= \delta g^{\mu\nu}R_{\mu\nu} + g^{\mu\nu}\delta R_{\mu\nu} \\
&= g^{\mu\nu}\left(\delta\Gamma^{\sigma}_{\mu\nu;\sigma} - \delta\Gamma^{\sigma}_{\mu\sigma;\nu}\right) - R^{\mu\nu}\delta g_{\mu\nu} \\
&= g^{\mu\nu}g^{\sigma\tau}(\delta g_{\mu\sigma;\nu\tau} - \delta g_{\mu\nu;\sigma\tau} - R^{\mu\nu}\delta g_{\mu\nu},
\end{aligned}
$$

$$
\delta g^{\mu\nu} = -g^{\mu\sigma}g^{\nu\tau}\delta g_{\sigma\tau}, \qquad \delta g^{1/2} = \frac{1}{2}g^{1/2}g^{\mu\nu}\delta g_{\mu\nu},
$$

where $\delta g_{\mu\nu}$ is an arbitrary variation in the metric tensor.

Applying these results to the action S_{G} and carrying out a covariant integration by parts,[3] we find

$$
\begin{aligned}
\delta S_{\mathrm{G}} &= \kappa\int \delta(g^{1/2}R)\mathrm{d}^4x + \lambda\int \delta g^{1/2}\mathrm{d}^4x \\
&= \kappa\int g^{1/2}\left[g^{\mu\nu}\left(\delta\Gamma^{\sigma}_{\mu\nu;\sigma} - \delta\Gamma^{\sigma}_{\mu\sigma;\nu}\right) - \left(R^{\mu\nu} - \frac{1}{2}g^{\mu\nu}R\right)\delta g_{\mu\nu}\right]\mathrm{d}^4x \\
&\quad + \frac{1}{2}\lambda\int g^{1/2}g^{\mu\nu}\delta g_{\mu\nu}\mathrm{d}^4x \\
&= -\int g^{1/2}\left[\kappa\left(R^{\mu\nu} - \frac{1}{2}g^{\mu\nu}R\right) - \frac{1}{2}\lambda g^{\mu\nu}\right]\delta g_{\mu\nu}\mathrm{d}^4x,
\end{aligned}
$$

whence it follows that

$$
\frac{\delta S_{\mathrm{G}}}{\delta g_{\mu\nu}} = -g^{1/2}\left[\kappa\left(R^{\mu\nu} - \frac{1}{2}g^{\mu\nu}R\right) - \frac{1}{2}\lambda g^{\mu\nu}\right].
$$

This yields *Einstein's gravitational field equations*:

$$
\kappa g^{1/2}\left(R^{\mu\nu} - \frac{1}{2}g^{\mu\nu}R\right) = \frac{1}{2}T^{\mu\nu} + \frac{1}{2}\lambda g^{1/2}g^{\mu\nu},
$$

---

[3] $\delta g_{\mu\nu}$ is assumed to have compact support in spacetime.

where

$$T^{\mu\nu} \equiv 2\frac{\delta S_{\mathrm{M}}}{\delta g_{\mu\nu}}.$$

$T^{\mu\nu}$, which is a tensor density, is known variously as the *stress–energy density*, *energy–momentum density*, or *energy–momentum–stress density* of the matter. The reason for this terminology does not become fully apparent until one has analyzed the dynamics of various kinds of bulk matter in some detail, but one can already make a beginning at understanding by examining what it looks like in the case of a free particle.

8.1 Energy–Momentum Density of a Free Particle

The action functional of the free particle is given by (5.1). To obtain the functional derivative of this action with respect to the metric tensor $g_{\mu\nu}$ we must subject $g_{\mu\nu}$ to a variation $\delta g_{\mu\nu}$. If the world line of the particle does not intersect the support of $\delta g_{\mu\nu}$, the action will remain unaffected. It is evident therefore that the functional derivative is going to involve a delta function $\delta(x, z)$ having as arguments the point x where the derivative is being taken and the location $z^\alpha(\lambda)$ of the particle. We shall use indices from the first part of the Greek alphabet to denote tensors taken at the point $z^\alpha(\lambda)$ and from the middle of the alphabet to denote tensors taken at the point x^μ. With this convention we may employ the abbreviations

$$g_{\mu\nu} = g_{\mu\nu}(x), \quad g_{\alpha\beta} = g_{\alpha\beta}(z(\lambda)).$$

We shall also need the identity

$$\frac{\delta g_{\alpha\beta}}{\delta g_{\mu\nu}} = \delta^{\mu\nu}_{\alpha\beta},$$

where

$$\delta^{\mu\nu}_{\alpha\beta} \equiv \frac{1}{2}(\delta^\mu_\sigma\delta^\nu_\tau + \delta^\nu_\sigma\delta^\mu_\tau)\,\delta(x,z)\big|_{\sigma=\alpha,\tau=\beta}.$$

$\delta^{\mu\nu}_{\alpha\beta}$ is a bitensor density of unit weight at the point x and zero weight at the point z. It satisfies

$$\delta^{\mu\nu}_{\alpha\beta;\nu} = -\frac{1}{2}(\delta^\mu_{\alpha;\beta} + \delta^\mu_{\beta;\alpha}), \quad \delta^\mu_\alpha \equiv \delta^\mu_\nu\,\delta(x,z)\big|_{\nu=\alpha},$$

as may be verified by passing to a coordinate system in which the derivatives of $g_{\mu\nu}$ vanish at x. Finally, for later use, we shall record here two other properties of the functional derivative:

- Functional differentiation is commutative (like ordinary differentiation).
- Functional differentiation commutes with ordinary differentiation with respect to coordinates or world line parameter λ (It does *not* commute with covariant differentiation!).

The computation of the energy–momentum density for the free particle is now elementary. We find

$$T^{\mu\nu} = 2\frac{\delta S}{\delta g_{\mu\nu}} = m\int \delta^{\mu\nu}_{\alpha\beta}\dot{z}^{\alpha}\dot{z}^{\beta}(-\dot{z}^2)^{-1/2}\mathrm{d}\lambda$$

$$\underset{\lambda\to\tau}{\longrightarrow} m\int \delta^{\mu\nu}_{\alpha\beta}\dot{z}^{\alpha}\dot{z}^{\beta}\mathrm{d}\tau = \int \delta^{\mu\nu}_{\alpha\beta}p^{\alpha}u^{\beta}\mathrm{d}\tau.$$

Let us look at the special form this expression takes in canonical coordinates in flat spacetime.[4]

$$\begin{aligned}
T^{\mu\nu} &= \int p^{\mu}(\tau)u^{\nu}(\tau)\delta(x - z(\tau))\mathrm{d}\tau \\
&= \int p^{\mu}(\tau)u^{\nu}(\tau)\delta(\boldsymbol{x} - \boldsymbol{z}(\tau))\delta\left(x^0 - z^0(\tau)\right)\frac{\mathrm{d}z^0(\tau)}{\dot{z}^0(\tau)} \\
&= \delta(\boldsymbol{x} - \boldsymbol{z})p^{\mu}\frac{u^{\nu}}{u^0}\bigg|_{z^0(\tau)=x^0},
\end{aligned}$$

so that

$$T^{\mu 0} = \delta(\boldsymbol{x} - \boldsymbol{z})p^{\mu}, \quad T^{\mu i} = \delta(\boldsymbol{x} - \boldsymbol{z})p^{\mu}v^{i}, \quad v^{i} = \frac{u^{i}}{u^{0}}.$$

The three-dimensional delta function appearing in these last equations displays like a beacon the pointlike character of the particle. T^{00} is clearly the particle's energy density: All the energy p^0 is located where the particle is! T^{i0} is just as clearly the momentum density. However, if one remembers the relativistic relation $p^i = p^0 v^i$ between momentum and energy, one can alternatively regard momentum as a *rate of transport of energy*. This permits T^{i0}, or T^{0i} if you like, to be interpreted also as a rate of flow of energy per unit area or *energy flux density*. In a similar vein, T^{ij} is to be regarded as a *momentum flux density*.

8.2 The Weak Field Approximation

We are now in a position to establish another point of contact between formalism and observation, a point of contact that will, in particular, enable us to determine

[4] In flat spacetime, the world line is of course straight, but we make no use of this at this point

the value of the constant κ. We again assume that the gravitational field is so weak that we may introduce a quasi-canonical coordinate system. We note that this forces us for the present to assume that the constant λ vanishes. For if it were not zero, spacetime could not be even approximately flat, in the large, even in the absence of matter. It would instead be forced to have a constant scalar curvature that is easily found by contracting Einstein's equations:

$$R = -2\lambda/\kappa.$$

To find the form that Einstein's equations take in the weak field approximation we first compute the Riemann tensor. For this purpose, we recall that we may use the expression obtained in (4.2) for the Riemann tensor in a coordinate system in which the first derivatives of the metric tensor vanish at the point of interest. We have seen that the terms that involve the first derivatives in an arbitrary coordinate system are, in a quasi-canonical coordinate system, of the second order in small quantities and hence may be dropped. Therefore we have used the notation and thereafter [see (6.8)],

$$R_{\mu\nu\sigma\tau} = -\frac{1}{2}(h_{\mu\sigma,\nu\tau} + h_{\nu\tau,\mu\sigma} - h_{\mu\tau,\nu\sigma} - h_{\nu\sigma,\mu\tau}).$$

This expression is known as the *linearized Riemann tensor*, and the weak field theory is often called the *linearized theory*. The linearized Riemann tensor has an important property: It is invariant under the approximate coordinate transformation law for the $h_{\mu\nu}$ given by (6.2). In this respect the linearized Riemann tensor is similar to the electromagnetic field tensor, which is invariant under gauge transformations,[5] and for this reason the approximate coordinate transformation law is often called a *gravitational gauge transformation*. The reason for the gauge invariance of the linearized Riemann tensor is not hard to see. When ξ^μ is small the functional form of the full Riemann tensor suffers a change that can be accurately expressed as a sum of terms of the form $-R_{\mu\nu\sigma\tau,\rho}\xi^\rho$, $-R_{\mu\nu\ \sigma\rho}\xi^\rho{}_{,\tau}$, etc. (infinitesimal coordinate transformation law). But as these terms are of the second order in small quantities they may be dropped in the weak field approximation. ($R_{\mu\nu\sigma\tau}$, unlike the metric tensor, is already of the first order in small quantities.) The gauge invariance of the linearized Riemann tensor may also, of course, be verified by direct computation.

Problem 30 Verify the gauge invariance of the linearized Riemann tensor by direct computation.

---

[5] The analogy goes deeper than this. The electromagnetic field tensor is a curl. The linearized Riemann tensor is a *double curl*. It is obtained by antisymmeterizing the second derivative $-h_{\mu\sigma,\nu\tau}/2$ in μ and ν and in σ and τ.

Solution 30 We have

$$\begin{aligned}\overline{R}_{\mu\nu\sigma\tau} &= R_{\mu\nu\sigma\tau} + \frac{1}{2}(\xi_{\mu,\sigma\nu\tau} + \xi_{\sigma,\mu\nu\tau} + \xi_{\nu,\tau\mu\sigma} + \xi_{\tau,\nu\mu\sigma} \\ &\quad -\xi_{\nu,\sigma\mu\tau} - \xi_{\sigma,\nu\mu\tau} - \xi_{\mu,\tau\nu\sigma} - \xi_{\tau,\mu\nu\sigma}) \\ &= R_{\mu\nu\sigma\tau}.\end{aligned}$$

The importance of the gauge invariance of the linearized Riemann tensor lies in the fact that the presence or absence of a real gravitational field is characterized by the presence or absence of a nonvanishing Riemann tensor. This tensor *represents* the gravitational field, and in the weak field approximation we have, in it, an *invariant* characterization of the field, i.e., an expression for the field that is *independent* of which quasi-canonical coordinate system we are using.

Let us now compute the linearized Ricci tensor and curvature scalar:

$$R_{\mu\nu} = \eta^{\sigma\tau} R_{\mu\sigma\nu\tau} = -\frac{1}{2}\eta^{\sigma\tau}\left(h_{\mu\nu,\sigma\tau} + h_{\sigma\tau,\mu\nu} - h_{\sigma\nu,\mu\tau} - h_{\mu\tau,\sigma\nu}\right),$$

$$R = \eta^{\mu\nu} R_{\mu\nu} = -h^{\mu}_{,\mu} + h^{\mu\nu}_{,\mu\nu}.$$

The linearized Einstein equations then follow immediately:

$$\begin{aligned}\frac{1}{2\kappa}T^{\mu\nu} &= -\frac{1}{2}\left(h^{\mu\nu\sigma}_{,\sigma} + h^{\mu\nu}_{,} - h^{\mu\sigma\nu}_{,\sigma} - h^{\nu\sigma\mu}_{,\sigma}\right) \\ &\quad + \frac{1}{2}\eta^{\mu\nu}\left(h^{\sigma}_{,\sigma} - h^{\sigma\tau}_{,\sigma\tau}\right) \\ &= -\frac{1}{2}\left(l^{\mu\nu\sigma}_{,\sigma} - \frac{1}{2}\eta^{\mu\nu}l^{\sigma}_{,\sigma} - l^{\mu\nu}_{,} - l^{\mu\sigma\nu}_{,\sigma} + \frac{1}{2}l^{\mu\nu}_{,}\right. \\ &\quad \left. - l^{\nu\sigma\mu}_{,\sigma} + \frac{1}{2}l^{\nu\mu}_{,} + \eta^{\mu\nu}l^{\sigma}_{,\sigma} + \eta^{\mu\nu}l^{\sigma\tau}_{,\sigma\tau} - \frac{1}{2}\eta^{\mu\nu}l^{\sigma}_{,\sigma}\right) \\ &= -\frac{1}{2}\left(l^{\mu\nu\sigma}_{,\sigma} - l^{\mu\sigma\nu}_{,\nu\sigma} - l^{\nu\sigma\mu}_{,\sigma} + \eta^{\mu\nu}l^{\sigma\tau}_{,\sigma\tau}\right).\end{aligned}$$

If we impose the supplementary condition $l^{\mu\nu}_{\nu} = 0$ (sometimes called choosing the *Lorentz, harmonic*, or *de Donder gauge*), these equations take the particularly simple form

$$\Box^2 l^{\mu\nu} = -\frac{1}{\kappa}T^{\mu\nu}.$$

Contact with observation is made by choosing for $T^{\mu\nu}$ the energy–momentum density of a point particle. There is, of course, a contradiction here. If the particle is a point the field $l^{\mu\nu}$ will become singular at the particle itself, thus violating the weak field approximation. Worse still, this will continue to be true even in the full theory, and hence the particle will have no geodesic to follow because the very notion of Riemannian manifold breaks down where the particle is. This means, of course, that the point particle picture is an idealization. We must smear the particle out. It is true that we can then no longer be absolutely sure that the point particle picture is a valid idealization but must check it later, after the fact, which we shall do.

Let us suppose that the particle has mass M and is located at the origin. Then its (idealized) energy–momentum density has only one nonvanishing component, viz., T^{00}, given by

$$T^{00} = M\delta(\boldsymbol{x}). \tag{8.3}$$

The particle being at rest, we may choose a stationary quasi-canonical coordinate system. Actually, this choice involves another assumption, namely that no gravitational waves are present. In the special gauge or coordinate system in which the supplementary condition $l^{\mu\nu}_{\ \ ,\nu} = 0$ is satisfied, $l^{\mu\nu}$ then has only one nonvanishing component l^{00}, which satisfies the equation

$$\nabla^2 l^{00} = -\frac{1}{\kappa}T^{00} = -\frac{1}{\kappa}M\delta(\boldsymbol{x}).$$

(We recall that the supplementary condition can also be imposed in a quasi-stationary coordinate system.) If more than one 'particle' is present, there will be additional terms on the right side of the equation. (This follows from the additivity of action functionals.) Owing to the interaction of the particles, they will no longer remain at rest. In order that it remain possible to keep the coordinates quasi-stationary, the density of the masses must satisfy $G\rho \ll (GM)^{-2}$ [see (6.6)] in accord with our previous statement that the particles must in reality be smeared out.

The solution of the above differential equation that satisfies

$$\lim_{|\boldsymbol{x}|\to\infty} l^{00} = 0,$$

and is to be taken seriously only for $|\boldsymbol{x}| \gg GM$, is

$$l^{00}(x) = \frac{M}{4\pi\kappa}\frac{1}{|\boldsymbol{x}|} = -\frac{1}{4\pi\kappa G}\Phi,$$

where Φ is the Newtonian potential of the particle. From this we get

$$l = -l^{00} = -l_{00}, \quad h_{00} = l_{00} + \frac{1}{2}l = \frac{1}{2}l_{00} = -\frac{1}{8\pi\kappa G}\Phi,$$

$$h_{ij} = l_{ij} - \frac{1}{2}\delta_{ij}l = \frac{1}{2}\delta_{ij}l_{00} = -\delta_{ij}\frac{1}{8\pi\kappa G}\Phi, \quad h_{0i} = l_{0i} = 0.$$

If we transform now to another quasi-stationary coordinate system using the transformation laws (6.3), h_{ij} and h_{0i} will assume other values, but h_{00} will remain unchanged.

In order that general relativity in the weak field approximation agree with Newtonian theory in its account of the motion of bodies under the action of gravitational forces, we have seen in (6.5) that we must make the identification $h_{00} = -2\Phi$. Comparing this with our present result we see that the constant κ must be given by

$$\kappa = \frac{1}{16\pi G}.$$

The full Einstein equations therefore take the form

$$g^{1/2}\left(R^{\mu\nu} - \frac{1}{2}g^{\mu\nu}R\right) = 8\pi G T^{\mu\nu} + 8\pi G\lambda g^{1/2}g^{\mu\nu}. \tag{8.4}$$

8.3 Energy–Momentum–Stress Density of a Gas at Equilibrium

If more than one particle is present their action functionals simply add together and the total energy–momentum density takes the form

$$T^{\mu\nu} = \sum_n \int \delta^{\mu\nu}_{\alpha_n\beta_n} p_n^{\alpha_n} u_n^{\beta_n} \mathrm{d}\tau_n = \sum_n \int \delta^{\mu\nu}_{\alpha_n\beta_n} p_n^{\alpha_n} u_n^{\beta_n} \frac{\mathrm{d}z_n^0}{u_n^0}.$$

If the particles are numerous enough to be described effectively by a continuous distribution function $f(x, \boldsymbol{p})$, the above summation may be replaced by an integration:

$$\begin{aligned} T^{\mu\nu} &= \int \mathrm{d}^3 z \int \mathrm{d}^3\boldsymbol{p} f(z,\boldsymbol{p}) \int \delta^{\mu\nu}_{\alpha\beta} p^\alpha u^\beta \frac{\mathrm{d}z^0}{u^0} \\ &= \int \mathrm{d}^4 z \int \frac{\mathrm{d}^3\boldsymbol{p}}{u^0} \delta^{\mu\nu}_{\alpha\beta} p^\alpha u^\beta f(z,\boldsymbol{p}) \\ &= \int p^\mu u^\nu f(x,\boldsymbol{p}) \frac{\mathrm{d}^3\boldsymbol{p}}{u^0} \\ &= 2\int p^\mu p^\nu F(x,p)\theta(p^0)\delta(p^2+m^2)\mathrm{d}^4 p, \end{aligned}$$

where $F(x, p)$ is an extension of $f(x, \boldsymbol{p})$ off the mass shell. Here the particles are assumed to be non-interacting (except through their averaged gravitational forces–Vlasov approximation). However, the expression obtained for $T^{\mu\nu}$ is also valid for interacting particles under the Boltzmann collision approximation, in particular for a gas at equilibrium in a spacetime with a timelike Killing vector ξ^μ. Choosing the distribution function in the form $F(x, p) = \Phi(\xi \cdot p)$ in this case, and defining

$$\Theta(y) = \int^y \mathrm{d}y' \int^{y'} \mathrm{d}y'' \Phi(y''),$$

we may write

$$\begin{aligned}T^{\mu\nu} &= 2\int p^\mu p^\nu \Phi(\xi\cdot p)\theta(p^0)\delta(p^2+m^2)\mathrm{d}^4p\\&= 2\frac{\partial^2}{\partial\xi_\mu\partial\xi_\nu}\left[g^{1/2}Z_m(\xi^2)\right]\\&= 4g^{1/2}\frac{\partial}{\partial\xi_\mu}[\xi^\nu Z_m'(\xi^2)]\\&= 4g^{1/2}\left[g^{\mu\nu}Z_m'(\xi^2)+2\xi^\mu\xi^\nu Z_m''(\xi^2)\right],\end{aligned}$$

where

$$g^{1/2}Z_m(\xi^2) = \int \Theta(\xi\cdot p)\theta(p^0)\delta(p^2+m^2)\mathrm{d}^4p.$$

It is customary to reexpress the energy–momentum–stress density of a gas in terms of the *pressure* p and the *proper energy density* w_0 defined by[6]

$$p = 4Z_m'(\xi^2),\quad w_0+p = 8(-\xi^2)Z_m''(\xi^2).$$

These definitions yield

$$\begin{aligned}T^{\mu\nu} &= g^{1/2}[(w_0+p)\overline{u}^\mu\overline{u}^\nu + pg^{\mu\nu}]\\&= g^{1/2}\left[w_0\overline{u}^\mu\overline{u}^\nu + p\overline{P}^{\mu\nu}\right].\end{aligned}\tag{8.5}$$

The appropriateness of these definitions may be checked by passing to a local canonical rest frame of the gas at any point. At the chosen point we then have $T^{i0}=0=T^{0i}$ and

$$w_0 = T^{00} = \int p^0 f(x,\mathbf{p})\mathrm{d}^3\mathbf{p} = \int p^0\Phi\left((-\xi^2)^{1/2}p^0\right)\mathrm{d}^3\mathbf{p},$$

$$p\delta_{ij} = T^{ij} = \int p^i v^j f(x,\mathbf{p})\mathrm{d}^3\mathbf{p} = \int p^i v^j\Phi\left((-\xi^2)^{1/2}p^0\right)\mathrm{d}^3\mathbf{p},$$

with $p^0=\sqrt{\mathbf{p}^2+m^2}$, in precise accord with our customary definitions of energy density and pressure.

---

6 Note that these equations allow one to determine the pressure and energy distribution in the gas directly from a knowledge of ξ^μ and Φ, and hence of the function Z_m.

Chapter 9
Conservation Laws

In the case of the free particle, we interpreted various components of the energy–momentum–stress density as fluxes of energy and momentum. This interpretation can obviously be extended also to particle ensembles and gases. When we speak of fluxes we usually think of quantities that are conserved. In special relativity, energy and momentum are conserved. In general relativity, they are no longer generally conserved, at least if we do not include the energy and momentum of the gravitational field itself. Nevertheless, their densities and fluxes satisfy a covariant generalization of a true conservation law, which is quite easy to obtain.

Consider the action functional for the matter, S_{M}. This functional is coordinate invariant. Therefore, if $\delta g_{\mu\nu}$ and $\delta\Phi^{\mathrm{A}}$ are the changes induced in the metric tensor and the matter dynamical variables by an infinitesimal coordinate transformation, we must have

$$0 \equiv \int \frac{\delta S_{\mathrm{M}}}{\delta g_{\mu\nu}} \delta g_{\mu\nu} \mathrm{d}^4 x + \frac{\delta S_{\mathrm{M}}}{\delta \Phi^A} \delta \Phi^A,$$

with implicit summation or integration over the index A. When the matter dynamical equations are satisfied, the second term vanishes. Therefore, writing

$$\delta g_{\mu\nu} = -\delta\xi_{\mu;\nu} - \delta\xi_{\nu;\mu},$$

assuming that $\delta\xi_\mu$ has compact support, and carrying out integration by parts, we have

$$\begin{aligned} 0 &= -\int \frac{\delta S_{\mathrm{M}}}{\delta g_{\mu\nu}} (\delta\xi_{\mu;\nu} + \delta\xi_{\nu;\mu}) \mathrm{d}^4 x \\ &= -\int T^{\mu\nu} \delta\xi_{\mu;\nu} \mathrm{d}^4 x = \int T^{\mu\nu}_{\ \ ;\nu} \delta\xi_\mu \mathrm{d}^4 x. \end{aligned}$$

B. Dewitt, *Bryce DeWitt's Lectures on Gravitation*,
Edited by Steven M. Christensen, Lecture Notes in Physics, 826,
DOI: 10.1007/978-3-540-36911-0_9, © Springer-Verlag Berlin Heidelberg 2011

As $\delta\xi_\mu$ is arbitrary, we have

$$T^{\mu\nu}_{;\nu} = 0,$$

whenever the matter dynamical equations are satisfied.

We emphasize that this last equation generally holds only when the matter dynamical equations are satisfied. In fact, in many cases, it is completely equivalent to the matter dynamical equations and can be used in place of them. This may be illustrated with the case of the free particle. We have

$$T^{\mu\nu}_{;\nu} = \int \delta^{\mu\nu}_{\alpha\beta;\nu} p^\alpha u^\beta \mathrm{d}\tau = -\int \delta^\mu_{\alpha;\beta} p^\alpha \dot{z}^\beta \mathrm{d}\tau$$
$$= -\int \dot{\delta}^\mu_\alpha p^\alpha \mathrm{d}\tau = \int \delta^\mu_\alpha \dot{p}^\alpha \mathrm{d}\tau.$$

Now let A_μ be an arbitrary covariant vector of compact support. Multiplying both sides of this equation by A_μ and integrating over spacetime. If $T^{\mu\nu}_{;\nu} = 0$, one gets

$$\int A_\alpha \dot{p}^\alpha \mathrm{d}\tau = 0.$$

As A_α is arbitrary, this implies $\dot{p}^\alpha = 0$.

Although $T^{\mu\nu}$ has vanishing covariant divergence, this does not imply a true conservation law. $T^{\mu\nu}$ accounts only for the energy and momentum of the matter. When a gravitational field is present (i.e., when spacetime is not flat), it can exchange energy and momentum with the matter. One might ask whether the energy and momentum of the gravitational field could be accounted for by treating the gravitational action functional S_G in the same way. It too is coordinate independent and hence satisfies

$$\left(\frac{\delta S_\mathrm{G}}{\delta g_{\mu\nu}}\right)_{;\nu} = 0.$$

In this case, however, the relation is an identity that holds whether or not the field equations are satisfied. Its explicit form is

$$0 \equiv -\frac{1}{16\pi G}\left\{g^{1/2}\left[\left(R^{\mu\nu} - \frac{1}{2}g^{\mu\nu}R\right) + 8\pi G\lambda g^{\mu\nu}\right]\right\}_{;\nu}.$$

The term in λ drops out right away, as also does the factor $g^{1/2}$, leaving

$$0 \equiv \left(R^{\mu\nu} - \frac{1}{2}g^{\mu\nu}R\right)_{;\nu}.$$

This is known as the *contracted Bianchi identity*.

Problem 31 Show that the identity

$$0 \equiv \left(R^{\mu\nu} - \frac{1}{2}g^{\mu\nu}R\right)_{;\nu}$$

can be obtained by contracting the Bianchi identity twice.

Solution 31 We have

$$\begin{aligned} 0 \equiv & R^{\mu\nu}_{\ \ \mu\nu;\sigma} + R^{\mu\nu}_{\ \ \nu\sigma;\mu} + R^{\mu\nu}_{\ \ \sigma\mu;\nu} \\ \equiv & R_{;\sigma} - R^{\mu}_{\sigma;\mu} - R^{\nu}_{\sigma;\nu} \equiv -2g_{\sigma\mu}\left(R^{\mu\nu} - \frac{1}{2}g^{\mu\nu}R\right)_{;\nu} . \end{aligned}$$

The contracted Bianchi identity imposes no constraint on the gravitational field. It does, however, impose a constraint on the matter through the Einstein equations, *forcing* $T^{\mu\nu}$ to have vanishing covariant divergence. In many cases, therefore, the matter dynamical equations are superfluous; the Einstein equations are sufficient. It is nice, however, to know that the equations obtained from the complete variational principle are at least consistent. They would *not* be consistent if S_{M} were not coordinate independent.

The presence of a gravitational field does not always mean that the matter variables satisfy no true conservation laws of their own. In special cases, conserved quantities *can* be built out of the $T^{\mu\nu}$, namely, when spacetime admits an isometric motion corresponding to a Killing vector ξ^{μ}. For we then have

$$\begin{aligned} (\xi_{\mu}T^{\mu\nu})_{;\nu} &= \xi_{\mu;\nu}T^{\mu\nu} + \xi_{\mu}T^{\mu\nu}_{\ \ ;\nu} \\ &= \frac{1}{2}(\xi_{\mu;\nu} + \xi_{\nu;\mu})T^{\mu\nu} = 0, \end{aligned}$$

implying the conservation of

$$\int_{\Sigma} \xi_{\mu}T^{\mu\nu}\mathrm{d}\Sigma_{\nu},$$

where Σ is any connected hypersurface that intersects completely the support of the vector density $\xi_{\mu}T^{\mu\nu}$, cutting all its flow lines an odd number of times.

Problem 32 Show that the stress tensor of a particle ensemble

$$T^{\mu\nu} = 2\int p^{\mu}p^{\nu}F(x,p)\theta(p^0)\delta(p^2+m^2)\mathrm{d}^4p,$$

has vanishing covariant divergence if the distribution function satisfies $F_{;\mu}p^{\mu} = 0$. In the special case of a dilute gas at equilibrium, in which F has the form $F(x, p) = \Phi(\xi\cdot\, p)$, identify the conserved quantity associated with the Killing vector ξ^{μ}.

Solution 32 The result is proven by introducing a coordinate system in which the derivatives $g_{\mu\nu,\sigma}$ vanish at x.

We identify the conserved quantity by

$$\xi_{\mu}T^{\mu\nu} = -g^{1/2}(-\xi^2)^{1/2}\overline{u}^{\nu}w_0,$$

whence

$$\int_{\Sigma} \xi_\mu T^{\mu\nu} \mathrm{d}\Sigma_\mu = -\int g^{1/2}\overline{u}^0 (-\xi^2)^{1/2} w_0 \mathrm{d}^3x = \text{ const.} \times \int \frac{w_0}{T} \mathrm{d}^3 V,$$

where $\mathrm{d}^3 V = g^{1/2}\overline{u}^0 \mathrm{d}^3 x$. Imagine spacetime to be divided into flow tubes parallel to ξ^μ. The conserved quantity may be defined as the sum of w_0/T over all the tubes, weighted by the 3-volume of the orthogonal section of each tube.

9.1 Energy, Momentum, Angular Momentum and Spin

When spacetime is flat, its group of motions is the full Poincaré group and there are correspondingly many conserved quantities. Strictly speaking, this case cannot be realized physically unless spacetime is empty, with nothing to be conserved (and with the constant λ equal to zero)! Spacetime can be flat only if dynamical behavior is withheld from the gravitational field and the geometry is "externally" imposed. Nevertheless, flatness is a highly accurate approximation in practice, except under extreme astrophysical conditions.

In canonical coordinates in flat spacetime, Killing's equation takes the form:

$$\xi_{\mu,\nu} + \xi_{\nu,\mu} = 0,$$

which has the general solution

$$\xi_\mu = a_\mu + l_{\mu\nu} x^\nu,$$

where the a_μ, $l_{\mu\nu}$ are constants with

$$l_{\mu\nu} = -l_{\nu\mu}.$$

The corresponding conserved quantity is

$$\int_{\Sigma} (a_\mu + l_{\mu\nu} x^\nu) T^{\mu\sigma} \mathrm{d}\Sigma_\sigma = a_\mu P^\mu - \frac{1}{2} l_{\mu\nu} J^{\mu\nu},$$

where

$$P^\mu = \int_{\Sigma} T^{\mu\nu} \mathrm{d}\Sigma_\nu \,, \qquad J^{\mu\nu} = \int_{\Sigma} (x^\mu T^{\nu\sigma} - x^\nu T^{\mu\sigma}) \mathrm{d}\Sigma_\sigma.$$

As the a_μ, $l_{\mu\nu}$ are arbitrary, it follows that P^μ and $J^{\mu\nu}$ are independently conserved. They are known respectively as the *total energy–momentum 4-vector* and *total angular momentum tensor* of the matter.

Although it is not easy to prove in general, the vector P^μ is always time-like[1] (and oriented to the future if the sign of S_M has been chosen correctly). Therefore, it may be used to define a *mean rest frame* for the matter as well as a total energy, or mass M, in that frame:

$$M^2 = -P^2.$$

The corresponding *mean 4-velocity* is given by

$$U^\mu = M^{-1} P^\mu.$$

Although the P^μ transform as the components of a vector under the full Poincaré group, the $J^{\mu\nu}$ transform as the components of a tensor only under the homogeneous Lorentz group. Under displacements

$$\bar{x}^\mu = x^\mu + \xi^\mu, \qquad \xi^\mu = \text{ constant},$$

they transform according to

$$\bar{J}^{\mu\nu} = J^{\mu\nu} + \xi^\mu P^\nu - \xi^\nu P^\mu.$$

A true tensor under the Poincaré group can be constructed out of $J^{\mu\nu}$ by passing to the mean rest frame of the matter or, equivalently, projecting $J^{\mu\nu}$ onto the corresponding hyperplane of simultaneity:

$$S^{\mu\nu} \equiv P^\mu_\sigma P^\nu_\tau J^{\sigma\tau}, \qquad P^{\mu\nu} = \eta^{\mu\nu} + U^\mu U^\nu.$$

In the mean rest frame, only the spatial components of $S^{\mu\nu}$ are nonvanishing and they then coincide with the spatial components of $J^{\mu\nu}$. As the mean 3-momentum vanishes in this frame, the contributions to the spatial components of $J^{\mu\nu}$ come only from the overall spin of the matter. $S^{\mu\nu}$ is, therefore, called the *spin angular momentum tensor*. The *orbital angular momentum tensor* may be defined as the difference between $J^{\mu\nu}$ and $S^{\mu\nu}$:

$$\begin{aligned} L^{\mu\nu} &= J^{\mu\nu} - S^{\mu\nu} \\ &= J^{\mu\nu} - J^{\mu\nu} - U^\mu U_\sigma J^{\sigma\nu} - U^\nu U_\tau J^{\mu\tau} - U^\mu U^\nu U_\sigma U_\tau J^{\sigma\tau} \\ &= (M^{-1} U_\sigma J^{\sigma\mu} + U^\mu \tau) P^\nu - (M^{-1} U_\sigma J^{\sigma\nu} + U^\nu \tau) P^\mu, \end{aligned}$$

where τ is an arbitrary parameter.

Strictly speaking, only the spatial components of $J^{\mu\nu}$ refer to angular momentum. If the hypersurface Σ is chosen to be the hyperplane $x^0 = t$, then the temporal components of $J^{\mu\nu}$ take the form:

$$J^{0i} = tP^i - X^i_{\mathrm{E}} P^0, \tag{9.1}$$

[1] See, however, the special case described on p. 175, in which P^μ is null in the eikonal approximation.

where X^i_{E} are the coordinates of the *center of energy* of the matter:

$$X^i_{\mathrm{E}} = \frac{1}{P^0}\int x^i T^{00} \mathrm{d}^3\vec{x}.$$

The conservation of the J^{0i} may be stated in the form:

$$\frac{\mathrm{d}X^i_{\mathrm{E}}}{\mathrm{d}t} = \frac{P^i}{P^0} = V^i, \qquad V^i = \frac{U^i}{U^0},$$

which says that the center of energy moves with the mean 3-velocity of the matter.

The concept of center of energy is not frame independent. To get an invariant concept, we must again pass to the mean rest frame. In this frame, we have $(U_\mu) = (-1, 0, 0, 0)$ and $P^0 = M$, and we may express the center of energy in the form:

$$\begin{aligned} X^i_{\mathrm{E}} &= \frac{1}{P^0}(-J^{0i} + P^i t) \\ &= \frac{1}{M}U_\nu J^{\nu i} + U^i t. \end{aligned}$$

If we also define, in this frame,

$$X^0_{\mathrm{E}} \equiv t = \frac{1}{M}U_\nu J^{\nu 0} + U^0 t,$$

and remember that rest-frame time t is equal to proper time τ, then we see that X^0_{E} and the X^i_{E} are equal to the components of the true 4-vector,

$$X^\mu \equiv \frac{1}{M}U_\nu J^{\nu\mu} + U^\mu \tau.$$

However, X^0_{E} and the X^i_{E} coincide with the components of X^μ only in the rest frame.

X^μ, which is really a linear function of the parameter τ, and hence a straight world line, may be called the *covariant center of energy*. In terms of it, the orbital angular momentum tensor may be reexpressed in the form (see above):

$$L^{\mu\nu} = X^\mu P^\nu - X^\nu P^\mu.$$

It will be noted that $L^{\mu\nu}$ is actually independent of τ. In fact, both $L^{\mu\nu}$ and $S^{\mu\nu}$ are separately conserved.

Chapter 10
Phenomenological Description of a Conservative Continuous Medium

The energy–momentum–stress density evidently plays the role of source for the gravitational field and is in many ways analogous to the electric charge in electromagnetic theory. However, because matter in *all* its forms is coupled to the gravitational field, this source can be much more complicated than electric charge. Many times we may wish to find the gravitational field produced by a certain material system without knowing or being able to write an action functional for the system. We then need a general description of the system and its dynamical behavior that will enable us to keep track of its energy and momentum content and obtain an energy–momentum–stress density for it without necessarily knowing its structure in all fundamental respects. An example of such a description, which covers a wide range of practical cases, is the phenomenological treatment of a conservative continuous medium. By "conservative" we mean that there are no irreversible dissipative processes at work. Once having found the energy–momentum–stress density for a conservative medium, we shall find it not difficult to introduce dissipative mechanisms either again phenomenologically or, if that is insufficient, by the use of distribution functions and all the paraphernalia of the Boltzmann and other types of transport equations.

We use the notation in Chap. 1 and thereafter, but we add two new elements, an orthonormal triad field n_a^μ defined *throughout* the medium and satisfying everywhere

$$n_a \cdot n_b = \delta_{ab}, \quad n_a \cdot u = 0, \quad u^\mu = \frac{\partial}{\partial \tau} x^\mu(\xi, \tau),$$

and a scalar field w_0 equal at each point to the *proper energy density* at that point, i.e., the density of total energy (rest mass as well as internal energy) as viewed in the local rest frame defined by the n_a^μ at that point. We do not impose any additional conditions on the n_a^μ, e.g., Fermi–Walker transport, beyond their orthonormality and orthogonality to u^μ.

B. DeWitt, *Bryce DeWitt's Lectures on Gravitation*,
Edited by Steven M. Christensen, Lecture Notes in Physics, 826,
DOI: 10.1007/978-3-540-36911-0_10,

We shall assume that the dynamical behavior of the medium is determined solely by its proper energy density and its internal stresses. The stresses, which will be analyzed purely phenomenologically, will be described presently. First, however, we devote attention to w_0. This density is defined in a local Cartesian rest frame of the medium. It will be convenient to re-express it also relative to the arbitrary curvilinear coordinates x^μ of spacetime, as well as in the internal coordinate system provided by the labels ξ^i.

The transformations between the local Cartesian rest frame and the (in general curvilinear) frame of the ξ^i are described by the transformation coefficients

$$A_{ai} \equiv n_{a\mu} x^\mu_{,i}$$

and their inverses

$$A_a^{-1i} = \xi^i_{,\mu} n^\mu_a.$$

We note that

$$A_{ai} A_{aj} = n_{a\mu} x^\mu_{,i} n_{a\nu} x^\nu_{,j} = P_{\mu\nu} x^\mu_{,i} x^\nu_{,j} = \gamma_{ij},$$

where $P_{\mu\nu}$ projects orthoognally to u^μ. Hence that[1]

$$\det(A_{ai}) = \gamma^{1/2}, \quad \text{where } \gamma = \det(\gamma_{ij}).$$

The proper energy density in the ξ coordinate system is therefore

$$w_\xi = \det(A_{ai}) w_0 = \gamma^{1/2} w_0.$$

Problem 33 Prove that A_a^{-1i} ($\equiv \xi^i_{,\mu} n^\mu_a$) is both a left and a right inverse of A_{ai} ($\equiv n_{a\mu} x^\mu_{,i}$). Prove also the following identities:

$$A_a^{-1i} A_a^{-1j} = \gamma^{ij}, \qquad \gamma_{ij} A_a^{-1i} A_b^{-1j} = \delta_{ab}, \qquad \gamma^{ij} A_{ai} A_{bj} = \delta_{ab},$$

where γ^{ij} is the contravariant proper metric tensor, inverse to γ_{ij}.

Solution 33 We have

$$A_a^{-1i} A_{aj} = \xi^i_{,\mu} n^\mu_a n_{a\nu} x^\nu_{,j} = \xi^i_{,\mu} (\delta^\mu_\nu + u^\mu u_\nu) x^\nu_{,j} = \delta^i_j,$$

$$A_{ai} A_b^{-1i} = n_{a\mu} x^\mu_{,i} \xi^i_{,\nu} n^\nu_b = n_{a\mu} (\delta^\mu_\nu - \dot{x}^\mu \tau_{,\nu}) n^\nu_b = \delta_{ab}.$$

Furthermore,

$$\gamma_{ik} A_a^{-1k} A_a^{-1j} = A_{bi} A_{bk} A_a^{-1k} A_a^{-1j} = A_{ai} A_a^{-1j} = \delta^j_i,$$

$$\gamma_{ij} A_a^{-1i} A_b^{-1j} = A_{ci} A_{cj} A_a^{-1i} A_b^{-1j} = \delta_{ca} \delta_{cb} = \delta_{ab},$$

[1] We assume the ξ axes to have the same relative orientation as the vectors n^μ_a.

$$\gamma^{ij} A_{ai} A_{bj} = A_c^{-1i} A_c^{-1j} A_{ai} A_{bj} = \delta_{ca}\delta_{cb} = \delta_{ab},$$

as required.

Let us also note that the matrices, formed from components of u^μ, n_a^μ satisfy

$$\begin{pmatrix} -u^\mu \\ n_a^\mu \end{pmatrix}^{\mathrm{tr}} \begin{pmatrix} u^\nu \\ n_a^\nu \end{pmatrix} = (-u^\mu u^\nu + n_a^\mu n_a^\nu) = (g^{\mu\nu}),$$

from which it may be inferred that

$$-\left[\det\begin{pmatrix} u^\mu \\ n_a^\mu \end{pmatrix}\right]^2 = \det(g^{\mu\nu}) = -g^{-1},$$

and hence, assuming u^μ, n_1^μ, n_2^μ, n_3^μ to have, respectively, the same relative orientation as positive displacements along the x^0, x^1, x^2, x^3 axes,

$$\det\begin{pmatrix} u^\mu \\ n_a^\mu \end{pmatrix} = g^{-1/2}.$$

From this and the fact that

$$\begin{pmatrix} -u_\mu \\ n_{a\mu} \end{pmatrix}^{\mathrm{tr}} = \begin{pmatrix} u^\mu \\ n_a^\mu \end{pmatrix}^{-1}, \qquad \det\begin{pmatrix} -u_\mu \\ n_{a\mu} \end{pmatrix} = g^{1/2},$$

it follows, by the theory of minors, that

$$\varepsilon_{abc} n_{a\mu} n_{b\nu} n_{c\sigma} = {}^{-1}\varepsilon_{\tau\mu\nu\sigma} g^{1/2} u^\tau,$$

and hence,

$$\begin{aligned} \varepsilon_{ijk}\gamma^{1/2} &= \varepsilon_{ijk}\det(A_{ab}) = \varepsilon_{abc} A_{ai} A_{bj} A_{ck} \\ &= \varepsilon_{abc} n_{a\mu} n_{b\nu} n_{c\sigma} x^\mu_{,i} x^\nu_{,j} x^\sigma_{,k} \\ &= {}^{-1}\varepsilon_{\tau\mu\nu\sigma} g^{1/2} u^\tau x^\mu_{,i} x^\nu_{,j} x^\sigma_{,k} \\ &= \varepsilon_{ijk} g^{1/2} \frac{\partial(x)}{\partial(\tau,\xi)}. \end{aligned}$$

This last relation enables us to write

$$w_\xi = \gamma^{1/2} w_0 = \frac{\partial(x)}{\partial(\tau,\xi)} w,$$

where w is the proper energy density of the medium relative to the coordinates x^μ:

$$w \equiv g^{1/2} w_0.$$

w_0 is a scalar under both transformations of the ξ (relabelling) and transformations of the x^μ, whereas w_ξ is a scalar under transformations of the x^μ but transforms as a density of unit weight under transformations of the ξ. The quantity w is a scalar under transformations of the ξ but transforms as a density of unit weight under transformations of the x^μ.

We now ask: How does the proper energy density vary with time? If the medium is *conservative*, which means that energy does not flow around by dissipative mechanisms, w_0 can vary only as a result of the action of forces on the component parts of the medium. These forces can be described phenomenologically by means of a *stress tensor*. Suppose for a moment that the coordinates x^μ have been chosen to be canonical at a certain point x, oriented in such a way that

$$\begin{pmatrix} u^\mu \\ n^\mu_a \end{pmatrix}$$

becomes the unit matrix at x, and adjusted in the neighbourhood of x so that the derivatives of the metric tensor vanish at x. Then the coordinates x^μ may be regarded as an extension of the local Cartesian (Minkowskian) frame (which strictly speaking has mathematical existence only in the tangent space) to a small neighborhood of x. Let $\mathrm{d}\Sigma_a$ be a directed surface element in this frame. Then, from simple continuity arguments, the material on the side of $\mathrm{d}\Sigma_a$ away from the direction in which $\mathrm{d}\Sigma_a$ points must exert on the material on the opposite side a force that depends linearly on $\mathrm{d}\Sigma_a$,:

$$\mathrm{d}F_a = t_{ab}\mathrm{d}\Sigma_b.$$

The coefficients t_{ab} of the linear dependence are called the *components of the stress tensor in the local Cartesian rest frame*.

The force $\mathrm{d}F_a$ is a contact force and, as such, must respect the law of action and reaction. This means that the material on the side of $\mathrm{d}\Sigma_a$ *toward* which $\mathrm{d}\Sigma_a$ points must exert a force $-\mathrm{d}F_a$ across $\mathrm{d}\Sigma_a$. As a consequence the total force experienced by a small volume V of the medium, as a result of the action of the surrounding medium, is given by

$$F_a = -\int_\Sigma \mathrm{d}F_a = -\int_\Sigma t_{ab}\mathrm{d}\Sigma_b = -\int_V t_{ab,b}\mathrm{d}^3\boldsymbol{x},$$

where Σ is the surface of V. Here, V is assumed to contain the point x and the derivative in the final integrand is taken with respect to the extended local coordinates. Because V is otherwise arbitrary, it is evident that the internal stresses which the tensor t_{ab} describes give rise to a net force density in the immediate vicinity of x given by

$$f_a = -t_{ab,b}.$$

Suppose the origin of the coordinates x^μ is taken at the point x. Then, lowering the spatial indices on the x^μ, we may express the torque, about x, exerted on V by the surrounding medium in the form

$$T_a = -\int_\Sigma \varepsilon_{abc} x_b \mathrm{d}F_c = -\varepsilon_{abc} \int_\Sigma x_b t_{cd} \mathrm{d}\Sigma_d$$
$$= -\varepsilon_{abc} \int_V (x_b t_{cd})_{,d} \mathrm{d}^3x = T_a^{\mathrm{I}} + T_a^{\mathrm{II}},$$

where

$$T_a^{\mathrm{I}} = \varepsilon_{abc} \int_V x_b f_c \mathrm{d}^3x, \qquad T_a^{\mathrm{II}} = \varepsilon_{abc} \int_V t_{bc} \mathrm{d}^3x.$$

T_a^{I}, whose value depends on the location of the origin, is what one would expect to get for the torque using the force density f_a. T_a^{II}, whose value is independent of the location of the origin, is an unexpected residual. We can argue that this residual must vanish, as follows. In the limit $V \to 0$, it may be expressed simply as

$$T_a^{\mathrm{II}} = V \varepsilon_{abc} t_{bc}.$$

On the other hand, the moment of inertia of V is of the order

$$I \sim w_0 V^{5/3}.$$

The residual torque therefore imparts a contribution to the angular acceleration of V given by

$$\dot{w}_a^{\mathrm{II}} = \frac{T_a^{\mathrm{II}}}{I} \sim V^{-2/3} w_0^{-1} \varepsilon_{abc} t_{bc},$$

which becomes infinite as $V \to 0$. But this is absurd. We must therefore conclude that

$$\varepsilon_{abc} t_{bc} = 0,$$

or, alternatively,

$$t_{ab} - t_{ba} = (\delta_{ac}\delta_{bd} - \delta_{ad}\delta_{bc}) t_{cd} = \varepsilon_{abe}\varepsilon_{ecd} t_{cd} = 0.$$

That is, *the stress tensor is necessarily symmetric.*

The symmetry of the stress tensor may be illustrated in the particularly simple case of a gas at equilibrium, where we obviously have

$$t_{ab} = p\delta_{ab},$$

p being the pressure. We note that p, like w_0, is a scalar.

The stress tensor, like the energy density, can be expressed not only in the local Cartesian rest frame, but also in the ξ coordinate system and in the general system of spacetime coordinates x^μ. When viewed in an arbitrary coordinate system, however, it is conveniently regarded as a tensor density, known as the *stress density*. The relevant definitions are then

$$t^{ij} = \gamma^{1/2} A^{-1i}_{\ a} A^{-1j}_{\ b} t_{ab}, \qquad t^{\mu\nu} = g^{1/2} n^{\mu}_{a} n^{\nu}_{b} t_{ab}.$$

We note that

$$t^{\mu\nu} u_{\nu} = 0.$$

Problem 34 Show that

$$t^{ij} = \frac{\partial(x)}{\partial(\tau, \xi)} \xi^{i}_{,\mu} \xi^{j}_{,\nu} t^{\mu\nu}, \qquad t^{\mu\nu} = \frac{\partial(\tau, \xi)}{\partial(x)} P^{\mu}_{\sigma} P^{\nu}_{\tau} x^{\sigma}_{,i} x^{\tau}_{,j} t^{ij}.$$

Solution 34 We have

$$\begin{aligned} t^{ij} &= \gamma^{1/2} A^{-1i}_{\ a} A^{-1j}_{\ b} t_{ab} = \gamma^{1/2} \xi^{i}_{,\mu} n^{\mu}_{a} \xi^{j}_{,\nu} n^{\nu}_{b} t_{ab} \\ &= \gamma^{1/2} g^{-1/2} \xi^{i}_{,\mu} \xi^{j}_{,\nu} t^{\mu\nu} = \frac{\partial(x)}{\partial(\tau, \xi)} \xi^{i}_{,\mu} \xi^{j}_{,\nu} t^{\mu\nu}, \end{aligned}$$

$$t_{ab} = \gamma^{-1/2} A_{ai} A_{bj} t^{ij} = \gamma^{-1/2} n_{a\mu} x^{\mu}_{,i} n_{b\nu} x^{\nu}_{,j} t^{ij},$$

whence

$$\begin{aligned} t^{\mu\nu} &= g^{1/2} n^{\mu}_{a} n^{\nu}_{b} t_{ab} = g^{1/2} n^{\mu}_{a} n^{\nu}_{b} \gamma^{-1/2} n_{a\sigma} x^{\sigma}_{,i} n_{b\tau} x^{\tau}_{,j} t^{ij} \\ &= \frac{\partial(\tau, \xi)}{\partial(x)} P^{\mu}_{\sigma} P^{\nu}_{\tau} x^{\sigma}_{,i} x^{\tau}_{,j} t^{ij}, \end{aligned}$$

as required.

Consider now three nonparallel infinitesimal displacements $\delta_i \xi^j$ that are fixed in the medium and have the same orientation as the vectors n^{μ}_{a}. Relative to the local Cartesian rest frame, these become

$$\delta_i x_a = A_{aj} \delta_i \xi^j,$$

and they define an infinitesimal parallelipiped whose volume is

$$\delta V = \det(\delta_i x_a) = \det(A_{ak}) \det(\delta_i \xi^j) = \gamma^{1/2} \det(\delta_i \xi^j).$$

The surface elements of the three pairs of opposite faces of this parallelipiped are $\pm \delta_i \Sigma_a$, where

$$\begin{aligned} \delta_i \Sigma_a &= \frac{1}{2} \varepsilon_{ijk} \varepsilon_{abc} \delta_j x_b \delta_k x_c \\ &= \frac{1}{2} \varepsilon_{ijk} \varepsilon_{abc} A_{bm} A_{cn} \delta_j \xi^m \delta_k \xi^n \\ &= \frac{1}{2} \gamma^{1/2} \varepsilon_{ijk} \varepsilon_{lmn} A^{-1l}_{\ \ a} \delta_j \xi^m \delta_k \xi^n. \end{aligned}$$

The forces exerted on these faces by the surrounding medium are $\pm\delta_i F_a$, where

$$\begin{aligned}\delta_i F_a &= -t_{ab}\delta_i\Sigma_b = -\frac{1}{2}\gamma^{1/2}\varepsilon_{ijk}\varepsilon_{lmn}A_b^{-1l}t_{ab}\delta_j\xi^m\delta_k\xi^n \\ &= -\frac{1}{2}\varepsilon_{ijk}\varepsilon_{lmn}A_{ar}t^{rl}\delta_j\xi^m\delta_k\xi^n.\end{aligned}$$

During an increment $d\tau$ of proper time, the faces of the parallelipiped will suffer displacements relative to its center given by

$$\pm\frac{1}{2}\left(\frac{d}{d\tau}\delta_i x_a\right)d\tau = \pm\frac{1}{2}\dot{A}_{aj}\delta_i\xi^j d\tau.$$

The rate of change of the energy density w_ξ with proper time may be computed by taking into account the work done by the forces $\pm\delta_i F_a$ on the faces of the parallelipiped as a result of these displacements:

$$\begin{aligned}\dot{w}_\xi \det(\delta_i\xi^j) &= \frac{d}{d\tau}\left[w_\xi \det(\delta_i\xi^j)\right] = \frac{d}{d\tau}(w_0\delta V) \\ &= \left(\frac{d}{d\tau}\delta_i x_a\right)\delta_i F_a = \dot{A}_{as}\delta_i\xi^s\delta_i F_a \\ &= -\frac{1}{2}\varepsilon_{ijk}\varepsilon_{lmn}\dot{A}_{as}A_{ar}t^{rl}\delta_i\xi^s\delta_j\xi^m\delta_k\xi^n.\end{aligned}$$

Factoring out the determinant, we get

$$\begin{aligned}\dot{w}_\xi &= -\dot{A}_{al}A_{ar}t^{rl} = -\frac{1}{2}(\dot{A}_{ai}A_{aj} + A_{ai}\dot{A}_{aj})t^{ij} \\ &= -\frac{1}{2}\dot{\gamma}_{ij}t^{ij} = -\frac{1}{2}\frac{\partial(x)}{\partial(\tau,\xi)}\dot{\gamma}_{ij}\xi^i{}_{,\mu}\xi^j{}_{,\nu}t^{\mu\nu} \quad \text{(see Problem 34)} \\ &= -\frac{1}{2}\frac{\partial(x)}{\partial(\tau,\xi)}r_{\mu\nu}t^{\mu\nu} = -\frac{\partial(x)}{\partial(\tau,\xi)}u_{\mu;\nu}t^{\mu\nu},\end{aligned} \tag{10.1}$$

where $r_{\mu\nu}$ is the rate-of-strain tensor [see (2.1) on p. 23]. However,

$$\begin{aligned}\dot{w}_\xi &= \frac{\partial}{\partial\tau}\left[\frac{\partial(x)}{\partial(\tau,\xi)}w\right] \\ &= \frac{\partial(x)}{\partial(\tau,\xi)}\left[\left(\frac{\partial\tau}{\partial x^\mu}\frac{\partial^2 x^\mu}{\partial\tau^2} + \frac{\partial\xi^i}{\partial x^\mu}\frac{\partial^2 x^\mu}{\partial\xi^i\partial\tau}\right)w + w_{,\mu}\dot{x}^\mu\right] \\ &= \frac{\partial(x)}{\partial(\tau,\xi)}\left[w\frac{\partial}{\partial x^\mu}(\dot{x}^\mu) + w_{,\mu}\dot{x}^\mu\right] = \frac{\partial(x)}{\partial(\tau,\xi)}(wu^\mu)_{;\mu}.\end{aligned} \tag{10.2}$$

Hence, finally,

$$(wu^\mu)_{;\mu} + u_{\mu;\nu}t^{\mu\nu} = 0,$$

or, alternatively,

$$-u_\mu(wu^\mu u^\nu + t^{\mu\nu})_{;\nu} = 0.$$

Having accounted for the energy balance in the medium, we have now to account for the momentum balance. This is much easier. Consider again the parallelipiped of volume δV. Its four-momentum is

$$p^\mu = w_0 u^\mu \delta V.$$

In the local (instantaneous) rest frame of the parallelipiped, the time rate of change of this momentum is equal to

$$n_{a\mu}\dot{p}^\mu = w_0 n_{a\mu}\dot{u}^\mu \delta V = w_0 a_a \delta V,$$

where the dot denotes the covariant proper time derivative and the a_a are the rest-frame components of the absolute acceleration of δV. This change of momentum can only be caused by the forces of stress which are

$$F_a = -g^{-1/2} n_{a\mu} t^{\mu\nu}_{;\nu} \delta V.$$

Equating F_a and $n_{a\mu}\dot{p}^\mu$, we get

$$\begin{aligned} 0 &= n_{a\mu}(w\dot{u}^\mu + t^{\mu\nu}_{;\nu}) \\ &= n_{a\mu}(wu^\mu_{;\nu}u^\nu + t^{\mu\nu}_{;\nu}) \\ &= n_{a\mu}(wu^\mu u^\nu + t^{\mu\nu})_{;\nu}, \end{aligned}$$

where orthogonality of $n_{a\mu}$ and u^μ has been used. This may be combined with the energy balance equation to yield finally

$$T^{\mu\nu}_{;\nu} = 0,$$

where $T^{\mu\nu}$ is the energy–momentum–stress density:

$$T^{\mu\nu} = wu^\mu u^\nu + t^{\mu\nu}.$$

It will be observed that this agrees completely with the result previously obtained for a gas at equilibrium [see (8.5) on p. 104] if we identify u^μ with $\overline{u}^\mu$ and remember that for a gas at equilibrium we have

$$t^{\mu\nu} = g^{1/2} p^{\mu\nu} p.$$

It is instructive to examine $T^{\mu\nu}$ in canonical coordinates in the case of flat spacetime in which one has the strictly conserved quantities

$$P^\mu = \int_\Sigma T^{\mu\nu} d\Sigma_\nu.$$

Separating P^μ into its energy and momentum components and choosing for Σ the hypersurface $x^0 = \text{constant}$, we have

$$P^0 = \int T^{00} \mathrm{d}^3 \boldsymbol{x}, \qquad P^i = \int T^{i0} \mathrm{d}^3 \boldsymbol{x}.$$

These expressions, together with the differential identities

$$T^{00}{}_{,0} + T^{0i}{}_{,i} = 0, \qquad T^{i0}{}_{,0} + T^{ij}{}_{,j} = 0,$$

allow one to make the identifications

- T^{00} = energy density,
- $T^{i0} = T^{0i}$ = momentum density = energy flux density,
- T^{ij} = momentum flux density, in complete agreement with our analysis of the more primitive case of the point particle (see p. 99).

In the case of the conservative medium we have (remembering that $g^{1/2} = 1$ in canonical coordinates)

$$T^{00} = w_0 u^0 u^0 + v_i v_j t_x^{ij}, \qquad T^{i0} = w_0 u^0 u^i + t^{ij} v_j,$$

in which we have used repeatedly

$$t^{\mu 0} = t_x^{\mu i} v_i, \qquad v_i = \frac{u_i}{u_0},$$

which follows from the constraint $t^{\mu\nu} u_\nu = 0$. The first terms on the right-hand sides of these equations are easy to understand. Because of Lorentz contraction the proper energy density, i.e., the total energy density of the medium in the local Cartesian rest frame becomes $w_0 u^0$ in an arbitrary Lorentz frame, and these terms evidently give the contributions to the densities of energy and momentum arising from the bulk motion of the matter. The remaining terms, however, are curious residuals arising from the internal stresses.

That the residuals are by no means unimportant and are, in fact, essential may be illustrated by the amusing example given in the following problem.

Problem 35 A battery B, an electric motor M, a paddle wheel W, and a tank containing a viscous liquid L are all mounted on a platform P that is supported,

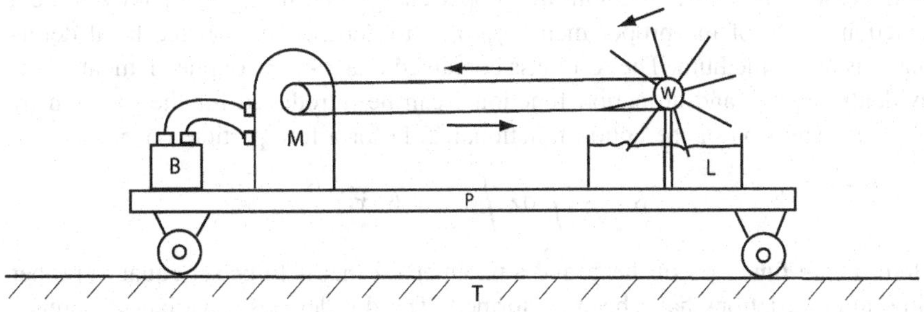

Fig. 10.1 The Battery Cart

through wheels with frictionless bearings, by a smooth table T at rest in the laboratory. The chemical energy of the battery drives the motor which, through a moving belt, turns the paddle wheel which stirs up and hence heats the viscous liquid. The platform is initially at rest in the laboratory. When the motor is turned on, energy leaves the battery and reappears in the form of heat in the liquid. Because of conservation of momentum, the center of energy of the device must remain motionless in the laboratory, and hence the platform must shift to the left. It is not possible to account for the energy transport by the mass motion of the belt, because as much mass is transported in one direction as the other. A term $t^{ij}v_j$ is needed for this purpose.

We may suppose the upper portion of the belt to be under tension and the lower portion to be experiencing no stresses. If the x^1 axis is taken in the direction of the belt then effectively the only nonvanishing component of t^{ij} in the upper portion of the belt will be t^{11}. Moreover, this component will be negative because the stress is one of tension. Other portions of the device are also under stress. For example, the region of the platform between the motor and the paddle wheel is under compression. This may be described by a single positive component t^{11} of t^{ij}.

Determine the *route* by which energy is transported from one end of the device to the other as seen in three different reference frames.

- The frame in which the platform is at rest.
- The laboratory frame in which the table is at rest.
- The frame in which the upper portion of the belt is at rest. Is the route Lorentz invariant?

Solution 35
- Energy flows to the right along the upper portion of the belt.
- Energy flows to the right along the upper portion of the belt and to the left, at a lesser rate, through the platform.
- Energy flows to the right through the platform. The route depends on the frame of reference.

10.1 The Elastic Medium

In the case of an elastic medium the proper energy density w_ξ is assumed to be a function solely of the proper metric γ_{ij}, i.e., to depend only on the local deformations of the medium. The rest mass contained in a proper volume element $d^3\xi$ is evidently $w_\xi d^3\xi$, and an action functional can be introduced for the medium by obvious extension of the action functional (5.1) for a free particle on p. 63:

$$S = -\int d\lambda \int d^3\xi w_\xi(-\dot{x}^2)^{1/2}.$$

Here, x^μ are functions of the ξ^i and a parameter λ that is to be set equal to proper time after variations have been performed. The dot denotes covariant differentiation with respect to λ. From the fact that

$$\dot{w}_\xi = \frac{\partial w_\xi}{\partial \gamma_{ij}} \dot{\gamma}_{ij},$$

we may infer that the stress tensor density in the ξ coordinate system is given by [see (10.1) on p. 117]

$$t^{ij} = -2 \frac{\partial w_\xi}{\partial \gamma_{ij}}. \tag{10.3}$$

Now we recall that

$$\gamma_{ij} = P_{\mu\nu} x^\mu_{,i} x^\nu_{,j}, \qquad P_{\mu\nu} = g_{\mu\nu} + u_\mu u_\nu, \qquad u^\mu = (-\dot{x}^2)^{-1/2} \dot{x}^\mu.$$

Hence, variation of the action will require some care to perform. Let us proceed systematically, using the covariant variation technique introduced on p. 64. First varying the dynamical variables of the medium, i.e., the x^μ, we have

$$\begin{aligned} \overline{\delta} u^\mu &= (-\dot{x}^2)^{-3/2} g_{\nu\sigma} \dot{x}^\nu \overline{\delta} \dot{x}^\sigma \dot{x}^\mu + (-\dot{x}^2)^{-1/2} \overline{\delta} \dot{x}^\mu \\ &= (-\dot{x}^2)^{-1/2} P^\mu_{\ \nu} \frac{\mathrm{D}}{\mathrm{D}\lambda} \delta x^\nu = P^\mu_{\ \nu} \delta x^\nu_{\ ;\sigma} u^\sigma, \end{aligned}$$

$$\overline{\delta} P_{\mu\nu} = u_\mu \overline{\delta} u_\nu + u_\nu \overline{\delta} u_\mu = (u_\mu P_{\nu\sigma} + u_\nu P_{\mu\sigma}) \delta x^\sigma_{\ ;\tau} u^\tau,$$

$$\begin{aligned} \delta w_\xi &= -\frac{1}{2} t^{ij} \delta \gamma_{ij} = -\frac{1}{2} t^{ij} \left(\overline{\delta} P_{\mu\nu} x^\mu_{,i} x^\nu_{,j} + 2 P_{\mu\nu} x^\mu_{,i} \overline{\delta} x^\nu_{,j} \right) \\ &= -u_\mu u^\tau P_{\sigma\nu} x^\mu_{,i} x^\nu_{,j} t^{ij} \delta x^\sigma_{\ ;\tau} - P_{\nu\mu} x^\mu_{,i} x^\sigma_{,j} t^{ij} \delta x^\nu_{\ ;\sigma} \\ &= -P^\mu_{\ \sigma} P^\nu_{\ \tau} x^\sigma_{,i} x^\tau_{,j} t^{ij} \delta x_{\mu;\nu} \\ &= -\frac{\partial(x)}{\partial(\lambda, \xi)} t^{\mu\nu} \delta x_{\mu;\nu} \qquad \text{(see Problem 34)}, \end{aligned}$$

where we have used

$$\begin{aligned} \overline{\delta} x^\nu_{,j} &= \delta x^\nu_{,j} + \Gamma^\nu_{\sigma\tau} x^\sigma_{,j} \delta x^\tau \\ &= (\delta x^\nu_{,\sigma} + \Gamma^\nu_{\tau\sigma} \delta x^\tau) x^\sigma_{,j} \\ &= \delta x^\nu_{\ ;\sigma} x^\sigma_{,j}, \end{aligned}$$

and where

$$\delta x_\mu = g_{\mu\nu} \delta x^\nu,$$

$$\begin{aligned} \delta(-\dot{x}^2)^{1/2} &= -(-\dot{x}^2)^{-1/2} g_{\mu\nu} \dot{x}^\nu \overline{\delta} \dot{x}^\mu = -u_\mu \frac{\mathrm{D}}{\mathrm{D}\lambda} \delta x^\mu \\ &= -(-\dot{x}^2)^{1/2} u^\mu u^\nu \delta x_{\mu;\nu}. \end{aligned}$$

On the other hand, if we vary the metric tensor we get

$$\delta u^\mu = \frac{1}{2}(-\dot{x}^2)^{-3/2}\dot{x}^\mu\dot{x}^\nu\dot{x}^\sigma\delta g_{\nu\sigma} = \frac{1}{2}u^\mu u^\nu u^\sigma \delta g_{\nu\sigma},$$

$$\delta u_\mu = u^\nu \delta g_{\mu\nu} + \frac{1}{2}u_\mu u^\nu u^\sigma \delta g_{\nu\sigma},$$

$$\begin{aligned}\delta P_{\mu\nu} &= \delta g_{\mu\nu} + u_\mu u^\sigma \delta g_{\nu\sigma} + u_\nu u^\sigma \delta g_{\mu\sigma} + u_\mu u_\nu u^\sigma u^\tau \delta g_{\sigma\tau} \\ &= P^\sigma_\mu P^\tau_\nu \delta g_{\sigma\tau},\end{aligned}$$

$$\delta w_\xi = -\frac{1}{2}t^{ij}P^\mu_\sigma P^\nu_\tau x^\sigma_{,i} x^\tau_{,j} \delta g_{\mu\nu} = -\frac{1}{2}\frac{\partial(x)}{\partial(\lambda,\xi)}t^{\mu\nu}\delta g_{\mu\nu},$$

$$\delta(-\dot{x}^2)^{1/2} = -\frac{1}{2}(-\dot{x}^2)^{-1/2}\dot{x}^\mu\dot{x}^\nu\delta g_{\mu\nu} = -\frac{1}{2}(-\dot{x}^2)^{1/2}u^\mu u^\nu \delta g_{\mu\nu}.$$

Under combined variation of the x^μ and the metric tensor, we therefore find

$$\begin{aligned}\delta S &= \int \mathrm{d}\lambda \int \mathrm{d}^3\xi \left[w_\xi(-\dot{x}^2)^{1/2}u^\mu u^\nu + \frac{\partial(x)}{\partial(\lambda,\xi)}t^{\mu\nu}\right]\left(\delta x_{\mu;\nu} + \frac{1}{2}\delta g_{\mu\nu}\right) \\ &\xrightarrow[\lambda\to\tau]{} \int T^{\mu\nu}s_{\mu\nu}\mathrm{d}^4x,\end{aligned}$$

where

$$T^{\mu\nu} = 2\frac{\delta S}{\delta g_{\mu\nu}(x)} = w u^\mu u^\nu + t^{\mu\nu},$$

$$s_{\mu\nu} = \frac{1}{2}(\delta x_{\mu;\nu} + \delta x_{\nu;\mu} + \delta g_{\mu\nu}).$$

$s_{\mu\nu}$ is known as the *invariant strain tensor*. It vanishes for any combined change in the x^μ and $g_{\mu\nu}$ that corresponds merely to an infinitesimal coordinate transformation.

We see that the energy–momentum–stress density has exactly the form that our phenomenological analysis found for it. Moreover, in this case the dynamical equations are identical with the divergence condition on $T^{\mu\nu}{}_{;\nu}$:

$$0 = \frac{\delta S}{\delta x^\mu(\lambda,\xi)}\xrightarrow[\lambda\to\tau]{} -\frac{\partial(x)}{\partial(\tau,\xi)}T^\nu_{\mu;\nu}.$$

10.2 The Viscous Thermally Conducting Gas

In a local Cartesian rest frame the energy–momentum–stress density of a conservative medium takes the block form

$$(T^{\mu\nu}) = \begin{pmatrix} w_0 & 0 \\ 0 & t_{ab} \end{pmatrix}.$$

In the case of a gas at equilibrium the form is actually diagonal:

$$(T^{\mu\nu}) = \begin{pmatrix} w_0 & 0 \\ 0 & p\delta_{ab} \end{pmatrix}.$$

This latter form is often used even for gases that are not at equilibrium. This corresponds to assuming that no matter how the state of the gas changes, locally at any point the gas always adjusts instantaneously to the equilibrium conditions appropriate to the proper particle number density ρ_0 and proper energy density w_0 at that point. Such instantaneous changes follow adiabatic curves and are reversible. The local entropy per particle at any point remains constant, and even though the temperature may vary from point to point, no energy transport by thermal conduction is allowed.

This approximation is adequate for the description of acoustic waves in a gas that is otherwise at equilibrium, but it fails in the presence of shock waves. Real gases are both thermally conducting and viscous. The viscosity alters the simple stress tensor $p\delta_{ab}$ and the thermal conduction produces an energy flux q_a even in the local rest frame, so that the energy–momentum–stress density takes the general form

$$(T^{\mu\nu}) = \begin{pmatrix} w_0 & q_b \\ q_a & t_{ab} \end{pmatrix},$$

or, in a general frame,

$$T^{\mu\nu} = wu^{\mu}u^{\nu} + u^{\mu}q^{\nu} + u^{\nu}q^{\mu} + t^{\mu\nu},$$

$$u \cdot g = 0, \qquad u_{\mu}t^{\mu\nu} = 0.$$

(We have dropped the bars over the u^{μ} that we had previously placed there in the case of gases to denote the mean four-velocity of the component particles.)

It is not difficult to decide on phenomenological expressions for q^{μ} and $t^{\mu\nu}$. $T_{;\mu} + u_{\mu;\nu}u^{\nu}T$ vanishes at equilibrium, so any departure from zero for this quantity must indicate a non-equilibrium situation, so we take

$$q^{\mu} = -\lambda g^{1/2} P^{\mu\nu}(T_{;\nu} + u_{\nu;\sigma}u^{\sigma}T), \quad \lambda \geq 0,$$

where λ is the *thermal diffusion coefficient* of the gas. It may depend on ρ_0 and w_0. For $t^{\mu\nu}$, we choose

$$t^{\mu\nu} = g^{1/2}\left[\left(p - \frac{1}{2}\mu_{\mathrm{B}}r\right)P^{\mu\nu} - \mu_{\mathrm{S}}\left(r^{\mu\nu} - \frac{1}{3}rP^{\mu\nu}\right)\right], \quad \mu_{\mathrm{B}} \geq 0, \quad \mu_{\mathrm{S}} \geq 0,$$

where μ_B and μ_S are respectively the *coefficients of bulk and shear viscosity*, and

$$r = r^\mu_\mu = P^{\mu\nu} r_{\mu\nu} = 2P^{\mu\nu} u_{\mu;\nu} = 2u^\mu_{;\mu},$$

$r_{\mu\nu}$ being the rate-of-strain tensor. μ_B and μ_S may also depend on ρ_0 and w_0.

Here, p and T are to be understood as the pressure and temperature the gas *would* have at any point if it were in a state of equilibrium appropriate to the values of ρ_0 and w_0 at that point. That is, p and T are to be taken as given by the *equilibrium equation of state* of the gas. A corresponding *entropy per particle*, s, may then be defined by the differential identity

$$\mathrm{d}s = \frac{\mathrm{d}e + p\mathrm{d}\rho_0^{-1}}{T},$$

where e is the *proper energy per particle*:

$$e = \frac{w_0}{\rho_0} = \frac{w}{\rho}, \qquad \rho = g^{1/2}\rho_0.$$

Evidently this phenomenological description of the gas will fail if the distribution function of the gas departs too markedly from an instantaneous equilibrium distribution.

Problem 36 [Taken from C. Eckart, Phys. Rev. **58**, 919 (1940).] The dynamical behavior of the viscous thermally conducting gas is completely determined by the divergence condition $T^{\mu\nu}_{;\nu} = 0$ together with the equilibrium equation of state and the equations that express the dependence (if any) of λ, μ_B, and μ_S on ρ_0 and w_0. It is worth checking, however, that the result is consistent with the elementary principles of thermodynamics. Consider a small element of the gas contained in a volume V in the local rest frame. If there were no heat flow out through the surface Σ of V, the entropy contained in V would be expected to increase as a result of irreversible processes. Heat flow, however, can reduce the entropy, at least when the flow is low, because of the term $\mathrm{d}e/T$ in the expression for $\mathrm{d}s$. If we choose the coordinates x^μ to match the local rest frame as closely as possible in V ($g_{\mu\nu,\sigma} = 0$), we therefore expect the following mathematical statement to hold:

$$\frac{\mathrm{d}}{\mathrm{d}x^0}\int_V \rho_0 s\,\mathrm{d}^3\boldsymbol{x} = -\int_\Sigma T^{-1} q^i \mathrm{d}\Sigma_i + \begin{matrix}\text{entropy increase due}\\ \text{to irreversible processes}\end{matrix},$$

or, equivalently,

$$\int_V (\rho s u^\mu + T^{-1} q^\mu)_{;\mu} \mathrm{d}^3\boldsymbol{x} \geq 0.$$

The final integrand being a covariant quantity, and the volume being arbitrary, we should have in any coordinate system

$$(\rho s u^\mu + T^{-1} q^\mu)_{;\mu} \geq 0.$$

Derive this inequality from the entropy equation together with the divergence conditions

$$T^{\mu\nu}_{;\nu} = 0, \qquad (\rho u^\mu)_{;\mu} = 0.$$

Hint: first prove the following:

$$(wu^\mu)_{;\mu} + q^\mu_{;\mu} + q^\mu u_{\mu;\nu} u^\nu + \frac{1}{2} t^{\mu\nu} r_{\mu\nu} = 0,$$

$$(wu^\mu)_{;\mu} = \rho e_{;\mu} u^\mu, \qquad (\rho^{-1})_{;\mu} u^\mu = \rho^{-1} u^\mu_{;\mu}. \tag{10.4}$$

Solution 36 We have

$$\rho e_{;\mu} u^\mu = (\rho e u^\mu)_{;\mu} = (wu^\mu)_{;\mu},$$

$$\begin{aligned} 0 = u_\mu T^{\mu\nu}_{;\nu} &= u_\mu (wu^\mu u^\nu + u^\mu q^\nu + u^\nu q^\mu + t^{\mu\nu})_{;\nu} \\ &= -(wu^\nu)_{;\nu} - q^\nu_{;\nu} + u_\mu u^\nu q^\mu_{;\nu} + u_\mu t^{\mu\nu}_{;\nu} \\ &= -(wu^\mu)_{;\mu} - q^\mu_{;\mu} - q^\mu u_{\mu;\nu} u^\nu - \frac{1}{2} t^{\mu\nu} r_{\mu\nu}, \end{aligned}$$

$$(\rho^{-1})_{;\mu} u^\mu = -\rho^{-2} \rho_{;\mu} u^\mu = \rho^{-1} u^\mu_{;\mu},$$

$$\begin{aligned} T(\rho s u^\mu + T^{-1} q^\mu)_{;\mu} &= T\rho s_{;\mu} u^\mu + T(T^{-1} q^\mu)_{;\mu} \\ &= \rho e_{;\mu} u^\mu + \rho p (\rho_0^{-1})_{;\mu} u^\mu + T(T^{-1} q^\mu)_{;\mu} \\ &= (wu^\mu)_{;\mu} + g^{1/2} p u^\mu_{;\mu} + T(T^{-1} q^\mu)_{;\mu} \\ &= -q^\mu_{;\mu} - q^\mu u_{\mu;\nu} u^\nu - \frac{1}{2} t^{\mu\nu} r_{\mu\nu} + \frac{1}{2} g^{1/2} p r + T(T^{-1} q^\mu)_{;\mu} \\ &= -T^{-1} q^\mu (T_{;\mu} + u_{\mu;\nu} u^\nu T) \\ &\quad + \frac{1}{2} g^{1/2} r_{\mu\nu} \left[\frac{1}{2} \mu_{\mathrm{B}} r P^{\mu\nu} + \mu_{\mathrm{S}} \left(r^{\mu\nu} - \frac{1}{3} P^{\mu\nu} r \right) \right] \\ &= \lambda g^{1/2} T^{-1} P^{\mu\nu} (T_{;\mu} + u_{\mu;\sigma} u^\sigma T)(T_{;\nu} + u_{\nu;\tau} u^\tau T) + \frac{1}{4} \mu_{\mathrm{B}} g^{1/2} r^2 \\ &\quad + \frac{1}{2} \mu_{\mathrm{S}} g^{1/2} \left(r_{\mu\nu} - \frac{1}{3} P_{\mu\nu} r \right) \left(r^{\mu\nu} - \frac{1}{3} P^{\mu\nu} r \right) \geq 0. \end{aligned}$$

Chapter 11
Solubility of the Einstein and Matter Equations

One of the essential conditions that must be satisfied by any system of dynamical equations is that they give rise to a well-posed initial value problem. There must be precisely enough equations to determine uniquely the full physical evolution of the system, once initial conditions have been specified. We now examine this problem for the Einstein and matter equations.

We consider the matter equations first, confining our attention to the cases in which they are either identical with or fully equivalent to the divergence equations $T^{\mu\nu}_{;\nu} = 0$. In the case of the elastic medium these equations must yield a set of world lines given by the functions $x^\mu(\xi, \lambda)$. These functions are not unique in a given physical situation, however, because the parameter λ is arbitrary. The action functional in the elastic medium is invariant under the group of diffeomorphisms of the one-dimensional λ-manifold. This corresponds to the fact that the functions $x^\mu(\xi, \lambda)$ do not really represent four degrees of freedom per component particle, but only three. Choose a space-like hypersurface Σ. Once one specifies the point of intersection of each world line with Σ (three pieces of information per particle) and its normalized tangent vector u^μ at that point (three more pieces of information per particle), the whole future history of the medium should be determined by the dynamical equations. Six pieces of information for each set of values for the labels ξ^i corresponds to three degrees of freedom per particle. The dynamical equations $T^{\mu\nu}_{;\nu} = 0$, however, are *four* equations per particle, and therefore the motion of the medium appears to be *underdetermined*.

In this case appearances are deceiving. The four equations $T^{\mu\nu}_{;\nu} = 0$ are actually not independent. They are connected by the identity

$$\begin{aligned}
-u_\mu T^{\mu\nu}_{;\nu} &\equiv -u_\mu (w u^\mu u^\nu + t^{\mu\nu})_{;\nu} \equiv (w u^\mu)_{;\mu} + u_{\mu;\nu} t^{\mu\nu} \\
&\equiv \frac{\partial(\tau, \xi)}{\partial(x)} \left(\dot{w}_\xi + \frac{1}{2} t^{ij} \dot{\gamma}_{ij} \right) \\
&\equiv 0,
\end{aligned}$$

B. DeWitt, *Bryce DeWitt's Lectures on Gravitation*,
Edited by Steven M. Christensen, Lecture Notes in Physics, 826,
DOI: 10.1007/978-3-540-36911-0_11, © Springer-Verlag Berlin Heidelberg 2011

where we have used (10.1) and (10.2) to go from the first to second lines, and the last line follows from the definition (10.3) of t^{ij}. In this case the motion of the particles determines how γ_{ij} varies with time, and w_ξ and t^{ij} are at all times completely determined by γ_{ij}. Therefore, the dynamical equations lead to a well-posed initial value problem, and the motion is completely determined.

The same is also true for gases. The motion of gases can also be described in terms of four functions $x^\mu(\xi, \lambda)$. Here, again the number of degrees of freedom per world line (really a mean flow line in this case) is only three because of the arbitrariness of the parameter λ. In the case of the viscous thermally conducting gas, however, the equations $T^{\mu\nu}_{;\nu} = 0$ do *not* satisfy an identity and are actually independent. This does not mean that the motion is overdetermined. The equation $u_\mu T^{\mu\nu}_{;\nu} = 0$ is needed in order to establish the time rate of change of the proper energy density w. Then w_ξ is no longer determined simply by γ_{ij} but depends on heat flow and the state of motion of neighboring parts of the gas. And w_ξ, or w, is needed, together with ρ, to determine p and T (through the equilibrium equation of state) and hence, together with $r_{\mu\nu}$, the stress tensor density $t^{\mu\nu}$. Only if the gas is viscousless and non-conducting, adjusting instantaneously to the equilibrium equation-of-state conditions, does $u_\mu T^{\mu\nu}_{;\nu}$ vanish identically. In this case the flow is isentropic and, once the entropy per particle has been set as a function of the ξ^i, it remains constant. w_ξ, which can be expressed in terms of ρ_0 and the entropy through the equation of state, then depends only on the γ_{ij} (on $\gamma^{1/2}$ really), and an action functional can be introduced which yields the equations $T^{\mu\nu}_{;\nu} = 0$ from a variational principle, just as for the elastic medium.

It will be noted that the dynamical equations satisfy an identity in precisely those cases in which an action functional can be introduced that is invariant under the actions of an infinite dimensional group (the diffeomorphism group of the λ-manifold in the above examples). This is no accident but is an illustration of a general rule. Other examples are the free particle ($m\ddot{z}^\alpha = 0$) for which the identity is $mg_{\alpha\beta}\dot{z}^\alpha\ddot{z}^\beta \equiv 0$, the electromagnetic field (Maxwell's equations, see Chap. 14) for which the identity is $F^{\mu\nu}_{;\nu\mu} \equiv 0$, and the gravitational field for which the identity is the contracted Bianchi identity. We now study the last of these examples.

The metric tensor $g_{\mu\nu}$ has ten components. Because the diffeomorphism group of spacetime is an invariance group for the theory, the particular values $g_{\mu\nu}$ assume in a given physical situation result partly from the accident of choosing a coordinate system. The specification of a diffeomorphism requires four functions over spacetime, and therefore only six of the $g_{\mu\nu}$ can be determined by the Einstein equations. It is often convenient to choose these six to be the 'spatial' components g_{ij} and to fix g_{00} and g_{0i} in advance as certain functions, or even functionals, of the g_{ij} and their 'time' derivatives $g_{ij,0}$ on each hypersurface $x^0 = \text{constant}$. It is always possible to fix g_{00} and g_{0i} in this way, as these four components merely determine the physical spacing between two hypersurfaces $x^0 = t$ and $x^0 = t + dt$, and the manner in which the coordinate mesh *in* each hypersurface distorts as one passes from such hypersurface to the next. That the six g_{ij} are not then overdetermined by the ten Einstein equations is due to the existence of the contracted

Bianchi identity, which actually consists of four equations. Of the ten Einstein equations only six are really independent. These may be taken to be the equations

$$\frac{\delta S_{\mathrm{G}}}{\delta g_{ij}} + \frac{1}{2} T_{ij} = 0.$$

The remaining Einstein equations are not, however, to be dropped from sight and disposed of just like that. It turns out that they impose *constraints* on the *initial data* g_{ij} and $g_{ij,0}$ on the initial hypersurface x^0 = constant. This is most easily seen by combining the contracted Bianchi identity with the matter equations and rewriting the combination in the form

$$\left(\frac{\delta S_{\mathrm{G}}}{\delta g_{\mu 0}} + \frac{1}{2} T^{\mu 0}\right)_{,0} = -\left(\frac{\delta S_{\mathrm{G}}}{\delta g_{\mu i}} + \frac{1}{2} T^{\mu i}\right)_{,i} - \Gamma^{\mu}_{\nu\sigma}\left(\frac{\delta S_{\mathrm{G}}}{\delta g_{\nu\sigma}} + \frac{1}{2} T^{\nu\sigma}\right).$$

The right-hand side of this equation contains no third 'time' derivatives of the g_{ij}. Neither does it contain any second time derivatives of g_{00} or the g_{0i}, as may be checked by referring to (4.2) and noting that, because of its symmetries, the Riemann tensor itself contains no second time derivatives of g_{00} or the g_{0i}. Therefore, even if g_{00} and g_{0i} are replaced by functionals of the g_{ij} and $g_{ij,0}$, the right-hand side of this equation still contains no third time derivatives of the g_{ij}.[1] The same must be true of the left-hand side. From this it follows that the four Einstein equations

$$\frac{\delta S_{\mathrm{G}}}{\delta g_{\mu 0}} + \frac{1}{2} T^{\mu 0} = 0$$

contain no second time derivatives of the g_{ij} but are constraints on the initial data.

Suppose these constraints are satisfied on one hypersurface x^0 = constant. Will the six other Einstein equations ensure that they remain satisfied on all the other hypersurfaces x^0 = constant? The answer is yes, as may be seen by noting that when the six dynamical equations are satisfied, the combination of the Bianchi identity and matter equations reduces to

$$\left(\frac{\delta S_{\mathrm{G}}}{\delta g_{\mu 0}} + \frac{1}{2} T^{\mu 0}\right)_{,0} = -\delta^{\mu}_{0}\left(\frac{\delta S_{\mathrm{G}}}{\delta g_{i0}} + \frac{1}{2} T^{i0}\right)_{,i} - 2\Gamma^{\mu}_{i0}\left(\frac{\delta S_{\mathrm{G}}}{\delta g_{i0}} + \frac{1}{2} T^{i0}\right)$$
$$- \Gamma^{\mu}_{00}\left(\frac{\delta S_{\mathrm{G}}}{\delta g_{00}} + \frac{1}{2} T^{00}\right),$$

so that the derivative of the left-hand side vanishes for all time.

Problem 37 In electromagnetic theory, one of Maxwell's equations is a constraint: $F^{0\mu}_{;\mu} - j^0 = 0$. It is shown that the other Maxwell equations $F^{i\mu}_{;\mu} - j^i = 0$

[1] This is true in practice even when $T^{\mu\nu}$ itself involves derivatives of the metric tensor.

together with the identity $(F^{\mu\nu}_{;\nu} - j^{\mu})_{;\mu} \equiv 0$ ensure that the constraint, when once imposed, holds forever.

Solution 37 We have

$$(F^{0\nu}_{;\nu} - j^{0})_{,0} \equiv -(F^{i\nu}_{;\nu} - j^{i})_{,i} - \Gamma^{\mu}_{\nu\mu}(F^{\nu\sigma}_{;\sigma} - j^{\nu})$$
$$= -\Gamma^{\mu}_{0\mu}(F^{0\nu}_{;\nu} - j^{0}),$$

as claimed.

We are now in a position to determine the number of physically distinct degrees of freedom possessed by the gravitational field. Because of the constraints, it is possible freely to specify only 8 independent combinations of the 12 initial data functions g_{ij} and $g_{ij,0}$. Not all of these eight combinations are physical, however. For example, we can carry out a transformation of coordinates *in* the initial hypersurface. If g_{00} and g_{0i} are defined in terms of the g_{ij} and $g_{ij,0}$ in a covariant way, so that g_{00} is a scalar and g_{0i} are the components of a covariant vector under such a transformation, then the geometry of the resulting spacetime (which is the only aspect of the gravitational field that is physical) will remain completely unchanged and so will the succeeding hypersurfaces $x^0 =$ constant. This means that at most five independent combinations of the initial data are physical.

Actually, there are only four. To see this suppose for example that spacetime is asymptotically flat. Then it is possible to give some absolute significance to the canonical coordinate systems at infinity because of the Poincaré group of isometric motions that holds sway there. But even if we choose one of these systems, we can extend the hypersurfaces $x^0 =$ constant into the nonflat region in many different ways. To each of these ways corresponds a different set of initial data. The mathematical expert on initial data problems accepts these as distinct, but we as physicists know better. The same conclusion also holds if 3-space is compact so that the 'universe' is finite. In this case there is no preferred coordinate system at infinity to which to tie the hypersurfaces $x^0 =$ constant. Furthermore, it makes no sense to speak of displacing the whole history of the gravitational field forward in time, thereby getting a physically distinct field, the way one can in the case of nongravitational fields in flat spacetime. Two universes having the same overall spacetime geometry are identical. One cannot be said to 'lie to the future' of the other. This means that the choice of initial hypersurface itself has no physical significance.

The freedom to choose the initial hypersurface and the freedom to select the coordinate system in this hypersurface together dispose of four freely specifiable functions, thus leaving only four independent combinations of the initial data that can affect the physical situation. The existence of these four indicates that the gravitational field has only two physically distinct degrees of freedom per point of 3-space. These two will make their appearance later in the form of the two independent states of polarization permitted to plane gravitational waves.

It is to be noted that the very existence of these degrees of freedom implies that the Einsteinian gravitational field is a dynamical entity in its own right, capable of propagating in the absence of sources. The gravitational field of Newton has no such freedom.

11.1 The Cosmological Constant

The term in λ in the gravitational action functional S_G gives rise to a term—$8\pi G\lambda g^{1/2}g^{\mu\nu}$ on the right-hand side of Einstein's equations (8.4), which adds to the matter 'source term' $8\pi GT^{\mu\nu}$. When $\lambda > 0$, this term corresponds to a medium having a proper energy density λ and a *negative* pressure $-\lambda$. When $\lambda < 0$, the pressure is positive but the energy density is negative. In either case it corresponds to a medium with unusual properties!

Because Einstein's theory with $\lambda = 0$ describes very well the gravitational dynamics of the solar system, the galaxy, and even clusters of galaxies, the term in λ can play a role only at the cosmological level. For this reason the λ term is known as the *cosmological term*, and λ itself is known as the *cosmological constant*. Moreover an upper bound can be placed on λ, which is roughly equal to the mean density of presently observed matter in the universe:

$$\lambda \lesssim 10^{-26}\,\text{kg/m}^3.$$

Because this density is so small and because we shall not be concerned with cosmological problems in these lectures, we shall from now on set $\lambda = 0$.

Chapter 12
Energy, Momentum and Stress in the Gravitational Field

12.1 Condensed Notation

In discussing the general theory of gravitation (and, indeed, other field theories as well), it is convenient on many occasions to employ a highly condensed notation. In brief, we shall replace the symbol $g_{\mu\nu}(x)$ by the symbol φ^i and denote functional differentiation by a comma, so that the Einstein and matter dynamical equations take the forms

$$\left.\begin{aligned} S_{\mathrm{G},i} + S_{\mathrm{M},i} &= 0 \\ S_{\mathrm{M},A} &= 0 \end{aligned}\right\}.$$

Here, the index i may be thought of as standing for the sextuplet μ, ν, x^0, x^1, x^2, x^3, and the symbol φ may be regarded as a replacement for g.

This notation, with φ replacing g, brings to mind the notation introduced to discuss realizations of continuous groups. The analogy is deliberate. The infinitesimal coordinate transformation law

$$\delta g_{\mu\nu} = -\delta\xi_{\mu;\nu} - \delta\xi_{\nu;\mu} = -\int (\delta_{\mu\sigma';\nu} + \delta_{\nu\sigma';\mu})\delta\xi^{\sigma'} d^4x'$$

yields a realization (a matrix representation, in fact) of the diffeomorphism group and is a special case of

$$\delta\varphi^i = \Phi^i_{\ a}\delta\xi^a \quad \text{(see Problem 14).}$$

Here the index a may be understood as standing for the quintuplet

$$\sigma, x'^0, x'^1, x'^2, x'^3,$$

and the symbol $\Phi^i_{\ a}$ as representing the bitensor density $-(\delta_{\mu\sigma';\nu} + \delta_{\nu\sigma';\mu})$.

B. DeWitt, *Bryce DeWitt's Lectures on Gravitation*,
Edited by Steven M. Christensen, Lecture Notes in Physics, 826,
DOI: 10.1007/978-3-540-36911-0_12,

With this notation, the coordinate invariance of the gravitational action functional may be expressed in the form

$$0 \equiv \delta S_{\mathrm{G}} \equiv S_{\mathrm{G},i}\delta\varphi^i \equiv S_{\mathrm{G},i}\Phi^i_{\ a}\delta\xi^a,$$

with summation or integration over repeated indices. Because $\Delta\xi^a$ is arbitrary, it follows that

$$S_{\mathrm{G},i}\Phi^i_{\ a} \equiv 0.$$

This is just the contracted Bianchi identity. In a similar manner, we may write the divergence condition satisfied by the energy–momentum–stress density:

$$T_i\Phi^i_{\ a} = 0 \quad \text{when} \quad S_{\mathrm{M},A} = 0,$$

where

$$T_i \equiv 2S_{\mathrm{M},i}.$$

12.2 Variation of the Action

Suppose, we have a solution of Einstein's empty space equations $S_{\mathrm{G}},i = 0$ (for example, flat spacetime). Let us denote this solution by φ^i_{B} in some coordinate system. Suppose, we introduce into the empty spacetime described by this solution a material system described by an action functional S_{M}. The introduction of the material system will cause the spacetime geometry to change. The new metric tensor may be expressed (in some coordinate system) in the form

$$\varphi^i = \varphi^i_{\mathrm{B}} + \phi^i,$$

where the field ϕ^i measures the deviation from the *background field* φ^i_{B} induced by the introduction of the material system. The field ϕ^i, together with the dynamical variables Φ^A of the material system, satisfies the dynamical equations

$$\left.\begin{aligned} S_{\mathrm{G},i}[\varphi_{\mathrm{B}} + \phi] + S_{\mathrm{M},i}[\Phi, \varphi_{\mathrm{B}} + \phi] &= 0 \\ S_{\mathrm{M,A}}[\Phi, \varphi_{\mathrm{B}} + \phi] &= 0 \end{aligned}\right\}.$$

We now note that if we subtract from the first equation the equation satisfied by the background field, namely

$$S_{\mathrm{G},i}[\varphi_{\mathrm{B}}] = 0,$$

the resulting pair of equations may be expressed in the form

$$\frac{\delta\bar{S}}{\delta\phi^i} = 0, \quad \frac{\delta\bar{S}}{\delta\varphi^A} = 0,$$

where $\bar{S}$ is a new action functional given by

$$\bar{S}[\varphi_{\mathrm{B}}, \phi, \Phi \equiv \bar{S}_{\mathrm{G}}[\varphi_{\mathrm{B}}, \phi] + S_{\mathrm{M}}[\Phi, \varphi_{\mathrm{B}} + \phi],$$

$$\bar{S}_{\mathrm{G}}[\varphi_{\mathrm{B}}, \phi] \equiv S_{\mathrm{G}}[\varphi_{\mathrm{B}} + \phi] - S_{\mathrm{G}}[\varphi_{\mathrm{B}}] - S_{\mathrm{G},i}[\varphi_{\mathrm{B}}]\phi^i.$$

The significance of the new action functional is that, using it, one may formally regard ϕ^i as being just an ordinary field like any other, and it (together with the material system) may be regarded as evolving in the presence of an externally imposed gravitational field φ^i_{B}. This may be seen in several ways. Firstly, a formal Taylor expansion of $\bar{S}$ yields

$$\bar{S}[\varphi_{\mathrm{B}}, \phi, \Phi = \frac{1}{2} S_{\mathrm{G},ij}[\varphi_{\mathrm{B}}]\phi^i\phi^j + \frac{1}{3!} S_{\mathrm{G},ijk}[\varphi_{\mathrm{B}}]\phi^i\phi^j\phi^k + \cdots$$
$$\cdots + S_{\mathrm{M}}[\varphi_{\mathrm{B}}, \Phi S_{\mathrm{M},i}[\varphi_{\mathrm{B}}, \Phi]\phi^i + \cdots.$$

The first term in the expansion is just the action functional for a field that propagates freely in the background φ^i_{B}, satisfying the linear field equations $S_{\mathrm{G},ij}[\varphi_{\mathrm{B}}]\phi^j = 0$. The remaining terms describe the coupling of this field to itself and to the material system.

Secondly, if we define

$$\mathcal{T}_i \equiv 2\frac{\delta \bar{S}_{\mathrm{G}}}{\delta \varphi^i_{\mathrm{B}}} \equiv 2\{S_{\mathrm{G},i}[\varphi_{\mathrm{B}} + \phi] - S_{\mathrm{G},i}[\varphi_{\mathrm{B}}] - S_{\mathrm{G},ij}[\varphi_{\mathrm{B}}]\phi^j\},$$

we may rewrite the field equations for ϕ^i in the form

$$S_{\mathrm{G},ij}[\varphi_{\mathrm{B}}]\phi^j = -\frac{1}{2}(T_i + \mathcal{T}_i), \qquad T_i \equiv 2\frac{\delta S_{\mathrm{M}}}{\delta \phi^i} \equiv 2\frac{\delta S_{\mathrm{M}}}{\delta \varphi^i_{\mathrm{B}}}, \tag{12.1}$$

which illustrates a *feedback principle* for the gravitational field: *The gravitational field partly produces itself.* The quantity $\mathcal{T}_i$ may be regarded as the energy–momentum–stress density carried by the field ϕ^i. It, together with the energy–momentum–stress density T_i of the material system, serves as a source for the field ϕ^i.

It is not hard to show that the total energy–momentum–stress density, $T_i + \mathcal{T}_i$, has vanishing covariant divergence relative to the background field φ^i_{B} when the dynamical equations $\delta\bar{S}/\Delta\phi^i = 0$ and $\delta\bar{S}/\delta\Phi^A = 0$ are satisfied. For this purpose, we have only to show that the action functional $\bar{S}$ is coordinate independent and then use arguments identical to those employed in showing that T_i (or $T^{\mu\nu}$) by itself has vanishing covariant divergence relative to the total field φ^i. The demonstration depends on the adoption of the following coordinate transformation laws:

$$\delta\varphi^i_{\mathrm{B}} = \Phi^i_{\ a}[\varphi_{\mathrm{B}}]\delta\xi^a, \quad \delta\varphi^i = \Phi^i_{\ a}[\varphi]\delta\xi^a = \Phi^i_{\ a}[\varphi_{\mathrm{B}} + \phi]\delta\xi^a,$$

$$\Delta\phi^i = \Delta\varphi^i - \Delta\varphi^i_{\mathrm{B}} = \Phi^i_{\ a,j}\phi^j\Delta\xi^a.$$

The final expression above follows from the fact that the metric tensor transforms linearly under the diffeomorphism group. We note that the tritensor density $\Phi^i_{a,j}$ is metric independent.

S_{M} is already invariant under the above transformations, so we have only to analyze $\bar{S}_{\mathrm{G}}$. Using the contracted Bianchi identity, we find that $\bar{S}_{\mathrm{G}}$ suffers the change

$$\begin{aligned}\delta\bar{S}_{\mathrm{G}}[\varphi_{\mathrm{B}},\phi] &\equiv S_{\mathrm{G},i}[\varphi_{\mathrm{B}}+\phi]\Phi^i_a[\varphi_{\mathrm{B}}+\phi]\delta\xi^a - S_{\mathrm{G},i}[\varphi_{\mathrm{B}}]\Phi^i_a[\varphi_{\mathrm{B}}]\delta\xi^a\\ &\quad - S_{\mathrm{G},ij}[\varphi_{\mathrm{B}}]\Phi^j_a[\varphi_{\mathrm{B}}]\phi^i\delta\xi^a - S_{\mathrm{G},j}[\varphi_{\mathrm{B}}]\Phi^j_{a,i}\phi^i\delta\xi^a\\ &\equiv -\frac{\delta\big(S_{\mathrm{G},j}[\varphi_{\mathrm{B}}]\Phi^j_a[\varphi_{\mathrm{B}}]\big)}{\delta\varphi^i_{\mathrm{B}}}\phi^i\delta\xi^a \equiv 0.\end{aligned}$$

That is, it suffers no change. We, therefore, have immediately

$$(T_i+\mathcal{T}_i)\Phi^i_a[\varphi_{\mathrm{B}}]=0 \quad \text{when} \quad \frac{\delta\bar{S}}{\delta\phi^i}=0 \text{ and } \frac{\delta\bar{S}}{\delta\Phi^A}=0.$$

It should be pointed out that this result is actually required for consistency. This may be seen by functionally differentiating the contracted Bianchi identity

$$S_{\mathrm{G},j}[\varphi_{\mathrm{B}}]\Phi^j_a[\varphi_{\mathrm{B}}]\equiv 0.$$

We have

$$S_{\mathrm{G},ij}[\varphi_{\mathrm{B}}]\Phi^j_a[\varphi_{\mathrm{B}}]+S_{\mathrm{G},j}[\varphi_{\mathrm{B}}]\Phi^j_{a,i}\equiv 0, \tag{12.2}$$

which reduces to

$$S_{\mathrm{G},ij}[\varphi_{\mathrm{B}}]\Phi^j_a[\varphi_{\mathrm{B}}]=0,$$

by virtue of the equation $S_{\mathrm{G},j}[\varphi_{\mathrm{B}}]=0$ satisfied by the background field. Therefore if $(T_i+\mathcal{T}_i)\Phi^i_a[\varphi_{\mathrm{B}}]$ did not already vanish by virtue of the dynamical equations, it would be forced to vanish by virtue of the equation

$$S_{\mathrm{G},ij}[\varphi_{\mathrm{B}}]\phi^j=\frac{1}{2}(T_i+\mathcal{T}_i)$$

alone.

In calling $(T_i+\mathcal{T}_i)\Phi^i_a[\varphi_{\mathrm{B}}]$ a covariant divergence, we are assuming that both T_i and $\mathcal{T}_i$ are contravariant tensor densities. We already know that T_i transforms as a contravariant tensor density under the infinitesimal coordinate transformation laws above. To show that $\mathcal{T}_i$ is a contravariant tensor density, we have only to show that it transforms similarly to T_i. Let $\delta\Phi^A$ be the change suffered by the matter dynamical variables under the infinitesimal coordinate transformation. Then, because of the coordinate invariance of S_{M} , we have

$$0\equiv\delta S_{\mathrm{M}}\equiv S_{\mathrm{M},A}\delta\Phi^A+S_{\mathrm{M},i}\Phi^i_a[\varphi_{\mathrm{B}}+\phi]\delta\xi^a.$$

Now $\delta\Phi^A$ is metric independent. Therefore, functional differentiation of this identity yields

$$0 \equiv S_{\mathrm{M},iA}\delta\Phi^A + S_{\mathrm{M},ij}\Phi^j_a[\varphi_\mathrm{B} + \phi]\delta\xi^a + S_{\mathrm{M},j}\Phi^j_{a,i}\delta\xi^a,$$

hence

$$\begin{aligned}\delta T_i &\equiv 2\delta S_{\mathrm{M},i} \equiv 2S_{\mathrm{M},iA}\delta\Phi^A + 2S_{\mathrm{M},ij}\Phi^j_a[\varphi_\mathrm{B} + \phi]\delta\xi^a \\ &\equiv -S_{\mathrm{M},j}\Phi^j_{a,i}\delta\xi^a \equiv -T_j\Phi^j_{a,i}\delta\xi^a.\end{aligned}$$

Similarly,

$$\begin{aligned}\delta\mathcal{T}_i &\equiv 2\delta\{S_{\mathrm{G},i}[\varphi_\mathrm{B} + \phi] - S_{\mathrm{G},i}[\varphi_\mathrm{B}] - S_{\mathrm{G},ij}[\varphi_\mathrm{B}]\phi^j\} \\ &\equiv 2\Big\{S_{\mathrm{G},ij}[\varphi_\mathrm{B} + \phi]\Phi^j_{\ a}[\varphi_\mathrm{B} + \phi] - S_{\mathrm{G},ij}[\varphi_\mathrm{B}]\Phi^j_{\ a}[\varphi_\mathrm{B}] \\ &2 - S_{\mathrm{G},ijk}[\varphi_\mathrm{B}]\Phi^j_{\ a}[\varphi_\mathrm{B}]\phi^k - S_{\mathrm{G},ij}[\varphi_\mathrm{B}]\Phi^j_{\ a,k}\phi^k\Big\}\delta\xi^a \\ &\equiv 2\Big\{-S_{\mathrm{G},j}[\varphi_\mathrm{B} + \phi]\Phi^j_{\ a,i} + S_{\mathrm{G},j}[\varphi_\mathrm{B}]\Phi^j_{\ a,i} + S_{\mathrm{G},jk}[\varphi_\mathrm{B}]\Phi^j_{\ a,i}\phi^k\Big\}\delta\xi^a \\ &\equiv -\mathcal{T}_j\Phi^j_{\ a,i}\delta\xi^a,\end{aligned}$$

where we have used (12.2) to reach the penultimate line. This proves the above claim that $\mathcal{T}_i$ transforms in the same way as T_i.

However, it is not convenient in practice to adopt the above coordinate transformation laws. This is because there is no unique way of splitting a given metric φ^i into a background metric φ^i_B and a remainder ϕ^i, *even when the background geometry is chosen a priori*. The background metric can still be chosen in any one of its infinity of equivalent presentations under the diffeomorphism group. Thus one cannot pretend that the field ϕ^i is well defined as a covariant tensor. In practice one must arbitrarily choose a particular form for φ^i_B, hold it fixed, and place the entire burden of coordinate transformation onto ϕ^i so that it transforms according to

$$\delta\phi^i = \delta\varphi^i = \Phi^i_{\ a}[\varphi]\delta\xi^a = \left(\Phi^i_{\ a}[\varphi_\mathrm{B}] + \Phi^i_{\ a,j}\phi^j\right)\delta\xi^a.$$

$T^{\mu\nu}$ continues to transform like a covariant tensor density under this transformation because S_M depends on ϕ^i only through the sum $\varphi^i_\mathrm{B} + \phi^i$. But the energy–momentum–stress density $\mathcal{T}^{\mu\nu}$ of the gravitational field no longer has a simple transformation law. Only under the isometry group of the background geometry, generated by Killing vectors $\delta\xi^a$ (if any) satisfying

$$\Phi^i_a[\varphi_\mathrm{B}]\delta\xi^a = 0,$$

does it transform like a contravariant tensor density. For this reason $\mathcal{T}^{\mu\nu}$ is often called a *pseudo-tensor density*.

Because $\mathcal{T}^{\mu\nu}$ does not transform like a tensor density under general coordinate transformations it is impossible to assign a definite location to the energy, momentum and stress in the gravitational field. Nevertheless, if 3-space is infinite and ϕ^i vanishes sufficiently rapidly at infinity, there is an absolutely conserved quantity, namely

$$\int_\Sigma \xi_\mu (T^{\mu\nu} + \mathcal{T}^{\mu\nu}) d\Sigma_\nu \quad (\Sigma \text{ spacelike}),$$

associated with every Killing vector ξ_μ that the background geometry possesses, which is well defined and *independent of the choice of coordinate system* despite the non-tensorial character of $\mathcal{T}^{\mu\nu}$. For consider any two distinct coordinate systems in the neighborhood of the spacelike hypersurface Σ. These may be extended into the future in such a way that they ultimately merge and become identical there. Because Σ itself may be displaced to this future region without altering the value of the integral, and because the integrals in the two coordinate systems are obviously identical there, it follows that they were always identical.

12.3 Asymptotic Stationary Gravitational Fields in the Full Nonlinear Theory

Suppose, we introduce into a flat empty spacetime a finite material system. The spacetime will then be only asymptotically flat and will depart more or less strongly from flatness near the material system, depending on its density. Such a combined system is conveniently described by choosing the Minkowski metric $\eta_{\mu\nu}$ as a background metric φ^i_{B}. The deviation ϕ^i of the true metric from Minkowskian is then simply the field $h_{\mu\nu}$ introduced in (6.1), and the gravitational field equations take the form

$$-\frac{1}{2}\left(l^{\mu\nu}{}_{,\sigma}{}^{\sigma} - l^{\mu\sigma}{}_{,\sigma}{}^{\nu} - l^{\nu\sigma}{}_{,\sigma}{}^{\mu} + \eta^{\mu\nu} l^{\sigma\tau}{}_{,\sigma\tau}\right) = 8\pi G(T^{\mu\nu} + \mathcal{T}^{\mu\nu}). \tag{12.3}$$

It will be convenient in what follows to choose units in which

$$G = 1.$$

These equations may then be reexpressed in the form

$$\frac{1}{16\pi} H^{\mu\sigma\nu\tau} = T^{\mu\nu} + \mathcal{T}^{\mu\nu},$$

where

$$H^{\mu\sigma\nu\tau} \equiv -(l^{\mu\nu}\eta^{\sigma\tau} + l^{\sigma\tau}\eta^{\mu\nu} - l^{\mu\tau}\eta^{\sigma\nu} - l^{\sigma\nu}\eta^{\mu\tau}).$$

$H^{\mu\sigma\nu\tau}$ has the same algebraic symmetries as the curvature tensor. In view of its antisymmetry in ν and τ, it follows that

$$H^{\mu\sigma\nu\tau}{}_{,\sigma\tau\nu} \equiv 0,$$

which is consistent with the conservation law

$$(T^{\mu\nu} + \mathcal{T}^{\mu\nu})_{,\nu} = 0.$$

This conservation law, which holds when the dynamical equations are satisfied, allows one to introduce the following absolutely conserved quantities:

$$P^{\mu} = \int\limits_{\Sigma} (T^{\mu\nu} + \mathcal{T}^{\mu\nu})\mathrm{d}\Sigma_{\nu},$$

$$J^{\mu\nu} = \int\limits_{\Sigma} [x^{\mu}(T^{\nu\sigma} + \mathcal{T}^{\nu\sigma}) - x^{\nu}(T^{\mu\sigma} + \mathcal{T}^{\mu\sigma})]\mathrm{d}\Sigma_{\sigma}.$$

These are respectively the total energy–momentum 4-vector and total angular momentum tensor of the combined matter–gravitational field system. Choosing Σ to be the hypersurface $x^0 = t$ and making use of the gravitational field equations, we may reexpress these quantities in the forms

$$\begin{aligned} P^{\mu} &= \int (T^{\mu 0} + \mathcal{T}^{\mu 0})\mathrm{d}^3\boldsymbol{x} \\ &= \frac{1}{16\pi}\int H^{\mu\sigma 0\tau}{}_{,\sigma\tau}\mathrm{d}^3\boldsymbol{x} = \frac{1}{16\pi}\int H^{\mu\sigma 0 i}{}_{,\sigma i}\mathrm{d}^3\boldsymbol{x} \\ &= \lim_{S\to\infty}\frac{1}{16\pi}\int_S H^{\mu\sigma 0 i}{}_{,\sigma}\mathrm{d}^2 S_i, \end{aligned}$$

$$\begin{aligned} J^{\mu\nu} &= \int [x^{\mu}(T^{\nu 0} + \mathcal{T}^{\nu 0}) - x^{\nu}(T^{\mu 0} + \mathcal{T}^{\mu 0})]\mathrm{d}^3\boldsymbol{x} \\ &= \frac{1}{16\pi}\int \left(x^{\mu}H^{\nu\sigma 0\tau}{}_{,\sigma\tau} - x^{\nu}H^{\mu\sigma 0\tau}{}_{,\sigma\tau}\right)\mathrm{d}^3\boldsymbol{x} \\ &= \frac{1}{16\pi}\int \left[\left(x^{\mu}H^{\nu\sigma 0\tau}{}_{,\sigma}\right)_{,\tau} - \left(x^{\nu}H^{\mu\sigma 0\tau}{}_{,\sigma}\right)_{,\tau} - H^{\nu\sigma 0\mu}{}_{,\sigma} + H^{\mu\sigma 0\nu}{}_{,\sigma}\right]\mathrm{d}^3\boldsymbol{x} \\ &= \frac{1}{16\pi}\int \Big[\left(x^{\mu}H^{\nu\sigma 0 i}{}_{,\sigma}\right)_{,i} - \left(x^{\nu}H^{\mu\sigma 0 i}{}_{,\sigma}\right)_{,i} \\ &\qquad - H^{\nu i 0\mu}{}_{,i} + H^{\mu i 0\nu}{}_{,i} - H^{\nu 0 0\mu}{}_{,0} + H^{\mu 0 0\nu}{}_{,0}\Big]\mathrm{d}^3\boldsymbol{x} \\ &= \lim_{S\to\infty}\frac{1}{16\pi}\int_S \left(x^{\mu}H^{\nu\sigma 0 i}{}_{,\sigma} - x^{\nu}H^{\mu\sigma 0 i}{}_{,\sigma} + H^{\mu i 0\nu} - H^{\nu i 0\mu}\right)\mathrm{d}^2 S_i, \end{aligned}$$

where S is a 2-surface homologous to a sphere and the instruction $\lim_{S\to\infty}$ means that S is to be expanded to infinity in all directions.

The presentation of P^{μ} and $J^{\mu\nu}$ as surface integrals at infinity makes it particularly easy to relate them to the asymptotical structure of the gravitational field. We shall in fact see that if the asymptotic field is stationary (in a certain Lorentz frame), it is completely determined by P^{μ} and $J^{\mu\nu}$. A stationary asymptotic field (no gravitational waves) is a weak field. In the asymptotic region, we may assume the coordinate system to be quasi-canonical and, in fact, quasi-stationary. We shall then restrict our coordinate transformations to those that preserve the quasi-stationary character of the coordinates.

We shall also assume that the field $h_{\mu\nu}$ vanishes at infinity at least as fast as $1/r$, where $r \equiv |\boldsymbol{x}|$. $\mathcal{T}^{\mu\nu}$ then vanishes at least as fast as $1/r^4$, for its dominant terms in the asymptotic region have forms like $h^{\sigma\tau}h^{\mu\nu}_{\sigma,\tau}$, and $h_{\sigma\tau,}{}^{\mu}h^{\sigma\tau,\nu}$. This asymptotic behavior, together with the finiteness of the material system, is entirely consistent with (12.3). In fact, it will be obvious later that no other assumption about asymptotic behavior would be consistent both with this equation and with the (assumed) finiteness of P^{μ} and $J^{\mu\nu}$.

Because $T^{\mu\nu}$ vanishes outside a finite region and $\mathcal{T}^{\mu\nu}$ vanishes so rapidly at infinity, the field in the asymptotic region effectively satisfies the equation

$$\begin{aligned} 0 &= l^{\mu\nu}{}_{,\sigma}{}^{\sigma} - l^{\mu\sigma}{}_{,\sigma}{}^{\nu} - l^{\nu\sigma}{}_{,\sigma}{}^{\mu} + \eta^{\mu\nu} l^{\sigma\tau}{}_{,\sigma\tau} \\ &= l^{\mu\nu i}{}_{,i} - l^{\mu i\nu}{}_{,i} - l^{\nu i}{}_{,i}{}^{\mu} + \eta^{\mu\nu} l^{ij}{}_{,ij}, \end{aligned}$$

the final form following from the stationarity of the asymptotic field. We now ask whether we can simplify this equation by imposing the supplementary condition $l^{\mu i}{}_{,i} = 0$, without violating the assumptions made thus far. Schematically (dropping indices), we have the asymptotic behavior

$$l \sim \frac{M}{r},$$

where M is some constant. In order to impose the supplementary condition we must carry out a gauge transformation (coordinate transformation) where the gauge parameter ξ^{μ} satisfies an equation of the form [see (6.4)]

$$\nabla^2 \xi = \nabla \cdot l.$$

This implies

$$\xi \sim M \ln \frac{r}{M}.$$

If we now choose r so big that both M/r and $(M/r)\ln(r/M)$ may be neglected in comparison with unity, then the quantities

$$l \nabla \xi \sim \left(\frac{M}{r}\right)^2, \quad \xi \nabla l \sim \left(\frac{M}{r}\right)^2 \ln \frac{r}{M}$$

may be neglected in comparison with l and $\nabla\xi$, and the gauge transformation may indeed by carried out consistently with asymptotic fall-off of order 1/r or faster. This defines the asymptotic region.

With the supplementary condition imposed (remember that coordinate transformations leave P^{μ} and $J^{\mu\nu}$ unaffected), we have only to solve the simple equation

$$\nabla^2 l_{\mu\nu} = 0.$$

The general solution of this equation, which has the required asymptotic behavior, is

$$l_{\mu\nu} = \frac{A_{\mu\nu}}{r} + \frac{B_{\mu\nu i}x^i}{r^3} + O(r^{-3}),$$

where $A_{\mu\nu}$ and $B_{\mu\nu i}$ are arbitrary constants. Some of the arbitrariness of these constants is removed by the supplementary condition. We have

$$0 = l_{\mu i,i} = -\frac{A_{\mu i}\hat{x}^i}{r^2} + \frac{B_{\mu ii} - 3B_{\mu ij}\hat{x}^i\hat{x}^j}{r^3} + O(r^{-4}),$$

where $\hat{x}^i \equiv x^i/r$, from which it may be inferred that

$$A_{\mu i} = 0, \quad B_{\mu ij} = B_\mu \delta_{ij} + C_{\mu ij},$$

where the B_μ and $C_{\mu ij}$ are certain constants, with the latter satisfying

$$C_{\mu ij} = -C_{\mu ji}.$$

By combining the antisymmetry of C_{ijk} in j and k with the symmetry of B_{ijk} in i and j, we can reexpress B_{ijk} completely in terms of B_i. We have

$$B_i\delta_{jk} + C_{ijk} = B_j\delta_{ik} + C_{jik},$$

or

$$C_{ijk} - C_{jik} = B_j\delta_{ik} - B_i\delta_{jk}, \quad C_{kji} - C_{jki} = B_j\delta_{ki} - B_k\delta_{ji},$$

$$C_{kij} - C_{ikj} = B_i\delta_{kj} - B_k\delta_{ij}.$$

Adding these last three equations and dividing by 2, we find

$$C_{ijk} = B_j\delta_{ik} - B_k\delta_{ij},$$

and hence,

$$B_{ijk} = B_i\delta_{jk} + B_j\delta_{ik} - B_k\delta_{ij}.$$

The asymptotic field now takes the form

$$l_{00} = \frac{A_{00}}{r} + \frac{B_{00i}x^i}{r^3} + O(r^{-3}), \quad l_{0i} = \frac{B_0 x^i + C_{0ij}x^j}{r^3} + O(r^{-3}),$$

$$l_{ij} = \frac{B_i x^j + B_j x^i - \delta_{ij}B_k x^k}{r^3} + O(r^{-3}).$$

This may be simplified by carrying out an additional gauge transformation with the gauge parameters given by

$$\xi_\mu = -\frac{B_\mu}{r},$$

in the appropriate region. These parameters satisfy

$$\nabla^2 \xi_\mu = 0 \quad \text{(in the asymptotic region)},$$

and hence the gauge transformation leaves the supplementary condition intact. In the new gauge, we have [see (6.3)]

$$\bar{l}_{00} = l_{00} - \xi_{i,i} = \frac{A_{00}}{r} + \frac{\bar{B}_{00i}x^i}{r^3} + O(r^{-3}), \quad \bar{B}_{00i} = B_{00i} - B_i,$$

$$\bar{l}_{0i} = l_{0i} - \xi_{0,i} = l_{0i} - \frac{B_0 x^i}{r^3} = \frac{C_{0ij}x^j}{r^3} + O(r^{-3}),$$

$$\begin{aligned}\bar{l}_{ij} &= l_{ij} - \xi_{i,j} - \xi_{j,i} + \delta_{ij}\xi_{k,k} \\ &= l_{ij} - \frac{B_i x^j + B_j x^i - \delta_{ij} B_k x^k}{r^3} = O(r^{-3}).\end{aligned}$$

From now on, we assume this transformation already to have been carried out so that we may drop the bars.

We are now ready to compute P^μ and $J^{\mu\nu}$. Because $H^{\mu\sigma\nu\tau}{}_{,\sigma\tau}$ vanishes in the asymptotic region, it does not matter what shapes the surface S assumes as it expands to infinity. For simplicity, we may assume it to be a sphere and write

$$d^2 S_i = \hat{x}^i r^2 d^2\Omega,$$

where $d^2\Omega$ is the element of solid angle subtended at the origin. We then have

$$\begin{aligned}P^0 &= \lim_{S\to\infty} \frac{1}{16\pi} \int_S H^{0i0j}{}_{,i} \mathrm{d}^2 S_j \\ &= -\lim_{S\to\infty} \frac{1}{16\pi} \int_S \left(l^{00}\eta^{ij} + l^{ij}\eta^{00} - l^{0j}\eta^{0i} - l^{0i}\eta^{0j}\right)_{,i} \mathrm{d}^2 S_j \\ &= \frac{A_{00}}{16\pi} \int_S \frac{x^j}{r^3} \mathrm{d}^2 S_j = \frac{A_{00}}{16\pi} \int_{4\pi} \mathrm{d}^2\Omega = \frac{1}{4} A_{00},\end{aligned}$$

$$\begin{aligned}P^i &= \lim_{S\to\infty} \frac{1}{16\pi} \int_S H^{ij0k}{}_{,j} \mathrm{d}^2 S_k \\ &= -\lim_{S\to\infty} \frac{1}{16\pi} \int_S \left(l^{i0}\eta^{jk} + l^{jk}\eta^{i0} - l^{ik}\eta^{j0} - l^{j0}\eta^{ik}\right)_{,j} \mathrm{d}^2 S_k \\ &= 0.\end{aligned}$$

In terms of the total rest mass (or rest energy)

$$M = (-P^2)^{1/2}$$

of the matter–gravitational field system, we have

$$A_{00} = 4M.$$

We note that the total 3-momentum vanishes, as befits a system that gives rise to a stationary asymptotic field. If the 3-momentum did not vanish, the asymptotic field would not be stationary.

The vanishing of P^i implies the vanishing of the total orbital angular momentum so that the spatial components of J^{ij} are just the components of the total spin angular momentum:

$$\begin{aligned}
S^{ij} &= J^{ij} \\
&= \lim_{S\to\infty} \frac{1}{16\pi} \int_S \left(x^i H^{jk0l}{}_{,k} - x^j H^{ik0l}{}_{,k} + H^{il0j} - H^{jl0i} \right) \mathrm{d}^2 S_l \\
&= -\lim_{S\to\infty} \frac{1}{16\pi} \int_S \Big[x^i \left(l^{j0}\eta^{kl} + l^{kl}\eta^{j0} - l^{jl}\eta^{k0} - l^{k0}\eta^{jl} \right)_{,k} \\
&\quad - x^j \left(l^{i0}\eta^{kl} + l^{kl}\eta^{i0} - l^{il}\eta^{k0} - l^{k0}\eta^{il} \right)_{,k} \\
&\quad + l^{i0}\eta^{lj} + l^{lj}\eta^{i0} - l^{ij}\eta^{l0} - l^{l0}\eta^{ij} \\
&\quad - l^{j0}\eta^{li} + l^{li}\eta^{j0} - l^{ji}\eta^{l0} - l^{l0}\eta^{ji} \Big] \mathrm{d}^2 S_l \\
&= \frac{1}{16\pi} \int_{4\pi} \Big[\hat{x}^i C_{0jk} \left(\delta_{kl} - 3\hat{x}^k \hat{x}^l \right) \hat{x}^l - \hat{x}^j C_{0ik} \left(\delta_{kl} - 3\hat{x}^k \hat{x}^l \right) \hat{x}^l \\
&\quad + C_{0ik} \hat{x}^k \hat{x}^j - C_{0jk} \hat{x}^k \hat{x}^i \Big] \mathrm{d}^2\Omega \\
&= \frac{3}{16\pi} \int_{4\pi} \left(C_{0ik} \hat{x}^k \hat{x}^j - C_{0jk} \hat{x}^k \hat{x}^i \right) \mathrm{d}^2\Omega \\
&= \frac{1}{4} \left(C_{0ij} - C_{0ji} \right) = \frac{1}{2} C_{0ij}.
\end{aligned} \tag{12.4}$$

As for the temporal components of $J^{\mu\nu}$, we have

$$\begin{aligned}
J^{0i} &= \lim_{S\to\infty} \frac{1}{16\pi} \int_S \left(t H^{ij0k}{}_{,j} - x^i H^{0j0k}{}_{,j} + H^{0k0i} - H^{ik00} \right) \mathrm{d}^2 S_k \\
&= \lim_{S\to\infty} \frac{1}{16\pi} \int_S \Big[x^i \left(l^{00}\eta^{jk} + l^{jk}\eta^{00} - l^{0k}\eta^{j0} - l^{j0}\eta^{0k} \right)_{,j} \\
&\quad - \left(l^{00}\eta^{ki} + l^{ki}\eta^{00} - l^{0i}\eta^{k0} - l^{k0}\eta^{0i} \right) \Big] \mathrm{d}^2 S_k \\
&= \lim_{r\to\infty} \frac{1}{16\pi} \int_{4\pi} \Big[\hat{x}^i \left(-A_{00} r \hat{x}^k + \bar{B}_{00k} - 3\bar{B}_{00j} \hat{x}^j \hat{x}^k \right) \hat{x}^k \\
&\quad - A_{00} r \hat{x}^i - \bar{B}_{00j} \hat{x}^j \hat{x}^i \Big] \mathrm{d}^2\Omega \\
&= -\frac{3}{16\pi} \int_{4\pi} \bar{B}_{00j} \hat{x}^j \hat{x}^i \mathrm{d}^2\Omega = -\frac{1}{4} \bar{B}_{00i}.
\end{aligned} \tag{12.5}$$

According to (9.1),

$$J^{0i} = tP^i - X_E^i P^0 = -X^i P^0 = -MX^i,$$

where X^i is the center of energy of the matter–gravitational field system in the rest frame (covariant center of energy). We may therefore make the identification

$$\bar{B}_{00i} = 4MX_i,$$

and we have finally

$$\left.\begin{aligned} l_{00} &= \tfrac{4M}{r} + \tfrac{4MX_i x^i}{r^3} + O(r^{-3}) \\ l_{0i} &= \tfrac{2S_{ij}x^j}{r^3} + O(r^{-3}) \\ l_{ij} &= O(r^{-3}) \end{aligned}\right\} \quad \text{in the asymptotic region.}$$

No particularly useful purpose is served by retaining the term in X_i above. This term arises from the fact that the origin of coordinates has not been placed at the center of energy.[1] It may be removed by carrying out the coordinate shift

$$\bar{x}^\mu = x^\mu + \xi^\mu,$$

where

$$\xi^0 = 0 \quad \text{and } \xi^i = -X^i.$$

We then have

$$\bar{P}^\mu = P^\mu, \quad \bar{J}^{ij} = J^{ij} + \xi^i P^j - \xi^j P^i = J^{ij},$$

$$\bar{J}^{0i} = J^{0i} + \xi^0 P^i - \xi^i P^0 = J^{0i} + MX^i = 0,$$

$$\begin{aligned} \bar{l}_{00} &= l_{00} - l_{00,\mu}\xi^\mu + O(r^{-3}) \\ &= l_{00} + l_{00,i}X^i + O(r^{-3}) \\ &= l_{00} - \frac{4MX_i x^i}{r^3} + O(r^{-3}) = \frac{4M}{r} + O(r^{-3}), \end{aligned}$$

$$\bar{l}_{0i} = l_{0i} - l_{0i,\mu}\xi^\mu + O(r^{-4}) = \frac{2S_{ij}x^j}{r^3} + O(r^{-3}),$$

$$\bar{l}_{ij} = l_{ij} - l_{ij,\mu}\xi^\mu + O(r^{-5}) = O(r^{-3}).$$

We shall henceforth assume this transformation already to have been carried out and drop the bars.

[1] If the source of the gravitational field is so dense that topological anomalies occur in the strong field region, the coordinate origin may not physically exist. The shift transformation is nevertheless valid.

It is now necessary to point out a slight inconsistency in the above derivations. In order to relate the asymptotic field to the total spin angular momentum it was necessary to retain not only terms of order $1/r$ but also terms of order $1/r^2$. Now the dominant spherically symmetric (monopole) term $4M/r$ in l_{00} gives rise to terms in $\mathcal{T}^{\mu\nu}$ that fall off asymptotically like $1/r^4$. When these terms are used as sources in the full nonlinear field equations (12.3), they give rise to corrections of order $1/r^2$ in the asymptotic fields, i.e., to terms falling off at the same rate as terms that have been retained. To be sure, these terms are of order $(M/r)^2$ in magnitude and hence are typically very much smaller than the terms that have been retained. However, we need to check whether they can affect the values that have been obtained for P^μ and $J^{\mu\nu}$, or, conversely, whether nonlinear corrections have to be introduced into the coefficients of the dominant asymptotic terms of $l_{\mu\nu}$ when they are expressed in terms of P^μ and $J^{\mu\nu}$ (actually M and S_{ij}).

Because the monopole term $4M/r$ is spherically symmetric, a coordinate system may be chosen in which the nonlinear corrections $\delta l_{\mu\nu}$ induced by this term are themselves spherically symmetric (no preferred directions). Explicitly,

$$\Delta l_{00} = \frac{W}{r^2}, \quad \Delta l_{0i} = X\frac{x^i}{r^3}, \quad \Delta l_{ij} = Y\frac{\delta_{ij}}{r^2} + Z\frac{x^i x^j}{r^4},$$

for some constants W, X, Y, Z. Being of order $1/r^2$, these terms are readily seen to have no effect on P^μ. Their effect on S^{ij} is determined by referring to (12.4). Noting that $\Delta l_{0i,i} = 0$, we have

$$\Delta S^{ij} = -\lim_{S\to\infty}\frac{1}{16\pi}\int_S \left(x^i \Delta l^{j0}{}_{,l} - x^j \Delta l^{i0}{}_{,l} + \Delta l^{i0}\eta^{lj} - \Delta l^{j0}\eta^{li}\right)\mathrm{d}^2 S_l.$$

Noting also that $\Delta l^{i0}{}_{,ll} = 0$, and hence

$$\left(x^i \Delta l^{j0}{}_{,l} - x^j \Delta l^{i0}{}_{,l} + \Delta l^{i0}\eta^{lj} - \Delta l^{j0}\eta^{li}\right)_{,l}$$
$$= \Delta l^{j0}{}_{,i} - \Delta l^{i0}{}_{,j} + \Delta l^{i0}{}_{,j} - \Delta l^{j0}{}_{,i} = 0,$$

we see that it does not matter how S is chosen in the computation of ΔS^{ij}. Choosing a sphere, we find

$$\Delta S^{ij} = -\frac{X}{16\pi}\int_{4\pi}\left[\hat{x}^i(\delta_{jl} - 3\hat{x}^j\hat{x}^l) - \hat{x}^j(\delta_{il} - 3\hat{x}^i\hat{x}^l) + \hat{x}^i\delta_{jl} - \hat{x}^j\delta_{il}\right]\hat{x}^l \mathrm{d}^2\Omega$$
$$= 0.$$

In the case of J^{0i}, we have from (12.5),

$$\Delta J^{0i} = \lim_{S\to\infty}\frac{1}{16\pi}\int_S \left(x^i \Delta l^{00}{}_{,k} - x^i \Delta l^{jk}{}_{,j} - \Delta l^{00}\eta^{ki} + \Delta l^{ki}\right)\mathrm{d}^2 S_k.$$

Here it *does* matter how S is chosen, for we have

$$\begin{aligned}\left(x^i\Delta l^{00}{}_{,k} - x^i\Delta l^{jk}{}_{,j} - \Delta l^{00}\eta^{ki} + \Delta l^{ki}\right)_{,k} &= \Delta l^{00}{}_{,i} + x^i\Delta l^{00}{}_{,kk} - \Delta l^{ji}{}_{,j} - x^i\Delta l^{jk}{}_{,jk} - \Delta l^{00}{}_{,i} \\ &\quad + \Delta l^{ki}{}_{,k} \\ &= x^i(\Delta l^{00}{}_{,kk} - \Delta l^{jk}{}_{,jk}),\end{aligned}$$

which is not generally equal to zero. The reason for the sensitivity of ΔJ^{0i} to the choice of S may be understood by remembering that J^{0i} gives the location of the center of energy of the matter–gravitational field system. ΔJ^{0i} comes from the asymptotic contribution to the integral $-\int x^i\mathcal{T}^{00}d^3\boldsymbol{x}$. Asymptotically, $\mathcal{T}^{00}$ is spherically symmetric, and hence one would say that the asymptotic contribution vanishes by symmetry. However, if one remembers that $\mathcal{T}^{00}$ falls off asymptotically like $1/r^4$, one sees that this is really an improper integral (logarithmically divergent). In order to make it well defined, one must specify precisely how its boundary is to tend to infinity. The symmetry argument is, of course, valid here. One should integrate over the region inside a sphere and then let the radius of the sphere tend to infinity.[2] This is equivalent to choosing the surface S above to be a sphere. With this choice and the replacement $d^2S_k = r^2\hat{x}^k d^2\Omega$, one readily finds that every term in the integrand for ΔJ^{0i} contains an odd number of unit vector components $\hat{x}^i$, and hence that

$$\Delta J^{0i} = 0,$$

by symmetry. The correction terms arising from the nonlinearity of the full field equations are therefore seen to have no effect on P^μ and $J^{\mu\nu}$. This does *not* mean that the gravitational field itself makes no contribution to these quantities. Indeed, in regions of strong curvature, $\mathcal{T}^{\mu\nu}$ can be even more important than $T^{\mu\nu}$. It means only that the *asymptotic* part of the field makes no contribution. The asymptotic field merely registers the *imprint* of P^μ and $J^{\mu\nu}$.

Summing up, we have the following theorem: For every asymptotically stationary gravitational field a coordinate system can be introduced in which the field takes the canonical asymptotic form

$$l_{00} \sim \frac{4M}{r} + \frac{W}{r^2} + O(r^{-3}), \quad l_{0i} \sim \frac{2S_{ij}x^j}{r^3} + X\frac{x^i}{r^3} + O(r^{-3}),$$

$$l_{ij} \sim Y\frac{\delta_{ij}}{r^2} + Z\frac{x^ix^j}{r^4} + O(r^{-3}),$$

where M is the total mass energy (gravitational as well as material) of the source, S_{ij} is the total spin angular momentum, and W, X, Y, Z are constants of order M^2.

---

[2] It does not matter where the sphere is centered. Provided only its center is held fixed while its radius tends to infinity, the final result will be invariant under displacements of the sphere.

Conversely, if a coordinate system is found in which the field takes the canonical asymptotic form one may immediately identify the total mass and spin of the source from the pertinent coefficients.

Computing

$$l = l^{\mu}_{\mu} = -l_{00} + l_{ii} \sim -\frac{4M}{r} - \frac{W - 3Y - Z}{r^2} + O(r^{-3}),$$

and remembering that

$$g_{\mu\nu} = \eta_{\mu\nu}\left(1 - \frac{1}{2}l\right) + l_{\mu\nu},$$

we find for the canonical asymptotic form of the metric tensor the expressions

$$g_{00} \sim -1 + \frac{2M}{r} + \frac{\bar{W}}{r^2} + O(r^{-3}), \quad g_{0i} \sim \frac{2S_{ij}x^j}{r^3} + \bar{X}\frac{x^i}{r^3} + O(r^{-3}),$$

$$g_{ij} \sim \left(1 + \frac{2M}{r}\right)\delta_{ij} + \bar{Y}\frac{\delta_{ij}}{r^2} + \bar{Z}\frac{x^i x^j}{r^4} + O(r^{-3}),$$

where

$$\bar{W} = \frac{1}{2}(W + 3Y + Z), \quad \bar{X} = X, \quad \bar{Y} = \frac{1}{2}(W - Y + Z), \quad \bar{Z} = Z.$$

The asymptotic form for the metric tensor of a uniformly moving source may be obtained from this by a Lorentz transformation.

Some concluding remarks are in order concerning the above results. We note first of all that the dominant (monopole) terms in the asymptotic metric are identical to those we previously calculated for a point particle in the weak field approximation. We have therefore justified the point particle idealization. Indeed, the justification goes farther than this. With a Minkowskian background metric, the center of energy of the combined matter–gravitational field system moves in a straight line. At a great distance from the material source, the dimensions of the source dwindle to insignificance and this line effectively marks the location of the source, i.e., it becomes the world line of the source. Now imagine that we have many such sources, all at large distances from one another. Collectively, they produce a background field that is no longer flat. Nevertheless, coordinate patches can be introduced around each source, which, near their outer edges, are canonical to high accuracy. Each patch may be regarded as marking the domain of a local flat background geometry. Each source moves in a straight line relative to a canonical coordinate system laid down in this local background. This implies that each source moves along a geodesic in the *global* background field produced collectively by all the other sources. For the geodesic is the only invariantly defined curve that has the appearance of a straight line ($d^2z^\alpha/d\tau^2 = 0$) in each locally canonical coordinate system ($\Gamma^\alpha_{\beta\gamma} = 0$) that is laid down along its length. Thus the point particle idealization is justified even to the extent of describing correctly the dynamical behavior of the sources. One must only take care to separate the 'self-field' of each source from the background geometry. Although such a separation cannot be unique close to the source, it is well defined asymptotically.

Suppose, finally, that the whole collection of sources is bounded and that even the global geometry ultimately becomes flat at great distances from the ensemble. Then a flat background can be introduced for the ensemble and a total energy, momentum and angular momentum defined. These quantities, which include contributions from the global curvature will be conserved if the motion of the sources is so slow that the global asymptotic field is quasi-stationary. We shall see later that these quantities will be conserved even if some of the sources are moving at high velocity, provided only that their mutual interactions are so weak that the acceleration of each is small relative to a coordinate system that is quasi-canonical in the field of all the other sources. If, on the other hand, the accelerations become appreciable (in some level of approximation) then gravitational radiation must be taken into account, and there will be a net outflow of energy, momentum and angular momentum from the ensemble, as we shall see.

12.4 Newtonian Approximation

It is of interest to compare the contributions which $\mathcal{T}^{\mu\nu}$ and $T^{\mu\nu}$ each make to the total energy when the conditions for the validity of the Newtonian approximation hold, namely, when the size R of each source is very much greater than its so-called *gravitational radius* GM:

$$R \gg GM.$$

Under these conditions, the maximum value that $\mathcal{T}^{00}$ can assume (in quasi-canonical coordinates) is, apart from factors of $1/16\pi$, etc., of order

$$\mathcal{T}^{00} \sim \frac{1}{G}(\nabla h)^2 \sim \frac{GM^2}{R^4}.$$

T^{00}, on the other hand, will be of order

$$T^{00} \sim \frac{M}{R^3}.$$

Therefore,

$$\frac{\mathcal{T}^{00}}{T^{00}} \sim \frac{GM}{R} \ll 1,$$

and $\mathcal{T}^{00}$ is seen to be negligible compared to T^{00}. This is true even though $\mathcal{T}^{00}$ is distributed in space around the source whereas T^{00} is confined to the material source itself. For the total contribution to P^0 from $\mathcal{T}^{00}$ is of order GM^2/R, while that from T^{00} is, of course, of order M.

It is also of interest to compare $\mathcal{T}^{00}$ with the other components of the energy–momentum–stress density. In the mean rest frame of the source the components T^{0i} typically arise from rotation and are of order

$$T^{0i} \sim vT^{00},$$

where v is the 'rim' rotation velocity. But the maximum value that v can attain without the source flying apart is, for astronomical objects,

$$v \sim \left(\frac{GM}{R}\right)^{1/2} \ll 1.$$

Therefore, $T^{0i} \ll T^{00}$. As for the components T^{ij}, these typically arise from internal pressure, which is of order

$$p \sim \frac{GM^2}{R^4} \ll \frac{M}{R^3}.$$

Therefore the T^{ij} too are negligible compared to T^{00}, and T^{00} is seen to dominate everything in the Newtonian approximation. This provides yet one more justification for the point particle idealization used in the weak field analysis, in which T^{00} was the only surviving component of $T^{\mu\nu}$ [see (8.3)].

Chapter 13
Measurement of Asymptotic Fields

The asymptotic field of a quasi-stationary source registers the imprint of the total mass and total spin angular momentum of the combined matter—gravitational field system. These quantities can therefore be determined by experiments carried out in the asymptotic region without ever going near the source. The total mass may be determined simply by examining the Keplerian orbits of test bodies and comparing their size with their periods, in the familiar manner of celestial mechanics.[1] In this chapter, we shall indicate how to measure (in principle) the spin angular momentum of the source by observations of the behavior of gyroscopes in the asymptotic region.

We shall see latter (Appendix A) that the spin angular momentum 4-vector S^μ of a gyroscope is Fermi–Walker transported along the gyroscope's world line. This means it satisfies the equations (see Problem 9 in Chap. 2 and Footnote 1 in Chap. 5)

$$\dot{S}^\mu = (S \cdot \dot{u})u^\mu, \quad S \cdot u = 0,$$

where u^μ is the gyroscope's velocity 4-vector ($u^2 = -1$) and the dot denotes covariant differentiation with respect to the proper time. In Problem 9 in Chap. 2, it was found that the vector S^μ precesses in the case of flat spacetime when $\dot{u}^\mu \neq 0$. This precession is known as *Thomas precession*. In a curved spacetime we shall see that there is a precession even when $\dot{u}^\mu = 0$.

In order to compute this precession, it is convenient to introduce a tetrad field e^α_μ satisfying

$$\eta_{\alpha\beta} e^\alpha_\mu e^\beta_\nu = g_{\mu\nu}, \quad e^\alpha_\mu e^\beta_\nu g^{\mu\nu} = \eta_{\alpha\beta}.$$

In the local canonical frame provided by the tetrad e^α_μ, the components of S^μ are given by

[1] The location of the center of energy relative to the asymptotic Cartesian coordinates is given by the focus of the Keplerian ellipses.

B. DeWitt, *Bryce DeWitt's Lectures on Gravitation*,

Edited by Steven M. Christensen, Lecture Notes in Physics, 826,
DOI: 10.1007/978-3-540-36911-0_13,

$$S^{\alpha} = e^{\alpha}_{\mu} S^{\mu}.$$

These components satisfy the differential equation

$$\begin{aligned}\dot{S}^{\alpha} &= \dot{e}^{\alpha}_{\mu} S^{\mu} + e^{\alpha}_{\mu} \dot{S}^{\mu} \\ &= e^{\alpha}_{\mu;\nu} S^{\mu} u^{\nu} + e^{\alpha}_{\mu} (S \cdot \dot{u}) u^{\mu} \\ &= e^{\alpha}_{\mu;\nu} e^{\mu}_{\beta} e^{\nu}_{\gamma} S^{\beta} u^{\gamma} + (S \cdot \dot{u}) u^{\alpha}\end{aligned} .$$

In order to describe precession, we need to relate S^{α} to a pure 3-vector. This we can do by carrying out a boost from the local tetrad frame to the local rest frame of the gyroscope. The boost transformation coefficients are those given in Problem 9 in Chap. 2:

$$(L^{\alpha}_{\beta}) = \begin{pmatrix} \gamma & -\gamma v \\ -\gamma v & \underline{1} + (\gamma - 1)\hat{v}\hat{v} \end{pmatrix}, \quad (L_{\beta}^{-1\alpha}) = \begin{pmatrix} \gamma & \gamma v \\ \gamma v & \underline{1} + (\gamma - 1)\hat{v}\hat{v} \end{pmatrix},$$

where

$$(u^{\alpha}) = (\gamma, \gamma v), \quad \hat{v} = \frac{v}{|v|}, \quad \gamma = \frac{1}{\sqrt{1 - v^2}}.$$

Defining

$$\overline{S}^{\alpha} = L^{\alpha}_{\beta} S^{\beta}, \quad \overline{u}^{\alpha} = L^{\alpha}_{\beta} u^{\beta},$$

we then have

$$\overline{S}^{0}, \quad (\overline{u}^{\alpha}) = (1, 0, 0, 0),$$

and

$$\begin{aligned}\dot{\overline{S}}^{\alpha} &= \dot{L}^{\alpha}_{\beta} S^{\beta} + L^{\alpha}_{\beta} \dot{S}^{\beta} \\ &= \dot{L}^{\alpha}_{\beta} L_{\gamma}^{-1\beta} \overline{S}^{\gamma} + L^{\alpha}_{\beta} e^{\beta}_{\mu;\nu} e^{\mu}_{\gamma} e^{\nu}_{\delta} L_{\varepsilon}^{-1\gamma} \overline{S}^{\varepsilon} u^{\delta} + (S \cdot \dot{u}) \overline{u}^{\alpha}\end{aligned} .$$

Taking only the spatial components of this equation and making use of the results of Problem 9 in Chap. 2 as well as of the fact that

$$L^{i}_{\alpha} \eta^{\alpha\beta} = \eta^{i\gamma} L_{\gamma}^{-1\beta} = L_{i}^{-1\beta},$$

we find

$$\dot{\overline{S}}^{i} = -\Omega_{ij} \overline{S}^{j},$$

where

$$\Omega_{ij} = -(\gamma - 1)\left(\hat{v}_i \dot{\hat{v}}_j - \hat{v}_j \dot{\hat{v}}_i\right) - e_{\alpha\mu;\nu} e^{\mu}_{\beta} e^{\nu}_{\gamma} L_{i}^{-1\alpha} L_{j}^{-1\beta} u^{\gamma}.$$

The first term on the right of this last equation may be rewritten in a form that distinguishes between contributions coming from the curvature of spacetime and contributions coming from the absolute acceleration of the gyroscope. We first write

$$v_i = \gamma^{-1} u^i, \quad \dot{v}_i = \gamma^{-1} \dot{u}^i - \gamma^{-2} \dot{\gamma} u^i,$$

and remember that $v^2\gamma^2 = \gamma^2 - 1$. This allows us to express the first term in the form

$$-\frac{1}{\gamma+1}(u^i \dot{u}^j - u^j \dot{u}^i).$$

We next compute

$$\begin{aligned} \dot{u}^\alpha &= \dot{e}^\alpha_\mu u^\mu + e^\alpha_\mu \dot{u}^\mu \\ &= e^\alpha_{\mu;\nu} u^\mu u^\nu + a^\alpha \\ &= e^\alpha_{\mu;\nu} e^\mu_\beta e^\nu_\gamma u^\beta u^\gamma + a^\alpha \end{aligned},$$

where the a^α are the components of the absolute acceleration in the local tetrad frame. The first term of Ω_{ij} then becomes

$$-\frac{\gamma}{\gamma+1}(v_i a_j - v_j a_i) - \frac{\gamma}{\gamma+1}(v_i e_{j\mu;\nu} - v_j e_{i\mu;\nu}) e^\mu_\beta e^\nu_\gamma u^\beta u^\gamma.$$

Some authors regard only the first of these two terms as giving rise to the Thomas precession, while others regard both as giving rise to it. That is, there is disagreement as to precisely what should be called the Thomas precession in general relativity. The first group of authors likes to regard gravitational forces as special, because these forces do not contribute to a_i. The second group likes to regard the (coordinate) accelerations produced by gravitational forces as no different from the accelerations produced by any other forces. Locally, of course, there is no way of distinguishing the contributions to the precession arising from the two kinds of forces, because all of the quantities appearing in the above expressions depend on an arbitrary choice of tetrad field. However, in a closed orbit in a stationary field there is a cumulative precession over each period, which is physically well defined and which can be split in a physically well-defined way into a part arising from absolute acceleration and a remainder determined by spacetime geometry.

Collecting the above results we now have

$$\Omega_{ij} = -\frac{\gamma}{\gamma+1}(v_i a_j - v_j a_i) - e_{\alpha\mu;\nu} e^\mu_\beta e^\nu_\gamma u^\gamma M^{\alpha\beta}_{ij},$$

where (using $e_{\alpha\mu;\nu}\, e^{\mu}_{\beta} = -\, e_{\beta\mu;\nu}\, e^{\mu}_{\alpha}$)

$$\begin{aligned} M^{\alpha\beta}_{ij} &= \frac{\gamma}{2(\gamma+1)}(v_i\delta^{\alpha}_{j}u^{\beta} - v_j\delta^{\alpha}_{i}u^{\beta} - v_i\delta^{\beta}_{j}u^{\alpha} + v_j\delta^{\beta}_{i}u^{\alpha}) \\ &+ \frac{1}{2}(L^{-1\alpha}_{i}L^{-1\beta}_{j} - L^{-1\alpha}_{j}L^{-1\beta}_{i}) \end{aligned}.$$

Explicitly, component by component, we find

$$M^{00}_{ij} = 0,$$

$$\begin{aligned} M^{0k}_{ij} &= -\frac{\gamma^2}{2(\gamma+1)}v_i\delta_{jk} + \frac{1}{2}\gamma v_i\left[\delta_{jk} + (\gamma-1)\hat{v}_j\hat{v}_k\right] - (i \leftrightarrow j) \\ &= \frac{\gamma}{2(\gamma+1)}(v_i\delta_{jk} - v_j\delta_{ik}) \end{aligned},$$

$$\begin{aligned} M^{kl}_{ij} &= \frac{\gamma^2}{2(\gamma+1)}(v_i\delta_{jk}v_l - v_i\delta_{jl}v_k) \\ &+ \frac{1}{2}[\delta_{ik} + (\gamma-1)\hat{v}_i\hat{v}_k]\left[\delta_{jl} + (\gamma-1)\hat{v}_j\hat{v}_l\right] - (i \leftrightarrow j), \\ &= \frac{1}{2}(\delta_{ik}\delta_{jl} - \delta_{jk}\delta_{il}) \end{aligned}$$

using

$$-\frac{\gamma^2}{\gamma+1} + \frac{\gamma-1}{v^2} = 0.$$

Hence finally,

$$\begin{aligned} \Omega_{ij} &= -\frac{\gamma}{\gamma+1}(v_ia_j - v_ja_i) - \frac{\gamma}{\gamma+1}(v_je^{\mu}_{i} - v_je^{\mu}_{i})e_{0\mu;\nu}e^{\nu}_{\alpha}u^{\alpha} \\ &- e_{i\mu;\nu}e^{\mu}_{j}e^{\nu}_{\alpha}u^{\alpha} \end{aligned}.$$

Suppose now that the gravitational field is weak and that the coordinate system is quasi-canonical so that $g_{\mu\nu} = \eta_{\mu\nu} + h_{\mu\nu}$, where $|h_{\mu\nu}| \ll 1$. Then in the weak field approximation, there is a natural choice for the tetrad field, namely

$$e_{\alpha\mu} = \eta_{\alpha\mu} + \frac{1}{2}h_{\alpha\mu} + O(h^2).$$

We then have

$$\begin{aligned} e_{\alpha\mu;\nu} &= e_{\alpha\mu,\nu} - \Gamma^{\sigma}_{\mu\nu}e_{\alpha\sigma} \\ &= \frac{1}{2}h_{\alpha\mu,\nu} - \frac{1}{2}(h_{\alpha\mu,\nu} + h_{\alpha\nu,\mu} - h_{\mu\nu,\alpha}) + O\left(h\frac{\partial h}{\partial x}\right), \\ &= \frac{1}{2}(h_{\mu\nu,\alpha} - h_{\alpha\nu,\mu}) + O\left(h\frac{\partial h}{\partial x}\right) \end{aligned}$$

and, dropping terms of order $h\partial h/\partial x$,

$$\begin{aligned}\Omega_{ij} &= -\frac{\gamma}{\gamma+1}(v_i a_j - v_j a_i) \\ &\quad - \frac{\gamma^2}{2(\gamma+1)}\left[v_i(h_{j0,0} - h_{00,j}) - v_j(h_{i0,0} - h_{00,i}) + v_i v_k(h_{jk,0} - h_{0k,j}) - v_j v_k(h_{ik,0} - h_{0k,i})\right] . \\ &\quad - \frac{1}{2}\gamma\left[h_{j0,i} - h_{i0,j} + v_k(h_{jk,i} - h_{ik,j})\right]\end{aligned}$$

In the special case of slow motion relative to a quasi-stationary coordinate system in a quasi-stationary gravitational field, this expression reduces to

$$\begin{aligned}\Omega_{ij} &= -\frac{1}{2}\left[v_i\left(a_j - \frac{1}{2}h_{00,j}\right) - v_j\left(a_i - \frac{1}{2}h_{00,i}\right)\right] \\ &\quad - \frac{1}{2}v_k(h_{jk,i} - h_{ik,j}) - \frac{1}{2}(h_{j0,i} - h_{i0,j})\end{aligned},$$

in which no distinction need be made between the 3-velocity v_i relative to the tetrad frame and the coordinate 3-velocity. (This assumes that a_i is not much greater than $h_{00,i}$ in order of magnitude.)

It will be observed that the Lense–Thirring field makes its appearance in the third (last) term of the final expression. When a Lense–Thirring field is present, the gyroscope will precess even when it is at rest. It will also be observed that the first and third terms are invariant under gauge transformations that preserve the quasi-stationary character of the coordinate system (see p. 72). The second term, however, is not. Under a gauge transformation it, and therefore Ω_{ij}, suffers the change

$$\Delta\Omega_{ij} = \frac{1}{2}v_k(\xi_{j,i} - \xi_{i,j})_{,k}.$$

This change is not physical but arises simply from the fact that the gauge transformation induces a change in the 'natural' tetrad field. The precession is being defined relative to the new local tetrad frames! Remembering that $v_k = \dot{x}_k$ (no distinction now between τ and t), we see that

$$\Delta\Omega_{ij} = \frac{1}{2}\frac{\mathrm{d}}{\mathrm{d}t}(\xi_{j,i} - \xi_{i,j})$$

and, hence over a closed orbit, $\Delta\Omega_{ij}$ produces no cumulative effect:

$$\oint \Delta\Omega_{ij}\mathrm{d}t = 0.$$

Let us now apply these results to a gyroscope moving in the asymptotic field of a quasi-stationary source. Referring to p. 146, we have

$$h_{00} = \frac{2M}{r}, \quad h_{0i} = \frac{2S^{\mathrm{M}}_{ij}x^j}{r^3}, \quad h_{ij} = \frac{2M}{r}\delta_{ij},$$

where M is the mass and S^{M}_{ij} is the spin angular momentum tensor of the source. Therefore,

$$h_{00,i} = -2M\frac{\hat{x}_i}{r^2}, \quad h_{ij,k} = -2M\delta_{ij}\frac{\hat{x}_k}{r^2},$$

$$h_{0i,j} = \frac{2}{r^3}\left(S^{\mathrm{M}}_{ij} - 3S^{\mathrm{M}}_{ik}\hat{x}_k\hat{x}_j\right), \quad \hat{x}_i = \frac{x^i}{r},$$

whence

$$\Omega_{ij} = -\frac{1}{2}(v_i a_j - v_j a_i) - \frac{3}{2}\frac{M}{r^2}(v_i\hat{x}_j - v_j\hat{x}_i) + \frac{1}{r^3}\left(2S^{\mathrm{M}}_{ij} - 3S^{\mathrm{M}}_{ik}\hat{x}_k\hat{x}_j + 3S^{\mathrm{M}}_{jk}\hat{x}_k\hat{x}_i\right).$$

Defining

$$\Omega_i = \frac{1}{2}\varepsilon_{ijk}\Omega_{jk}, \quad S^{\mathrm{M}}_i = \frac{1}{2}\varepsilon_{ijk}S^{\mathrm{M}}_{jk},$$

and noting that

$$\begin{aligned}-3\varepsilon_{ijk}S^{\mathrm{M}}_{jl}\hat{x}_l\hat{x}_k &= -3\varepsilon_{ijk}\varepsilon_{jlm}S^{\mathrm{M}}_m\hat{x}_l\hat{x}_k \\ &= 3(\delta_{il}\delta_{km} - \delta_{im}\delta_{kl})S^{\mathrm{M}}_m\hat{x}_l\hat{x}_k\,, \\ &= 3\hat{x}_i\hat{x}_j S^{\mathrm{M}}_j - 3S^{\mathrm{M}}_i\end{aligned}$$

we find that we may write the gyroscope precession equation in the form

$$\frac{\mathrm{d}\overline{\boldsymbol{S}}}{\mathrm{d}t} = \Omega \times \overline{\boldsymbol{S}},$$

where, in standard units with the constants G and c restored, the gyroscope precession frequency Ω is given by

$$\Omega = -\frac{1}{2c^2}\boldsymbol{v} \times \boldsymbol{a} - \frac{3GM}{2c^2r^2}\boldsymbol{v} \times \hat{\boldsymbol{x}} - \frac{G}{c^2r^3}(\underline{1} - 3\hat{\boldsymbol{x}}\hat{\boldsymbol{x}}) \cdot \boldsymbol{S}^{\mathrm{M}}.$$

Assuming that $\boldsymbol{a}$ and M are known, we see finally that an observation of Ω yields a determination of $\boldsymbol{S}^{\mathrm{M}}$.

It will be observed that the contribution to Ω arising from $\boldsymbol{S}^{\mathrm{M}}$ causes the gyroscope axis to turn in the same direction as the source, i.e., Ω parallel to $\boldsymbol{S}^{\mathrm{M}}$, in polar regions ($\hat{\boldsymbol{x}}$ parallel or antiparallel to $\boldsymbol{S}^{\mathrm{M}}$) and in the opposite direction in equatorial regions ($\hat{\boldsymbol{x}}$ perpendicular to $\boldsymbol{S}^{\mathrm{M}}$). This effect has an analog in hydrodynamics. Consider a rotating solid sphere immersed in a viscous fluid. As it rotates, the sphere will drag the fluid along with it. At various points in the fluid, one may imagine little rods, free to rotate about a central pivot but otherwise held fast. Near the poles, the fluid will rotate the rods in the same direction as the sphere rotates. But near the equator, because the fluid is dragged more rapidly at small radii than at large, the end of a rod closest to the sphere experiences a stronger dragging force than the end farthest from the sphere. Consequently the rod rotates in a direction opposite to that of the sphere. We may therefore say that a rotating source of gravitation affects its surroundings as if it were dragging some kind of a medium along with it.

Problem 38 In the case of the earth the asymptotic field region begins already at the earth's surface. Moreover the contribution of the gravitational field to M and $\boldsymbol{S}^{\mathrm{M}}$ is negligible, so that M may be regarded as simply the mechanical mass of the earth, and $\boldsymbol{S}^{\mathrm{M}}$ may be expressed as $I\boldsymbol{\omega}$ where I is the earth's mechanical moment of inertia and $\boldsymbol{\omega}$ is its rotational angular frequency vector. Suppose the gyroscope is at rest on the surface of the earth at a latitude α. Using

$$I = 0.334MR^2,$$

where R is the radius of the earth and the factor 0.334 has been determined from geological and astronomical studies, obtain an expression for the average $\langle \Omega \rangle$ of the precession frequency over one sidereal day (86 164 s) in terms of G, c, M, R, ω and α. Using the values

$$G = 6.67 \times 10^{-11}\,\mathrm{kg}^{-1}\mathrm{m}^3\mathrm{s}^{-2}, \quad M = 5.98 \times 10^{24}\,\mathrm{kg}, \quad R = 6.38 \times 10^6\,\mathrm{m},$$

find the magnitude of $\langle \Omega \rangle$ at the equator. Express your answer in seconds of arc per year.

Now suppose the gyroscope is in circular orbit about the earth. Let r be the radius of the orbit and $\boldsymbol{n}$ the unit vector perpendicular to the orbit. Obtain an expression for $\langle \Omega \rangle$ (orbital average) in terms of G, c, M, I, R, $\boldsymbol{\omega}$ and $\boldsymbol{n}$. Show that when the orbit is polar the contribution to $\langle \Omega \rangle$ from the orbital motion is at right angles to that arising from the Lense–Thirring field produced by $\boldsymbol{S}^{\mathrm{M}}$. Compute the magnitudes of these two contributions for the case in which the orbit is at an altitude of 500 km above the earth's surface.

Solution 38 We have

$$\boldsymbol{a} = \frac{GM}{R^2}\hat{\boldsymbol{x}}, \quad \Omega = -\frac{2GM}{c^2R^2}\boldsymbol{v} \times \hat{\boldsymbol{x}} - \frac{GI}{c^2R^3}(\underline{1} - 3\hat{\boldsymbol{x}}\hat{\boldsymbol{x}}) \cdot \boldsymbol{\omega},$$

$$\boldsymbol{v} = R\boldsymbol{\omega} \times \hat{\boldsymbol{x}}, \quad \boldsymbol{v} \times \hat{\boldsymbol{x}} = R(\boldsymbol{\omega} \times \hat{\boldsymbol{x}}) \times \hat{\boldsymbol{x}} = -R(\underline{1} - \hat{\boldsymbol{x}}\hat{\boldsymbol{x}}) \cdot \boldsymbol{\omega},$$

$$\begin{aligned}\Omega &= \frac{2GM}{c^2R}(\underline{1} - \hat{\boldsymbol{x}}\hat{\boldsymbol{x}}) \cdot \boldsymbol{\omega} - \frac{0.334GM}{c^2R}(\underline{1} - 3\hat{\boldsymbol{x}}\hat{\boldsymbol{x}}) \cdot \boldsymbol{\omega} \\ &= \frac{GM}{c^2R}(1.67\underline{1} - 1.00\hat{\boldsymbol{x}}\hat{\boldsymbol{x}}) \cdot \boldsymbol{\omega}\end{aligned},$$

whence

$$\langle \Omega \rangle = \frac{GM}{c^2R}(1.67 - 1.00 \sin^2 \alpha)\boldsymbol{\omega}$$

Now,

$$|\boldsymbol{\omega}| = \frac{2\pi}{86164}\,\mathrm{s}^{-1} = 7.30 \times 10^{-5}\,\mathrm{rad\ s}^{-1},$$

and

$$1\,\mathrm{arcsec\ yr}^{-1} = \frac{2\pi}{360 \times 3600 \times 3.16 \times 10^{7}} = 1.54 \times 10^{-13}\,\mathrm{rad\ s}^{-1}.$$

At the equator ($\alpha = 0$), we have

$$\begin{aligned} |\langle \Omega \rangle| &= \frac{1.67GM}{c^2 R}|\omega| \\ &= \frac{1.67 \times 6.67 \times 10^{-11} \times 5.98 \times 10^{24}}{9 \times 10^{16} \times 6.38 \times 10^{6}} \times 7.30 \times 10^{-5} \\ &= 8.47 \times 10^{-14}\,\mathrm{rad\ s}^{-1} \\ &= \frac{8.47 \times 10^{-14}}{1.54 \times 10^{-13}} = 0.550\,\mathrm{arcsec\ yr}^{-1} \end{aligned}$$

In orbit we, have $a = 0$ and

$$\frac{v^2}{r} = \frac{GM}{r^2}, \quad |\boldsymbol{v}| = \left(\frac{GM}{r}\right)^{1/2}, \quad \boldsymbol{v} \times \hat{\boldsymbol{x}} = -\left(\frac{GM}{r}\right)^{1/2} \boldsymbol{n},$$

$$\Omega = \frac{3}{2}\frac{(GM)^{3/2}}{r^{5/2}}\boldsymbol{n} - \frac{GI}{c^2 r^3}(\underline{1} - 3\hat{\boldsymbol{x}}\hat{\boldsymbol{x}}) \cdot \boldsymbol{\omega}, \quad \langle \hat{\boldsymbol{x}}\hat{\boldsymbol{x}} \rangle = \frac{1}{2}(\underline{1} - \boldsymbol{n}\boldsymbol{n}),$$

so that

$$\langle \Omega \rangle = \frac{3(GM)^{3/2}}{2\ c^2 r^{5/2}}\boldsymbol{n} + \frac{GI}{2c^2 r^3}(\underline{1} - 3\boldsymbol{n}\boldsymbol{n}) \cdot \boldsymbol{\omega}$$

In polar orbit we have $\boldsymbol{n} \cdot \boldsymbol{\omega} = 0$ and this reduces to

$$\langle \Omega \rangle = \langle \Omega^{\mathrm{orb}} \rangle + \langle \Omega^{\mathrm{LT}} \rangle$$

where

$$\langle \Omega^{\mathrm{orb}} \rangle = \frac{3(GM)^{3/2}}{2\ c^2 r^{5/2}}\boldsymbol{n}, \quad \langle \Omega^{\mathrm{LT}} \rangle = \frac{GI}{2c^2 r^3}\boldsymbol{\omega}$$

Now,

$$\begin{aligned} GM &= 6.67 \times 10^{-11} \times 5.98 \times 10^{24} = 3.99 \times 10^{14}\,\mathrm{m}^3\,\mathrm{s}^{-2}, \\ (GM)^{3/2} &= 7.97 \times 10^{21}\,\mathrm{m}^{9/2}\,\mathrm{s}^{-3}, \end{aligned}$$

$$r = 6.38 \times 10^6 + 0.50 \times 10^6 = 6.88 \times 10^6\,\mathrm{m},$$
$$r^{5/2} = 1.24 \times 10^{17}\,\mathrm{m}^{5/2},$$
$$I = 0.334MR^2 = 0.334 \times 5.98 \times 10^{24} \times (6.38)^2 \times 10^{12} = 8.13 \times 10^{37}\,\mathrm{kg\ m}^2.$$

Therefore,

$$\begin{aligned}|\langle \Omega^{\mathrm{orb}} \rangle| &= \frac{3}{2}\frac{7.97 \times 10^{21}}{9 \times 10^{16} \times 1.24 \times 10^{17}} = 1.07 \times 10^{-12}\,\mathrm{rad\ s}^{-1} \\ &= \frac{1.07 \times 10^{-12}}{1.54 \times 10^{-13}} = 6.96\,\mathrm{arcsec\ yr}^{-1}\end{aligned}$$

and

$$\begin{aligned}|\langle \Omega^{\mathrm{LT}} \rangle| &= \frac{6.67 \times 10^{-11} \times 8.13 \times 10^{37}}{2 \times 9 \times 10^{16} \times (6.88)^3 \times 10^{18}} \times 7.30 \times 10^{-5} \\ &= 6.76 \times 10^{-15}\,\mathrm{rad\ s}^{-1} \\ &= \frac{5.95 \times 10^{-15}}{1.54 \times 10^{-13}} = 0.439\,\mathrm{arcsec\ yr}^{-1}\end{aligned}$$

Chapter 14
The Electromagnetic Field

The action functional for the free electromagnetic field in a curved spacetime is a straightforward generalization of that in canonical coordinates in a flat spacetime:

$$S_{\mathrm{E}} = -\frac{1}{16\pi}\int g^{1/2} F_{\mu\nu}F^{\mu\nu}\,\mathrm{d}^4x,$$

$$\begin{aligned} F_{\mu\nu} &\equiv A_{\nu;\mu} - A_{\mu;\nu} = A_{\nu,\mu} - \Gamma^{\sigma}_{\nu\mu}A_{\sigma} - A_{\mu,\nu} + \Gamma^{\sigma}_{\mu\nu}A_{\sigma} \\ &= A_{\nu,\mu} - A_{\mu,\nu}. \end{aligned}$$

In a locally canonical coordinate system, one may make the usual identifications

$$(A_{\mu}) = (-\phi, \boldsymbol{A}),$$

$$\begin{aligned} E_i &= F_{i0} = A_{0,i} - A_{i,0}, \quad \boldsymbol{E} = -\nabla\phi - \boldsymbol{A}_{,0}, \\ H_i &= \frac{1}{2}\varepsilon_{ijk}F_{jk} = \varepsilon_{ijk}A_{k,j}, \quad \boldsymbol{H} = \nabla\times\boldsymbol{A}, \end{aligned}$$

where ϕ is the scalar potential, $\boldsymbol{A}$ is the 3-vector potential, $\boldsymbol{E}$ is the electric field vector, and $\boldsymbol{H}$ is the magnetic field vector. We note that the electromagnetic field tensor satisfies the covariant generalization of a familiar identity:

$$\begin{aligned} F_{\mu\nu;\sigma} + F_{\nu\sigma;\mu} + F_{\sigma;\mu\nu} &= A_{\nu;\mu\sigma} - A_{\mu;\nu\sigma} + A_{\sigma;\nu\mu} - A_{\nu;\sigma\mu} + A_{\mu;\sigma\nu} - A_{\sigma;\mu\nu} \\ &= A_{\tau}(R^{\tau}_{\ \nu\mu\sigma} + R^{\tau}_{\ \mu\sigma\nu} + R^{\tau}_{\ \sigma\nu\mu}) = 0. \end{aligned}$$

The electromagnetic field tensor, and hence the action functional itself, is invariant under gauge transformations:

$$\overline{A}_{\mu} = A_{\mu} + \xi_{;\mu} = A_{\mu} + \xi_{,\mu} \quad (\xi \text{ is a scalar}).$$

Denoting by δS_{E} the change induced in S_{E} by an infinitesimal gauge transformation generated by an infinitesimal gauge parameter $\delta\xi$, we may write

B. DeWitt, *Bryce DeWitt's Lectures on Gravitation*,
Edited by Steven M. Christensen, Lecture Notes in Physics, 826,
DOI: 10.1007/978-3-540-36911-0_14,

$$
\begin{aligned}
0 \equiv \delta S_{\mathrm{E}} &\equiv \int \frac{\delta S_{\mathrm{E}}}{\delta A_\mu} \delta A_\mu \mathrm{d}^4 x \equiv \int \frac{\delta S_{\mathrm{E}}}{\delta A_\mu} \delta \xi_{;\mu} \mathrm{d}^4 x \\
&\equiv - \int \left(\frac{\delta S_{\mathrm{E}}}{\delta A_\mu} \right)_{;\mu} \delta \xi \mathrm{d}^4 x.
\end{aligned}
$$

Because of the arbitrariness of $\delta \xi$, this implies

$$
\left(\frac{\delta S_{\mathrm{E}}}{\delta A_\mu} \right)_{;\mu} \equiv 0.
$$

But under a general variation δA_μ, we have

$$
\begin{aligned}
\delta S_{\mathrm{E}} &= -\frac{1}{8\pi} \int g^{1/2} F^{\mu\nu} (\delta A_{\nu;\mu} - \delta A_{\mu;\nu}) \mathrm{d}^4 x \\
&= -\frac{1}{4\pi} \int g^{1/2} F^{\mu\nu}_{;\nu} \delta A_\mu \mathrm{d}^4 x,
\end{aligned}
$$

whence

$$
\begin{aligned}
\frac{\delta S_{\mathrm{E}}}{\delta A_\mu} &\equiv -\frac{1}{4\pi} g^{1/2} F^{\mu\nu}_{;\nu} \\
&\equiv -\frac{1}{4\pi} (g^{1/2} F^{\mu\nu})_{;\nu} \\
&\equiv -\frac{1}{4\pi} \left[(g^{1/2} F^{\mu\nu})_{,\nu} + \Gamma^{\mu}_{\sigma\nu} g^{1/2} F^{\sigma\nu} \right] \\
&\equiv -\frac{1}{4\pi} (g^{1/2} F^{\mu\nu})_{,\nu}.
\end{aligned}
$$

The above differential identity may now be verified directly. Because $\delta S_{\mathrm{E}} / \delta A_\mu$ is a vector density, we have

$$
\left(\frac{\delta S_{\mathrm{E}}}{\delta A_\mu} \right)_{;\mu} \equiv \left(\frac{\delta S_{\mathrm{E}}}{\delta A_\mu} \right)_{,\mu} \equiv -\frac{1}{4\pi} (g^{1/2} F^{\mu\nu})_{,\nu\mu} \equiv 0.
$$

Alternatively,

$$
\begin{aligned}
F^{\mu\nu}_{;\nu\mu} &\equiv \frac{1}{2} (F^{\mu\nu}_{;\nu\mu} - F^{\mu\nu}_{;\mu\nu}) \\
&= \frac{1}{2} (F^{\sigma\nu} R^{\mu}_{\sigma\nu\mu} + F^{\mu\sigma} R^{\nu}_{\sigma\nu\mu}) \\
&= \frac{1}{2} (F^{\sigma\nu} R_{\sigma\nu} - F^{\mu\sigma} R_{\sigma\mu}) \equiv 0.
\end{aligned}
$$

The electromagnetic energy–momentum–stress density is readily computed by first writing S_{E} in the form

$$
S_{\mathrm{E}} \equiv -\frac{1}{16\pi} \int g^{1/2} g^{\mu\sigma} g^{\nu\tau} F_{\mu\nu} F_{\sigma\tau} \mathrm{d}^4 x,
$$

and remembering that $F_{\mu\nu}$ is metric independent when its indices are downstairs. Using $\delta g^{\mu\nu} = -g^{\mu\sigma} g^{\nu\tau} \delta g_{\sigma\tau}$, we find

$$T_{\mathrm{E}}^{\mu\nu} \equiv 2\frac{\delta S_{\mathrm{E}}}{\delta g_{\mu\nu}} \equiv \frac{1}{4\pi} g^{1/2} \left(F^{\mu}_{\ \sigma} F^{\nu\sigma} - \frac{1}{4} g^{\mu\nu} F_{\sigma\tau} F^{\sigma\tau} \right).$$

The combined action functional for the electromagnetic and gravitational fields in the presence of charged matter is given by

$$S \equiv S_{\mathrm{G}}[g_{\mu\nu}] + S_{\mathrm{E}}[g_{\mu\nu}, A_{\mu}] + S_{\mathrm{M}}[g_{\mu\nu}, A_{\mu}, \Phi^{A}],$$

where S_{M} is the action functional for the material system and the Φ^A are the matter dynamical variables. (Here we are treating the electromagnetic field separately from the material system.) Note that S_{M} is a functional of the $g_{\mu\nu}$ and the A_μ as well as of the Φ^A. In quantum field theory the Φ^A are the components of a complex tensor (or spinor) field φ whose coupling to the electromagnetic field is determined by the *principle of minimal coupling* (akin to the strong equivalence principle): Every ordinary derivative $\varphi_{,\mu}$ appearing in the matter field Lagrangian when no electromagnetic field is present, is replaced by the combination $\varphi_{,\mu} - \mathrm{i}\, eA_\mu\varphi$, where e is the unit charge of the matter field quanta (typically, the charge on the electron). Because ordinary derivatives become covariant derivatives when a gravitational field is present, this means that in general relativity matter field derivatives occur only in the combination

$$\mathrm{D}_\mu \varphi, \quad \text{where } \mathrm{D}_\mu = \frac{\partial}{\partial x^\mu} + G^{\nu}_{\sigma} \Gamma^{\sigma}_{\ \nu\mu} - \mathrm{i} e A_\mu.$$

D_μ may be regarded as a kind of generalization of the covariant derivative. It is invariant under electromagnetic gauge transformations provided the matter field is understood as gauge transforming according to the law

$$\overline{\varphi} = \mathrm{e}^{\mathrm{i}e\xi} \varphi.$$

In the non-quantal description of bulk matter, the coupling to the electromagnetic field is usually described by an explicit coupling term in the matter action functional S_{M}. For example, the action functional for a point particle bearing a charge e becomes

$$S_{\mathrm{M}} \equiv -m \int (-\dot{z}^2)^{1/2} \mathrm{d}\lambda + e \int A_\alpha \dot{z}^\alpha \mathrm{d}\lambda,$$

where the new second term is the electromagnetic coupling term. It is always important to check that the coupling term is gauge invariant. For the point particle we have

$$\delta S_{\mathrm{M}} \equiv e \int \delta\xi_{;\alpha} \dot{z}^\alpha \mathrm{d}\lambda \equiv e \int \frac{\mathrm{d}}{\mathrm{d}\lambda} \delta\xi(z) \mathrm{d}\lambda \equiv 0,$$

in which it is assumed that $\delta\xi$ has compact support.

The dynamical equations for the combined gravitational–electromagnetic–charged matter system are

$$0 = \frac{\delta S}{\delta g_{\mu\nu}} \equiv \frac{1}{16\pi G}\left[-g^{1/2}\left(R^{\mu\nu} - \frac{1}{2}g^{\mu\nu}R\right) + 8\pi G(T_{\mathrm{E}}^{\mu\nu} + T_{\mathrm{M}}^{\mu\nu})\right],$$

$$0 = \frac{\delta S}{\delta A_\mu} \equiv \frac{1}{4\pi}(-g^{1/2}F^{\mu\nu}_{;\nu} + 4\pi j^\mu) \quad \text{(Maxwell's equations)},$$

$$0 = \frac{\delta S}{\delta \Phi^A} \equiv \frac{\delta S_{\mathrm{M}}}{\delta \Phi^A},$$

where

$$T_{\mathrm{M}}^{\mu\nu} \equiv 2\frac{\delta S_{\mathrm{M}}}{\delta g_{\mu\nu}}, \quad j^\mu \equiv \frac{\delta S_{\mathrm{M}}}{\delta A_\mu},$$

and j^μ is the *charge current density*. Because of the dynamical equations themselves, we have the following covariant divergence laws:

$$(T_{\mathrm{E}}^{\mu\nu} + T_{\mathrm{M}}^{\mu\nu})_{;\nu} = 0, \quad j^\mu_{;\mu} = 0.$$

The former, which assures consistency with the contracted Bianchi identity, has already been proved (see Chap. 9). The latter, which assures consistency with the identity $F^{\mu\nu}_{;\nu\mu} \equiv 0$, can be proved analogously. Let $\delta\Phi^A$ be the change (if any) induced in the Φ^A by an infinitesimal gauge transformation $\delta\xi$. Then because of the gauge invariance of S_{M}, we have

$$0 \equiv \delta S_{\mathrm{M}} \equiv \int \frac{\delta S_{\mathrm{M}}}{\delta A_\mu}\delta\xi_{;\mu}\mathrm{d}^4x + \frac{\delta S_{\mathrm{M}}}{\delta \Phi^A}\delta\Phi^A = -\int j^\mu_{;\mu}\delta\xi\mathrm{d}^4x.$$

The arbitrariness of $\delta\xi$ leads at once to $j^\mu_{;\mu} = 0$. If the Φ^A are unaffected by gauge transformations then the covariant divergence of j^μ vanishes ibecause dentically whether the dynamical equations are satisfied or not. The point particle provides an example of this. In this case we have

$$j^\mu = e\int \delta^\mu_\alpha \dot{z}^\alpha \mathrm{d}\lambda,$$

and

$$j^\mu_{;\mu} = e\int \delta^\mu_{\alpha;\mu}\dot{z}^\alpha\mathrm{d}\lambda = -e\int \delta_{;\alpha}(x,z)\dot{z}^\alpha\mathrm{d}\lambda$$
$$= -e\int \frac{\mathrm{d}}{\mathrm{d}\lambda}\delta(x,z)\mathrm{d}\lambda = 0,$$

the final integral vanishing because the world line of the particle is constrained to be timelike, and hence $z(\lambda)$ ultimately becomes infinitely remote from any spacetime point x.

Because the covariant divergence of a vector density is the same as the ordinary divergence, the condition $j^{\mu}_{;\mu} = 0$, unlike $(T^{\mu\nu}_{\mathrm{E}} + T^{\mu\nu}_{\mathrm{M}})_{;\nu} = 0$, is a true conservation law. The conserved quantity in this case is the *total charge* Q:

$$Q = \int_{\Sigma} j^{\mu} \mathrm{d}\Sigma_{\mu}.$$

In the case of the point particle, we may evaluate this integral by choosing Σ to be a hypersurface $x^0 =$ constant:

$$\begin{aligned} Q &= e \int_{\Sigma} \mathrm{d}\Sigma_{\mu} \int \mathrm{d}\lambda \delta^{\mu}_{\alpha} \dot{z}^{\alpha} = e \int \mathrm{d}^3 \boldsymbol{x} \int \mathrm{d}\lambda \delta(x, z) \dot{z}^0 \\ &= e \int \mathrm{d}^3 \boldsymbol{x} \int \mathrm{d}z^0 \delta(x, z) = e. \end{aligned}$$

Problem 39 In the case of the point particle, show that the divergence law

$$(T^{\mu\nu}_{\mathrm{E}} + T^{\mu\nu}_{\mathrm{M}})_{;\nu} = 0,$$

when combined with Maxwell's equations $g^{1/2} F^{\mu\nu}_{;\nu} = 4\pi\, j^{\mu}$, implies the equations of motion $m\ddot{z}^{\alpha} = eF^{\alpha}_{\beta}\dot{z}^{\beta}$.

Solution 39 We have

$$\begin{aligned} 0 &= (T^{\mu\nu}_{\mathrm{E}} + T^{\mu\nu}_{\mathrm{M}})_{;\nu} \\ &= \left[\frac{1}{4\pi} g^{1/2} \left(F^{\mu}_{\sigma} F^{\nu\sigma} - \frac{1}{4} g^{\mu\nu} F_{\sigma\tau} F^{\sigma\tau} \right) + m \int \delta^{\mu\nu}_{\alpha\beta} \dot{z}^{\alpha} \dot{z}^{\beta} \mathrm{d}\tau \right]_{;\nu} \\ &= \frac{1}{8\pi} g^{1/2} (F^{\mu}_{\sigma;\nu} + F^{\mu}_{\nu\mu;\sigma} + F^{\mu}_{\sigma\nu;}) F^{\nu\sigma} + \frac{1}{4\pi} g^{1/2} F^{\mu}_{\sigma} F^{\nu\sigma}_{;\nu} \\ &\quad + m \int \delta^{\mu\nu}_{\alpha\beta;\nu} \dot{z}^{\alpha} \dot{z}^{\beta} \mathrm{d}\tau \\ &= -F^{\mu}_{\sigma} j^{\sigma} - m \int \delta^{\mu}_{\alpha;\beta} \dot{z}^{\alpha} \dot{z}^{\beta} \mathrm{d}\tau \\ &= -eF^{\mu}_{\nu} \int \delta^{\nu}_{\alpha} \dot{z}^{\alpha} \mathrm{d}\tau - m \int \dot{z}^{\alpha} \frac{\mathrm{D}}{\mathrm{D}\tau} \delta^{\mu}_{\alpha} \mathrm{d}\tau \\ &= \int \delta^{\mu}_{\alpha} (-eF^{\alpha}_{\beta} \dot{z}^{\beta} + m\ddot{z}^{\alpha}) \mathrm{d}\tau, \end{aligned}$$

whence

$$m\ddot{z}^{\alpha} = eF^{\alpha}_{\beta}\dot{z}^{\beta}.$$

14.1 Electromagnetic Waves

In studying special solutions of Maxwell's equations it is convenient to impose the so-called *Lorentz supplementary condition* on the vector potential:

$$A^{\mu}_{;\mu} = 0.$$

If this condition does not already hold it can be made to hold by carrying out the gauge transformation

$$\overline{A}_{\mu} = A_{\mu} + \xi_{;\mu},$$

where the gauge parameter ξ is a solution of the inhomogeneous curved spacetime wave equation

$$\xi^{;\mu}_{;\mu} = -A^{\mu}_{;\mu}.$$

Solutions of this equation can be found with the aid of appropriate biscalar Green's functions. From now on we shall assume the gauge transformation already to have been carried out and drop the bar over the A_{μ}.

In this section we shall study solutions of Maxwell's equations in the absence of charged matter ($S_{\mathrm{M}} = 0$). When the supplementary condition holds these equations take the form

$$\begin{aligned} 0 = -F^{\mu\nu}_{;\nu} &= -A^{\nu\mu}_{;\nu} + A^{\mu\nu}_{;\nu} \\ &= A^{\mu\nu}_{;\nu} - A^{\sigma}R^{\nu\mu}_{\sigma\nu} = A^{\mu\nu}_{;\nu} - R^{\mu}_{\nu}A^{\nu}. \end{aligned}$$

We seek a solution of these equations in the so-called *eikonal approximation*. This approximation is based on an expansion of the form

$$A_{\mu} = \left(\overset{0}{A}_{\mu} + \overset{2}{A}_{\mu} + \cdots\right)\cos\phi + \left(\overset{1}{A}_{\mu} + \overset{3}{A}_{\mu} + \cdots\right)\sin\phi.$$

The solution is assumed to be that of a locally plane monochromatic wave having a propagation vector k_{μ} at any point equal to the gradient of the so-called *eikonal function* ϕ at that point:

$$k_{\mu} = \phi_{;\mu}.$$

The coefficients $\overset{n}{A}_{\mu}$ are assumed to vary slowly compared to ϕ, and their magnitudes are assumed to decrease in such a way that none of the products $k_{\mu}\overset{n}{A}_{\nu}$ are bigger in order of magnitude than the biggest of the derivatives $\overset{n-1}{A}_{\mu;\nu}$. Moreover, the components of the curvature tensor are assumed to be of the second order in smallness compared to the products $k_{\mu}k_{\nu}$.

Substituting the expansion into the supplementary condition and into Maxwell's equations and setting terms of like order independently equal to zero, we obtain the following equations:

$$-\overset{0}{A}{}^{\mu}k_{\mu}=0,$$

$$\overset{1}{A}{}^{\mu}k_{\mu}+\overset{0}{A}{}^{\mu}_{;\mu}=0,$$

$$-\overset{2}{A}{}^{\mu}k_{\mu}+\overset{1}{A}{}^{\mu}_{;\mu}=0,$$

$$\overset{3}{A}{}^{\mu}k_{\mu}+\overset{2}{A}{}^{\mu}_{;\mu}=0,$$

and so on, together with

$$-\overset{0}{A}{}^{\mu}k^2=0,$$

$$-\overset{1}{A}{}^{\mu}k^2-2\overset{0}{A}{}^{\mu}_{;\nu}k^{\nu}-\overset{0}{A}{}^{\mu}k^{\nu}_{;\nu}=0,$$

$$-\overset{2}{A}{}^{\mu}k^2+2\overset{1}{A}{}^{\mu}_{;\nu}k^{\nu}+\overset{1}{A}{}^{\mu}k^{\nu}_{;\nu}+\overset{0}{A}{}^{\mu\nu}_{;\nu}-R^{\mu}_{\nu}\overset{0}{A}{}^{\nu}=0,$$

$$-\overset{3}{A}{}^{\mu}k^2-2\overset{2}{A}{}^{\mu}_{;\nu}k^{\nu}-\overset{2}{A}{}^{\mu}k^{\nu}_{;\nu}+\overset{1}{A}{}^{\mu\nu}_{;\nu}-R^{\mu}_{\nu}\overset{1}{A}{}^{\nu}=0,$$

and so on. We have immediately

$$k^2=0,$$

and hence,

$$k^{\nu}k_{\nu;\mu}=0,$$

which, together with the solution

$$k_{\mu;\nu}=k_{\nu;\mu}$$

that follows from k_{μ} being the gradient of a scalar, implies

$$k_{\mu;\nu}k^{\nu}=0.$$

That is, the lines orthogonal to the wave fronts $\phi=$ constant are null geodesics. These lines are the rays of geometrical optics, and we see that light (electromagnetic radiation) really does travel with the velocity of light!

If we now define

$$a^2\equiv\overset{0}{A}_{\mu}\overset{0}{A}{}^{\mu},\quad e_{\mu}\equiv\overset{0}{f}_{\mu},\quad \overset{n}{f}_{\mu}\equiv a^{-1}\overset{n}{A}_{\mu},$$

we then have

$$e^2=1,\quad k\cdot e=0,\quad k\cdot\overset{n}{f}=(-1)^n a^{-1}(a\overset{n-1}{f}{}^{\mu})_{;\mu},\quad n\geq 1,$$

$$(a^2k^{\mu})_{;\mu}=0,\quad 2a_{;\mu}k^{\mu}+ak^{\mu}_{;\mu}=0,\quad e^{\mu}_{;\nu}k^{\nu}=0,$$

$$\overset{n}{f}{}^{\mu}_{\nu}k^{\nu} = \frac{1}{2}(-1)^{n}\left[a^{-1}(a\overset{n-1}{f}{}^{\mu})_{;\nu}^{\nu} - R^{\mu}_{\nu}\overset{n-1}{f}{}^{\nu}\right], \quad n \geq 1.$$

The eikonal approximation consists in writing

$$A_{\mu} \approx a\, e_{\mu} \cos\phi,$$

and in recognizing a as the amplitude and e_{μ} as the polarization vector of the wave. The equation $(a^{2}k^{\mu})_{;\mu} = 0$ is a conservation law for the wave, and the equation $e^{\mu}_{;\nu}k^{\nu} = 0$ says that the wave is linearly polarized with the polarization vector being transported in a parallel fashion along the null rays orthogonal to the wave fronts. The coefficient vectors $\overset{n}{f}{}^{\mu}$ represent small corrections to the eikonal approximation. They may all (including e^{μ}) be chosen tangent to some initial spacelike hypersurface, with components satisfying the initial constraints

$$k \cdot \overset{n}{f} = (-1)^{n}a^{-1}(a\overset{n-1}{f}{}^{\mu})_{;\mu}.$$

Their values elsewhere may then be obtained by integrating each of the equations for $\overset{n}{f}{}^{\mu}_{;\nu}k^{\nu}$ (in succession) along the null rays. The integration automatically preserves the constraint equations, as may be verified by the following computation:

$$\begin{aligned}
0 &= (k\cdot\overset{n}{A})_{;\nu}k^{\nu} - (k\cdot\overset{n}{A})_{;\nu}k^{\nu}\\
&= (-1)^{n}\overset{n-1}{A}{}^{\mu}{}_{;\mu\nu}k^{\nu} + \frac{1}{2}k\cdot\overset{n}{A}k^{\nu}_{;\nu} - \frac{1}{2}(-1)^{n}k_{\mu}\left(\overset{n-1}{A}{}^{\mu\,\nu}_{;\nu} - R^{\mu}_{\nu}\overset{n-1}{A}{}^{\nu}\right)\\
&= (-1)^{n}\overset{n-1}{A}{}^{\mu}{}_{;\nu\mu}k^{\nu} + (-1)^{n}\overset{n-1}{A}{}^{\sigma}R^{\mu}_{\sigma\mu\nu}k^{\nu} + \frac{1}{2}(-1)^{n}\overset{n-1}{A}{}^{\mu}{}_{;\mu}k^{\nu}_{;\nu}\\
&\quad - \frac{1}{2}(-1)^{n}\left(k_{\mu}\overset{n-1}{A}{}^{\mu}{}_{;\nu}\right)^{\nu}_{;} + \frac{1}{2}(-1)^{n}k^{\nu}_{\mu;}\overset{n-1}{A}{}^{\mu}{}_{;\nu} + \frac{1}{2}(-1)^{n}k_{\mu}R^{\mu}_{\nu}\overset{n-1}{A}{}^{\nu}\\
&= (-1)^{n}\left(\overset{n-1}{A}{}^{\mu}{}_{;\nu}k^{\nu}\right)_{;\mu} - \frac{1}{2}(-1)^{n}\overset{n-1}{A}{}^{\mu}{}_{;\nu}k^{\nu}_{;\mu} - \frac{1}{2}(-1)^{n}k_{\mu}R^{\mu}_{\nu}\overset{n-1}{A}{}^{\nu}\\
&\quad + \frac{1}{2}(-1)^{n}\overset{n-1}{A}{}^{\mu}{}_{;\mu}k^{\nu}_{;\nu} - \frac{1}{2}(-1)^{n}\left(k\cdot\overset{n-1}{A}\right)^{\nu}_{;\nu} + \frac{1}{2}(-1)^{n}\left(k_{\nu;\mu}\overset{n-1}{A}{}^{\mu}\right)^{\nu}_{;}\\
&= -\frac{1}{2}(-1)^{n}\left(\overset{n-1}{A}{}^{\mu}k^{\nu}_{;\nu}\right)_{;\mu} - \frac{1}{2}\left(\overset{n-2}{A}{}^{\mu\,\nu}_{;\nu} - R^{\mu}_{\nu}\overset{n-2}{A}{}^{\nu}\right)_{;\mu} - \frac{1}{2}(-1)^{n}k_{\mu}R^{\mu}_{\nu}\overset{n-1}{A}{}^{\nu}\\
&\quad + \frac{1}{2}(-1)^{n}\overset{n-1}{A}{}^{\mu}{}_{;\mu}k^{\nu}_{;\nu} + \frac{1}{2}\overset{n-2}{A}{}^{\mu\ \nu}_{;\mu\nu} + \frac{1}{2}(-1)^{n}k^{\nu}_{\nu;\mu}\overset{n-1}{A}{}^{\mu}\\
&= -\frac{1}{2}(-1)^{n}\overset{n-1}{A}{}^{\mu}k^{\sigma}R^{\nu}_{\sigma\nu\mu} - \frac{1}{2}\overset{n-2}{A}{}^{\mu\ \nu}_{;\nu\mu} - \frac{1}{2}\overset{n-2}{A}{}^{\sigma}{}_{;\nu}R^{\mu\nu}_{\sigma\mu} - \frac{1}{2}\overset{n-2}{A}{}^{\mu}{}_{;\sigma}R^{\sigma\nu}_{\nu\mu}\\
&\quad + \frac{1}{2}\left(R^{\mu}_{\nu}\overset{n-2}{A}{}^{\nu}\right)_{;\mu} - \frac{1}{2}(-1)^{n}k_{\mu}R^{\mu}_{\nu}\overset{n-1}{A}{}^{\nu} + \frac{1}{2}\overset{n-2}{A}{}^{\mu}{}_{;\nu\mu}{}^{\nu} + \frac{1}{2}\left(\overset{n-2}{A}{}^{\sigma}R^{\mu}_{\sigma\mu\nu}\right)^{\nu}_{;}\\
&= 0.
\end{aligned}$$

The coefficients $\overset{n}{f}{}^{\mu}$ in the eikonal expansion for a given wave are not uniquely determined, for we may always carry out a gauge transformation of the form

$$\overline{A}_{\mu} = A_{\mu} + \xi_{;\mu},$$

where the gauge parameter ξ satisfies the homogeneous wave equation

$$\xi^{\mu}_{;\mu} = 0.$$

If we write an eikonal expansion for ξ,

$$\xi = a(\overset{0}{a} + \overset{2}{a} + \cdots)\sin\phi - a(\overset{1}{a} + \overset{3}{a} + \cdots)\cos\phi,$$

we find that the coefficients $\overset{n}{a}$ must satisfy

$$\overset{0}{a}_{;\mu}k^{\mu} = 0, \quad \overset{n}{a}_{;\mu}k^{\mu} = \frac{1}{2}(-1)^{n}a^{-1}\left(a\overset{n-1}{a}\right)^{\mu}_{;\mu}, \quad n \geq 1,$$

and that the coefficients $\overset{n}{f}_{\mu}$ suffer the gauge transformations

$$\overline{e}_{\mu} = e_{\mu} + \overset{0}{a}k_{\mu}, \quad \overline{\overset{n}{f}}_{\mu} = \overset{n}{f}_{\mu} + \overset{n}{a}k_{\mu} - (-1)^{n}a^{-1}\left(a\overset{n-1}{a}\right)_{;\mu}, \quad n \geq 1.$$

These transformations of course leave the relations satisfied by the $\overset{n}{f}{}^{\mu}$ unchanged:

$$\overline{e}^{2} = 1, \quad k\cdot\overline{e} = 0,$$

$$\begin{aligned}
k\cdot\overline{\overset{n}{f}} - (-1)^{n}a^{-1}\left(a\overline{\overset{n-1}{f}{}^{\mu}}\right)_{;\mu} &= k\cdot\overset{n}{f} - (-1)^{n}a^{-1}\left(a\overset{n-1}{a}\right)_{;\mu}k^{\mu} - (-1)^{n}a^{-1}\left(a\overset{n-1}{f}{}^{\mu}\right)_{;\mu} \\
&= -(-1)^{n}a^{-1}\left[a\overset{n-1}{a}k^{\mu} - (-1)^{n-1}\left(a\overset{n-2}{a}\right)^{\mu}_{;}\right]_{;\mu} \\
&= -(-1)^{n}\left[2\overset{n-1}{a}_{;\mu}k^{\mu} - (-1)^{n-1}a^{-1}\left(a\overset{n-2}{a}\right)^{\mu}_{;\mu}\right] = 0,
\end{aligned}$$

$$\begin{aligned}
&\overline{\overset{n}{f}}{}^{\mu}_{;\nu}k^{\nu} - \frac{1}{2}(-1)^{n}\left[a^{-1}\left(a\overline{\overset{n-1}{f}{}^{\mu}}\right)^{\nu}_{;\nu} - R^{\mu}_{\nu}\overline{\overset{n-1}{f}{}^{\nu}}\right]\overset{n}{f}{}^{\mu}_{;\nu}k^{\nu} + \left(\overset{n}{a}k^{\mu}\right)_{;\nu}k^{\nu} - (-1)^{n}\left[a^{-1}\left(a\overset{n-1}{a}\right)^{\mu}_{;}\right]_{;\nu}k^{\nu} \\
&\quad - \frac{1}{2}(-1)^{n}\left\{a^{-1}\left[a\overset{n-1}{f}{}^{\mu} + a\overset{n-1}{a}k^{\mu} - (-1)^{n-1}\left(a\overset{n-2}{a}\right)^{\mu}_{;}\right]^{\nu}_{;\nu}\right. \\
&\quad \left. - R^{\mu}_{\nu}\left[\overset{n-1}{f}{}^{\nu} + \overset{n-1}{a}k^{\nu} - (-1)^{n-1}a^{-1}\left(a\overset{n-2}{a}\right)^{\nu}_{;}\right]\right\}
\end{aligned}$$

$$
\begin{aligned}
&= \frac{1}{2}(-1)^n a^{-1}\left(a\overset{n-1}{a}\right)_{;\nu}^{\;\;\nu} k^\mu - \frac{1}{2}(-1)^n a^{-1}\left(a\overset{n-1}{a}\right)_{;}^{\;\mu} k^\nu_{;\nu} \\
&\quad - (-1)^n a^{-1}\left[\left(a\overset{n-1}{a}\right)_{;\nu} k^\nu\right]_{;}^{\;\mu} + (-1)^n a^{-1}\left(a\overset{n-1}{a}\right)_{;\nu} k^{\nu\mu}_{;} \\
&\quad - \frac{1}{2}(-1)^n a^{-1}\left(a\overset{n-1}{a}\right)_{;\nu}^{\;\;\nu} k^\mu - (-1)^n a^{-1}\left(a\overset{n-1}{a}\right)_{;\nu} k^{\mu\nu}_{;} \\
&\quad - \frac{1}{2}(-1)^n \overset{n-1}{a} k^{\mu\nu}_{;\nu} - \frac{1}{2} a^{-1}\left(a\overset{n-2}{a}\right)_{;\nu}^{\;\;\nu} \\
&\quad + \frac{1}{2}(-1)^n R^\mu_\nu \left[\overset{n-1}{a} k^\nu - (-1)^{n-1} a^{-1}\left(a\overset{n-2}{a}\right)_{;}^{\;\nu}\right] \\
&= -\frac{1}{2}(-1)^n a^{-1}\left(a\overset{n-1}{a}\right)_{;}^{\;\mu} k^\nu_{;\nu} + \frac{1}{2}(-1)^n a^{-1}\left(a\overset{n-1}{a} k^\nu_{;\nu}\right)_{;}^{\;\mu} \\
&\quad + \frac{1}{2} a^{-1}\left(a\overset{n-2}{a}\right)_{;\nu}^{\;\;\nu\mu} - \frac{1}{2}(-1)^n \overset{n-1}{a} k^{\mu\nu}_{\nu;} - \frac{1}{2} a^{-1}\left(a\overset{n-2}{a}\right)_{;\nu}^{\;\;\nu\mu} \\
&\quad - \frac{1}{2} a^{-1}\left(a\overset{n-2}{a}\right)_{;\sigma} R^{\sigma\mu\nu}_{\;\;\;\;\;\nu} + \frac{1}{2}(-1)^n R^\mu_\nu \left[\overset{n-1}{a} k^\nu - (-1)^{n-1} a^{-1}\left(a\overset{n-2}{a}\right)_{;}^{\;\nu}\right] \\
&= \frac{1}{2}(-1)^n \overset{n-1}{a} k^\sigma R^{\nu\mu}_{\sigma\nu} + \frac{1}{2}(-1)^n R^\mu_\nu \overset{n-1}{a} k^\nu = 0.
\end{aligned}
$$

Now suppose the wave sweeps over a test particle of charge e and mass m. Suppose further that the wave is very weak so that we may treat the disturbance $\delta\xi^\alpha$ in the particle coordinates, produced by the wave, as an infinitesimal. In the absence of the wave, the particle follows a geodesic:

$$
\frac{\mathrm{D}}{\mathrm{D}\lambda}\left[\left(-\dot{z}^2\right)^{-1/2}\dot{z}^\alpha\right] = 0.
$$

The equation for $\delta\xi^\alpha$ is therefore

$$
\begin{aligned}
eF^\alpha_\beta \dot{z}^\beta &= m\delta\frac{\mathrm{D}}{\mathrm{D}\lambda}\left[\left(-\dot{z}^2\right)^{-1/2}\dot{z}^\alpha\right] = m\bar{\delta}\frac{\mathrm{D}}{\mathrm{D}\lambda}\left[\left(-\dot{z}^2\right)^{-1/2}\dot{z}^\alpha\right] \\
&\underset{\lambda\to\tau}{\longrightarrow} m\ddot{\eta}^\alpha - mR^\alpha_{\beta\gamma\delta}u^\beta u^\gamma \eta^\delta,
\end{aligned}
$$

(see Chap. 5), where u^α $(= \dot{z}^\alpha)$ is the 4-velocity the particle would have in the absence of the wave and

$$
\eta^\alpha = P^\alpha_\beta \delta z^\beta.
$$

Now the Maxwell field that appears in this equation has the form

$$
\begin{aligned}
F_{\alpha\beta} &= A_{\beta;\alpha} - A_{\alpha;\beta} \\
&= -\left[a(k_\alpha e_\beta - k_\beta e_\alpha) + \cdots\right]\sin\phi \\
&\quad + \left[(ae_\beta)_{;\alpha} - (ae_\alpha)_{;\beta} + a(k_\alpha \overset{1}{f}_\beta - k_\beta \overset{1}{f}_\alpha) + \cdots\right]\cos\phi.
\end{aligned}
$$

It is convenient to use the gauge flexibility and choose the polarization vector so that it is perpendicular to the undisturbed 4-velocity:

$$e \cdot u = 0.$$

If this equation is not already satisfied we have only to carry out a gauge transformation with a gauge parameter ξ whose coefficient $\overset{0}{a}$, in its eikonal expansion, is given along the world line of the particle by

$$\overset{0}{a} = -\frac{e \cdot u}{k \cdot u}.$$

The equation for η^α then becomes

$$\ddot{\eta}^\alpha = \frac{ea}{m} e^\alpha (k \cdot u) \sin\phi + R^\alpha_{\beta\gamma\delta} u^\beta u^\gamma \eta^\delta + \cdots .$$

If we assume that there is no appreciable gravitational radiation present, so that the term in the curvature tensor varies slowly compared to $\sin\phi$, and if we remember that a, e^α, and $k{\cdot}u$ also vary slowly compared to $\sin\phi$, then we may write effectively

$$\ddot{\eta}^\alpha = \frac{ea}{m} \frac{\mathrm{D}}{\mathrm{D}\tau}(e^\alpha \cos\phi) = -\frac{e}{m} \frac{\mathrm{D}^2}{\mathrm{D}\tau^2}\left(\frac{a\, e^\alpha}{k \cdot u} \sin\phi\right),$$

which has the solution

$$\eta^\alpha = -\frac{ea}{m} \frac{e^\alpha}{k \cdot u} \sin\phi.$$

Along the (undisturbed) world line of the particle, $\sin\phi$ oscillates with angular frequency

$$\omega = -k \cdot u.$$

The electromagnetic wave is seen to cause the test particle to oscillate with this same frequency about its undisturbed position. The amplitude of the oscillation is $ea/m\omega$.

Waves of nearly monochromatic electromagnetic radiation need not be linearly polarized. It is possible to superpose two waves that have the same eikonal function but are 90° out of phase, obtaining

$$A_\mu = a_{\mathrm{I}}(e_{\mathrm{I}\mu} + \cdots) \cos\phi - a_{\mathrm{II}}(e_{\mathrm{II}\mu} + \cdots) \sin\phi.$$

If $a_{\mathrm{I}} = a_{\mathrm{II}}$ and the unit vectors e^μ_{I} and e^μ_{II} are orthogonal, forming a right handed system with $e^\mu_{\mathrm{III}} \equiv w^{-1} P^{\mu\nu} k_\nu$ (k^μ assumed pointing to the future), then this wave is circularly polarized with a right handed helicity. If $a_{\mathrm{I}} = -a_{\mathrm{II}}$, the wave is circularly polarized with a left handed helicity. The motion of a test particle in such a field is given by

$$\eta^\alpha = \frac{e}{m\omega}\left(a_{\rm I} e_{\rm I}^\alpha \sin\phi + a_{\rm II} e_{\rm II}^\alpha \cos\phi\right).$$

14.2 Energy, Momentum, and Angular Momentum in Electromagnetic Waves

The energy–momentum–stress density in an electromagnetic wave is easily obtained in the eikonal approximation. For a linearly polarized wave, we have

$$\begin{aligned} F_{\mu\nu} &= -a(k_\mu e_\nu - k_\nu e_\mu)\sin\phi, \\ F^\mu_\sigma F^{\nu\sigma} &= a^2(k^\mu e_\sigma - k_\sigma e^\mu)(k^\nu e^\sigma - k^\sigma e^\nu)\sin^2\phi \\ &= a^2 k^\mu k^\nu \sin^2\phi, \end{aligned}$$

and

$$F_{\sigma\tau}F^{\sigma\tau} = 0,$$

so that

$$T^{\mu\nu} = \frac{g^{1/2}a^2}{4\pi}k^\mu k^\nu \sin^2\phi.$$

This approximation rigorously satisfies the divergence law $T^{\mu\nu}_{;\nu} = 0$, as may immediately be verified by using the relations $k^2 = 0$, $k^\mu_{;\nu}k^\nu = 0$ and $(a^2k^\nu)_{;\nu} = 0$. It is also frequently useful to introduce the mean value of $T^{\mu\nu}$ averaged over a wavelength:

$$\langle T^{\mu\nu}\rangle = \frac{g^{1/2}a^2}{8\pi}k^\mu k^\nu.$$

It too satisfies the divergence law:

$$\langle T^{\mu\nu}\rangle_{;\nu} = 0.$$

In order to discuss angular momentum in the wave, it is necessary to include the next term in the eikonal expansion of $F_{\mu\nu}$. For simplicity we confine the analysis to the case of a wave packet in flat spacetime for which the propagation vector is constant, i.e., $k_{\mu;\nu} = 0$. In canonical coordinates the following equations then hold:

$$k^2 = 0, \quad k_{\mu,\nu} = 0, \quad a_{,\mu}k^\mu = 0,$$

$$\overset{1}{f}{}^\mu_{,\nu}k^\nu = -\frac{1}{2}a^{-1}(ae^\mu)^{\nu}_{,\nu}, \quad k\cdot \overset{1}{f} = -a^{-1}(ae^\mu)_{,\mu},$$

$$e^\mu_{,\nu}k^\nu = 0, \quad k\cdot e = 0.$$

If we specialize to the case of a circularly polarized wave for which the polarization vectors e_{I}^{μ} and e_{II}^{μ} are both orthogonal and constant, we then have

$$e_{\mathrm{I,II},\nu}^{\mu}=0,\quad k\cdot e_{\mathrm{I,II}}=0,\quad k\cdot \overset{1}{f}_{\mathrm{I,II}}=-a^{-1}a_{,\mu}e_{\mathrm{I,II}}^{\mu},$$

$$\overset{1}{f}{}_{\mathrm{I,II},\nu}^{\mu}k^{\nu}=-\frac{1}{2}a^{-1}a_{,\nu}^{\ \nu}e_{\mathrm{I,II}}^{\mu}.$$

If the helicity of the packet is right handed ($a_{\mathrm{I}}=a_{\mathrm{II}}=a$), the electromagnetic field tensor takes the form

$$\begin{aligned}F_{\mu\nu}=&-a\left[(k_{\mu}e_{\mathrm{I}\nu}-k_{\nu}e_{\mathrm{I}\mu})\sin\phi+(k_{\mu}e_{\mathrm{II}\nu}-k_{\nu}e_{\mathrm{II}\mu})\cos\phi\right]\\&+\left[a_{,\mu}e_{\mathrm{I}\nu}-a_{,\nu}e_{\mathrm{I}\mu}+a\left(k_{\mu}\overset{1}{f}_{\mathrm{I}\nu}-k_{\nu}\overset{1}{f}_{\mathrm{I}\mu}\right)\right]\cos\phi\\&-\left[a_{,\mu}e_{\mathrm{II}\nu}-a_{,\nu}e_{\mathrm{II}\mu}+a\left(k_{\mu}\overset{1}{f}_{\mathrm{II}\nu}-k_{\nu}\overset{1}{f}_{\mathrm{II}\mu}\right)\right]\sin\phi,\end{aligned}$$

and we have

$$\begin{aligned}F_{\sigma}^{\mu}F^{\nu\sigma}=&\,a^{2}(k^{\mu}e_{\mathrm{I}\sigma}-k_{\sigma}e_{\mathrm{I}}^{\mu})(k^{\nu}e_{\mathrm{I}}^{\sigma}-k^{\sigma}e_{\mathrm{I}}^{\nu})\sin^{2}\phi\\&+a^{2}(k^{\mu}e_{\mathrm{I}\sigma}-k_{\sigma}e_{\mathrm{I}}^{\mu})(k^{\nu}e_{\mathrm{II}}^{\sigma}-k^{\sigma}e_{\mathrm{II}}^{\nu})\sin\phi\cos\phi\\&+a^{2}(k^{\mu}e_{\mathrm{II}\sigma}-k_{\sigma}e_{\mathrm{II}}^{\mu})(k^{\nu}e_{\mathrm{I}}^{\sigma}-k^{\sigma}e_{\mathrm{I}}^{\nu})\sin\phi\cos\phi\\&+a^{2}(k^{\mu}e_{\mathrm{II}\sigma}-k_{\sigma}e_{\mathrm{II}}^{\mu})(k^{\nu}e_{\mathrm{II}}^{\sigma}-k^{\sigma}e_{\mathrm{II}}^{\nu})\cos^{2}\phi\\&-a(k^{\mu}e_{\mathrm{I}\sigma}-k_{\sigma}e_{\mathrm{I}}^{\mu})\left[a_{,}^{\nu}e_{\mathrm{I}}^{\sigma}-a_{,}^{\sigma}e_{\mathrm{I}}^{\nu}+a\left(k^{\nu}\overset{1}{f}{}_{\mathrm{I}}^{\sigma}-k^{\sigma}\overset{1}{f}{}_{\mathrm{I}}^{\nu}\right)\right]\sin\phi\cos\phi\\&+a(k^{\mu}e_{\mathrm{I}\sigma}-k_{\sigma}e_{\mathrm{I}}^{\mu})\left[a_{,}^{\nu}e_{\mathrm{II}}^{\sigma}-a_{,}^{\sigma}e_{\mathrm{II}}^{\nu}+a\left(k^{\nu}\overset{1}{f}{}_{\mathrm{II}}^{\sigma}-k^{\sigma}\overset{1}{f}{}_{\mathrm{II}}^{\nu}\right)\right]\sin^{2}\phi\\&-a(k^{\mu}e_{\mathrm{II}\sigma}-k_{\sigma}e_{\mathrm{II}}^{\mu})\left[a_{,}^{\nu}e_{\mathrm{I}}^{\sigma}-a_{,}^{\sigma}e_{\mathrm{I}}^{\nu}+a\left(k^{\nu}\overset{1}{f}{}_{\mathrm{I}}^{\sigma}-k^{\sigma}\overset{1}{f}{}_{\mathrm{I}}^{\nu}\right)\right]\cos^{2}\phi\\&+a(k^{\mu}e_{\mathrm{II}\sigma}-k_{\sigma}e_{\mathrm{II}}^{\mu})\left[a_{,}^{\nu}e_{\mathrm{II}}^{\sigma}-a_{,}^{\sigma}e_{\mathrm{II}}^{\nu}+a\left(k^{\nu}\overset{1}{f}{}_{\mathrm{II}}^{\sigma}-k^{\sigma}\overset{1}{f}{}_{\mathrm{II}}^{\nu}\right)\right]\sin\phi\cos\phi\\&+\begin{array}{l}\text{four more terms like the last four}\\\text{but with the indices }\mu\text{ and }\nu\text{ interchanged}\end{array}\\=&\,a^{2}k^{\mu}k^{\nu}\\&+a\Big[-k^{\mu}a_{,}^{\nu}-k^{\nu}a_{,}^{\mu}+k^{\mu}e_{\mathrm{I}}^{\nu}e_{\mathrm{I}}^{\sigma}a_{,\sigma}+k^{\nu}e_{\mathrm{I}}^{\mu}e_{\mathrm{I}}^{\sigma}a_{,\sigma}\\&-2k^{\mu}k^{\nu}(e_{\mathrm{I}}\cdot\overset{1}{f}_{\mathrm{I}})a-k^{\nu}e_{\mathrm{I}}^{\mu}e_{\mathrm{I}}^{\sigma}a_{,\sigma}-k^{\mu}e_{\mathrm{I}}^{\nu}e_{\mathrm{I}}^{\sigma}a_{,\sigma}\\&+k^{\mu}a_{,}^{\nu}+k^{\nu}a_{,}^{\mu}-k^{\mu}e_{\mathrm{II}}^{\nu}e_{\mathrm{II}}^{\sigma}a_{,\sigma}-k^{\nu}e_{\mathrm{II}}^{\mu}e_{\mathrm{II}}^{\sigma}a_{,\sigma}\\&+2k^{\mu}k^{\nu}(e_{\mathrm{II}}\cdot\overset{1}{f}_{\mathrm{II}})a+k^{\nu}e_{\mathrm{II}}^{\mu}e_{\mathrm{II}}^{\sigma}a_{,\sigma}+k^{\mu}e_{\mathrm{II}}^{\nu}e_{\mathrm{II}}^{\sigma}a_{,\sigma}\Big]\sin\phi\cos\phi\end{aligned}$$

$$+a\left[-k^\mu e_{\mathrm{II}}^\nu e_{\mathrm{I}}^\sigma a_{,\sigma} - k^\nu e_{\mathrm{II}}^\mu e_{\mathrm{I}}^\sigma a_{,\sigma} + 2k^\mu k^\nu (e_{\mathrm{I}}\cdot \overset{1}{f}_{\mathrm{II}})a + k^\nu e_{\mathrm{I}}^\mu e_{\mathrm{II}}^\sigma a_{,\sigma} + k^\mu e_{\mathrm{I}}^\nu e_{\mathrm{II}}^\sigma a_{,\sigma}\right]\sin^2\phi$$
$$+a\left[k^\mu e_{\mathrm{I}}^\nu e_{\mathrm{II}}^\sigma a_{,\sigma} + k^\nu e_{\mathrm{I}}^\mu e_{\mathrm{II}}^\sigma a_{,\sigma} - 2k^\mu k^\nu (e_{\mathrm{II}}\cdot \overset{1}{f}_{\mathrm{I}})a - k^\nu e_{\mathrm{II}}^\mu e_{\mathrm{I}}^\sigma a_{,\sigma} - k^\mu e_{\mathrm{II}}^\nu e_{\mathrm{I}}^\sigma a_{,\sigma}\right]\cos^2\phi.$$

In the present case, with constant k^μ and $e^\mu_{\mathrm{I,II}}$ in flat spacetime, it is possible to choose $\overset{1}{f}{}^\mu_{\mathrm{I,II}}$ so that

$$e_{\mathrm{I}}\cdot\overset{1}{f}_{\mathrm{I}} = e_{\mathrm{II}}\cdot\overset{1}{f}_{\mathrm{II}},\quad e_{\mathrm{I}}\cdot\overset{1}{f}_{\mathrm{II}} = e_{\mathrm{II}}\cdot\overset{1}{f}_{\mathrm{I}} = 0.$$

The above expansion then reduces to

$$F^\mu_{\ \sigma}F^{\nu\sigma} = a^2k^\mu k^\nu + ak^\mu(e_{\mathrm{I}}^\nu e_{\mathrm{II}}^\sigma - e_{\mathrm{II}}^\nu e_{\mathrm{I}}^\sigma)a_{,\sigma} + ak^\nu(e_{\mathrm{I}}^\mu e_{\mathrm{II}}^\sigma - e_{\mathrm{II}}^\mu e_{\mathrm{I}}^\sigma)a_{,\sigma}.$$

This yields

$$F_{\sigma\tau}F^{\sigma\tau} = 0,$$

and[1]

$$T^{\mu\nu} = \frac{a^2}{4\pi}k^\mu k^\nu + \frac{1}{8\pi}\left[k^\mu(e_{\mathrm{I}}^\nu e_{\mathrm{II}}^\sigma - e_{\mathrm{II}}^\nu e_{\mathrm{I}}^\sigma) + k^\nu(e_{\mathrm{I}}^\mu e_{\mathrm{II}}^\sigma - e_{\mathrm{II}}^\mu e_{\mathrm{I}}^\sigma)\right](a^2)_{,\sigma}. \tag{14.1}$$

To compute the energy, momentum, and angular momentum, it is convenient to choose a gauge in which $e_{\mathrm{I}}^0 = 0 = e_{\mathrm{II}}^0$ in the coordinate system one is working in. Then, choosing Σ to be a hypersurface $x^0 =$ constant, we find

$$\begin{aligned} P^\mu &= \int_\Sigma T^{\mu\nu}\mathrm{d}\Sigma_\nu \\ &= k^\mu\int\frac{\omega a^2}{4\pi}\mathrm{d}^3\boldsymbol{x} + \frac{\omega}{8\pi}(e_{\mathrm{I}}^\mu e_{\mathrm{II}}^i - e_{\mathrm{II}}^\mu e_{\mathrm{I}}^i)\int(a^2)_{,i}\mathrm{d}^3\boldsymbol{x} \\ &= N\hbar k^\mu, \end{aligned}$$

where $\omega = k^0$ and

$$N \equiv \frac{1}{4\pi\hbar}\int\omega a^2\mathrm{d}^3\boldsymbol{x} = \frac{1}{4\pi\hbar}\int_\Sigma a^2k^\mu\mathrm{d}\Sigma_\mu.$$

[1] Note that this energy–momentum–stress density satisfies the conservation law

$$T^{\mu\nu}_{\ \ ,\nu} = \frac{1}{8\pi}k^\mu(e_{\mathrm{I}}^\nu e_{\mathrm{II}}^\sigma - e_{\mathrm{II}}^\nu e_{\mathrm{I}}^\sigma)(a^2)_{,\nu\sigma} = 0.$$

N is the number of coherent photons out of which, in the quantum theory, the electromagnetic wave is built. This number is conserved in the eikonal approximation because of the divergence law $(a^2 k^\mu)_{;\mu} = 0$.

The angular momentum tensor is computed in a similar fashion:

$$
\begin{aligned}
J^{\mu\nu} &= \int_\Sigma (x^\mu T^{\nu\sigma} - x^\nu T^{\mu\sigma}) \mathrm{d}\Sigma_\sigma \\
&= \int \frac{\omega a^2}{4\pi} (x^\mu k^\nu - x^\nu k^\mu) \mathrm{d}^3\boldsymbol{x} \\
&\quad + \int \frac{\omega}{8\pi} \left[x^\mu (e_{\mathrm{I}}^\nu e_{\mathrm{II}}^i - e_{\mathrm{II}}^\nu e_{\mathrm{I}}^i) - x^\nu (e_{\mathrm{I}}^\mu e_{\mathrm{II}}^i - e_{\mathrm{II}}^\mu e_{\mathrm{I}}^i) \right] (a^2)_{,i} \mathrm{d}^3\boldsymbol{x} \\
&= X_{\mathrm{E}}^\mu P^\nu - X_{\mathrm{E}}^\nu P^\mu + N\hbar (e_{\mathrm{I}}^\mu e_{\mathrm{II}}^\nu - e_{\mathrm{II}}^\mu e_{\mathrm{I}}^\nu),
\end{aligned}
$$

where X_{E}^μ is the center of energy:

$$
X_{\mathrm{E}}^\mu = (P^0)^{-1} \int x^\mu T^{00} \mathrm{d}^3\boldsymbol{x} = \frac{1}{4\pi N\hbar} \int x^\mu \omega a^2 \mathrm{d}^3\boldsymbol{x}.
$$

Because P^μ is a null vector in the present case, it is not possible to pass to a mean rest frame of the radiation in which to define a spin component of the angular momentum. We adopt instead an alternative procedure. We define the 4-vector

$$
K^\mu \equiv \frac{1}{2} {}^1\varepsilon^{\mu\nu\sigma\tau} P_\nu J_{\sigma\tau},
$$

which is both gauge invariant and independent of the location of the origin of coordinates. We note that

$$
K \cdot P = 0,
$$

and because $\det(\eta_{\mu\nu}) = -1$,

$$
\begin{aligned}
K^2 &= -\frac{1}{4} {}^1\varepsilon^{\mu\nu\sigma\tau} \, {}^{-1}\varepsilon_{\mu\rho\kappa\lambda} P_\nu J_{\sigma\tau} P^\rho J^{\kappa\lambda} \\
&= -\frac{1}{4} \Big(\delta_\rho^\nu \delta_\kappa^\sigma \delta_\lambda^\tau + \delta_\kappa^\nu \delta_\lambda^\sigma \delta_\rho^\tau + \delta_\lambda^\nu \delta_\rho^\sigma \delta_\kappa^\tau \\
&\quad - \delta_\rho^\nu \delta_\lambda^\sigma \delta_\kappa^\tau + \delta_\lambda^\nu \delta_\kappa^\sigma \delta_\rho^\tau - \delta_\kappa^\nu \delta_\rho^\sigma \delta_\lambda^\tau \Big) P_\nu J_{\sigma\tau} P^\rho J^{\kappa\lambda} \\
&= P_\mu J^{\mu\sigma} P^\nu J_{\nu\sigma} = (P \cdot X_{\mathrm{E}})^2 P^2 = 0.
\end{aligned}
$$

That is, K^μ is a null vector orthogonal to P^μ. Because P^μ is itself a null vector this implies that K^μ is parallel to P^μ:

$$
K^\mu = S P^\mu.
$$

The constant of proportionality S is defined to be the spin angular momentum of the field. It is easily computed by setting $\mu = 0$:

$$N\hbar\omega S = P^0 S = K^0 = \frac{1}{2}\varepsilon_{ijk} P_i J_{jk}$$
$$= N\hbar\varepsilon_{ijk} P_i e_{\mathrm{I}j} e_{\mathrm{II}k} = (N\hbar)^2 \omega \varepsilon_{ijk} \hat{k}_i e_{\mathrm{I}j} e_{\mathrm{II}k},$$

where $\hat{k}_i = k_i/|\boldsymbol{k}|$. Assuming that $\boldsymbol{e}_{\mathrm{I}}$, $\boldsymbol{e}_{\mathrm{II}}$, $\hat{\boldsymbol{k}}$ (in that order) form a right handed orthonormal system (like the coordinates), we have

$$S = N\hbar.$$

Another way of computing S, and one that gives the rationale for calling it the spin angular momentum, is to introduce at any instant (x^0 = constant) a coordinate system with origin at the center of energy ($X^i_{\mathrm{E}} = 0$) and then to compute the angular momentum 3-vector

$$J_i = \frac{1}{2}\varepsilon_{ijk} J_{jk}.$$

It is easy to see that one gets $J_i = S\hat{k}_i$. Because the center of energy is at the origin, this angular momentum is pure spin. It is seen to be parallel to the propagation vector. The magnitude $N\hbar$ found for S may be interpreted as indicating that the N photons that make up the wave all have spin angular momentum $\hbar$ and that all the spins are pointing in the same direction. When the wave has right handed (positive) helicity, the spins all point parallel to the propagation vector. When the wave has left handed (negative) helicity, S is equal to $-N\hbar$, and the spins all point antiparallel to the propagation vector. In a linearly polarized wave, half the photons have their spins parallel to $\hat{\boldsymbol{k}}$ and the other half have their spins antiparallel to $\hat{\boldsymbol{k}}$, and S is zero. The spins of massless quanta can only be found in the parallel and antiparallel configurations.

Chapter 15
Gravitational Waves

Consider a region of spacetime free of matter in which Einstein's empty space equations hold:

$$0 = \frac{\delta S_G}{\delta g_{\mu\nu}} \equiv -\frac{1}{16\pi} g^{1/2} \left(R^{\mu\nu} - \frac{1}{2} g^{\mu\nu} R \right) \quad (G = 1).$$

Suppose the curvature tensor in this region is separable into two components, a slowly varying component and a rapidly varying component. Stated more precisely, suppose there exists a family of coordinate patches in which the components of the curvature may be expressed as the sum of two terms, one that is slowly varying and has magnitude of order $1/\mathcal{R}^2$ and one that varies rapidly over distances of order $\lambda \ll \mathcal{R}$. Then we may say that we have a situation in which *gravitational waves* of wavelengths $\sim \lambda$ are propagating in a smooth background geometry having a curvature characterized by the length $\mathcal{R}$, which may be called a *mean radius of curvature*. We may express the separation between background and waves by writing the full metric tensor, which we shall denote by $g^{\text{tot}}_{\mu\nu}$, as the sum of two terms, a tensor $g_{\mu\nu}$ representing the background geometry and a tensor $h_{\mu\nu}$ representing the waves:

$$g^{\text{tot}}_{\mu\nu} = g_{\mu\nu} + h_{\mu\nu}.$$

We have seen that in the completely general theory such a separation has no physical meaning. Even when spacetime is asymptotically flat, it generally has only a global meaning. Here, however, because of the existence of the special coordinate systems (special family of patches) in which the curvature separates into two parts, we shall find that the corresponding separation of the metric has quasi-local physical meaning. We shall consider two kinds of coordinate transformations. Under the first, both $g_{\mu\nu}$ and $h_{\mu\nu}$ transform like tensors. But in order to remain within the family of special coordinate systems, the transformation coefficients $\partial\bar{x}^\mu/\partial x^\nu$ must be smooth functions varying appreciably only over distances

B. DeWitt, *Bryce DeWitt's Lectures on Gravitation*,
Edited by Steven M. Christensen, Lecture Notes in Physics, 826,
DOI: 10.1007/978-3-540-36911-0_15,

much greater than λ. Moreover, they must nowhere correspond to a boost[1] that Doppler shifts the wavelengths of any of the gravitational waves to magnitudes large compared to λ.

Under the second kind of coordinate transformation, the transformation coefficients $\partial\bar{x}^\mu/\partial x^\nu$ may vary rapidly, but in this case the $g_{\mu\nu}$ are held fixed and the full burden of the transformation is placed on $h_{\mu\nu}$. In order to remain within the family of special coordinate systems, transformations of this kind, which may be called gauge transformations, must involve displacement of the coordinate mesh through distances of order much smaller than λ. This will prove to be the basic reason why it makes good physical sense to speak of the distribution of energy, momentum, and stress carried by gravitational waves, and to assign to each contribution to the total energy, momentum, and stress a location in a region having dimensions d of the order of a few λ.

In order to simplify the problem of analyzing gravitational waves, we shall assume that they are weak enough so that we may neglect their interaction with each other and apply the superposition principle as a valid approximation. Explicitly, we assume that in any one of the special coordinate systems that is locally canonical in the background geometry, we have

$$h_{\mu\nu} \sim a, \text{ where } a \ll 1.$$

It is then appropriate to expand Einstein's equations about the background geometry:

$$0 = \frac{\delta S_\mathrm{G}[g^\mathrm{tot}]}{\delta g^\mathrm{tot}_{\mu\nu}} = \frac{\delta S_\mathrm{G}[g]}{\delta g_{\mu\nu}} + \left[\delta\frac{\delta S_\mathrm{G}[g]}{\delta g_{\mu\nu}} + \frac{1}{2}\delta^2\frac{\delta S_\mathrm{G}[g]}{\delta g_{\mu\nu}} + \cdots\right]_{\delta g_{\alpha\beta}=h_{\alpha\beta}},$$

where

$$\delta\frac{\delta S_\mathrm{G}}{\delta g_{\mu\nu}} \equiv \int \frac{\delta^2 S_\mathrm{G}}{\delta g_{\mu\nu}\delta g_{\alpha'\beta'}}\delta g_{\alpha'\beta'}\mathrm{d}^4x',$$

$$\delta^2\frac{\delta S_\mathrm{G}}{\delta g_{\mu\nu}} \equiv \int \mathrm{d}^4x' \int \mathrm{d}^4x'' \frac{\delta^3 S_\mathrm{G}}{\delta g_{\mu\nu}\delta g_{\alpha'\beta'}\delta g_{\gamma''\delta''}}\delta g_{\alpha'\beta'}\delta g_{\gamma''\delta''},$$

and so on.

Our first (and most tedious) task is to compute these first and second variations of the Einstein equations. The first variation is readily calculated with the aid of the results assembled in Chap. 8. We find

[1] We shall see that gravitational waves propagate with the speed of light.

$$
\begin{aligned}
16\pi\delta\frac{\delta S_{\mathrm{G}}}{\delta g_{\mu\nu}} &= -\delta\left[g^{1/2}\left(g^{\mu\sigma}g^{\nu\tau}-\frac{1}{2}g^{\mu\nu}g^{\sigma\tau}\right)R_{\sigma\tau}\right] \\
&= g^{1/2}\left[-\frac{1}{2}\left(R^{\mu\nu}-\frac{1}{2}g^{\mu\nu}R\right)g^{\alpha\beta}+g^{\mu\alpha}R^{\beta\nu}+g^{\nu\alpha}R^{\mu\beta}-\frac{1}{2}g^{\mu\alpha}g^{\beta\nu}R-\frac{1}{2}g^{\mu\nu}R^{\alpha\beta}\right]_{\alpha\beta} \\
&\quad +\frac{1}{2}g^{1/2}\big[-g^{\mu\alpha}g^{\nu\beta}g^{\gamma\delta}(\delta g_{\alpha\gamma;\beta\delta}+\delta g_{\beta\gamma;\alpha\delta}-\delta g_{\alpha\beta;\gamma\delta}-\delta g_{\gamma\delta;\alpha\beta}) \\
&\quad +g^{\mu\nu}g^{\alpha\beta}g^{\gamma\delta}(\delta g_{\alpha\gamma;\beta\delta}-\delta g_{\alpha\beta;\gamma\delta})\big].
\end{aligned}
$$

Before beginning the computation of the second variation, let us make more precise the nature of the splitting of the total metric in $g_{\mu\nu}$ and $h_{\mu\nu}$, and the kind of approximation scheme we are envisaging. We shall require that

$$
\left\langle g_{\mu\nu}^{\mathrm{tot}}\right\rangle = g_{\mu\nu}, \quad \langle h_{\mu\nu}\rangle = 0,
$$

where the angle brackets denote the following averaging process. Choose one of the special coordinate systems. Make it as canonical as possible (relative to the background geometry) in as big a region as possible around the point x at which the average is desired. The domain of effective canonicity will have dimensions much larger than λ, for significant departures from canonicity will occur only over distances of order $\mathcal{R}$. Let $f(x-x')$ be a smooth non-negative function of the coordinate differences $x^\mu - x'^\mu$, which vanishes for $|x^\mu - x'^\mu| > d$, where d is of the order of a few λ, and that satisfies the normalization condition

$$
\int f(x)\mathrm{d}^4x = 1
$$

in these coordinates. Then for any tensor field ϕ, we define its average at x by

$$
\langle\phi(x)\rangle = \int f(x-x')\phi(x')\mathrm{d}^4x'.
$$

Once having defined the average in this nearly canonical coordinate system, we then treat the average as a tensor of the same type, in transforming to any other coordinate system.

As covariant derivatives become ordinary derivatives in the domain of effective canonicity (about x) of the above coordinate system, we have

$$
\begin{aligned}
\langle\phi(x)\rangle_{;\mu} = \langle\phi(x)\rangle_{,\mu} &= \int\frac{\partial}{\partial x^\mu}f(x-x')\phi(x')\mathrm{d}^4x' = -\int\left[\frac{\partial}{\partial x'^\mu}f(x-x')\right]\phi(x')\mathrm{d}^4x' \\
&= \int f(x-x')\phi_{,\mu'}(x')\mathrm{d}^4x' = \int f(x-x')\phi_{;\mu'}(x')\mathrm{d}^4x' \\
&= \langle\phi_{;\mu}(x)\rangle + O(d/\mathcal{R}^2)\langle\phi(x)\rangle,
\end{aligned}
$$

where covariant differentiation is to be understood in this section as defined with the background metric. We also have $\partial f/\partial x^\mu \sim f/d$ and hence

$$\langle \phi_{;\mu} \rangle \sim \frac{1}{d} \langle \phi \rangle.$$

This has the consequence that

$$\langle \phi_{;\mu} \psi \rangle \sim - \langle \phi \psi_{;\mu} \rangle + 0 \left(\frac{1}{d} \langle \phi \psi \rangle \right),$$

where ϕ and ψ are any two tensor fields. Finally, we note that

$$\left\langle g_{\text{tot}}^{1/2} \phi \right\rangle = g^{1/2} \langle \phi \rangle, \quad \left\langle g_{\mu\nu}^{\text{tot}} \phi \right\rangle = g_{\mu\nu} \langle \phi \rangle,$$

$$\left\langle R_{\mu\nu\sigma\tau}^{\text{tot}} \phi \right\rangle = R_{\mu\nu\sigma\tau} \langle \phi \rangle,$$

and so on.

To discuss the approximation scheme of this section, it will be convenient to employ the condensed notation introduced in Chap. 12. For $g_{\mu\nu}^{\text{tot}}$, $g_{\mu\nu}$, and $h_{\mu\nu}$, we shall write φ_{tot}^i, φ^i, and ϕ^i, respectively:

$$\varphi_{\text{tot}}^i = \varphi^i + \phi^i, \quad \langle \varphi_{\text{tot}}^i \rangle = \varphi^i, \quad \langle \phi^i \rangle = 0.$$

Einstein's equations take the form

$$0 = S_{\text{G},i}[\varphi + \phi] = S_{\text{G},i} + S_{\text{G},ij} \phi^j + \frac{1}{2} S_{\text{G},ijk} \phi^j \phi^k + \frac{1}{3!} S_{\text{G},ijkl} \phi^j \phi^k \phi^l + \cdots,$$

in which the functional derivative coefficients in the expansion are to be understood as evaluated with the background metric. The term $S_{\text{G},ij}\phi^j$ is just the first variation that we computed in present chapter, evaluated with $\delta g_{\alpha\beta} = h_{\alpha\beta}$. Using the averaging rules obtained above, one readily sees that

$$\langle S_{\text{G},ij} \phi^j \rangle = S_{\text{G},ij} \langle \phi^j \rangle = 0.$$

Therefore, performing the averaging operation on Einstein's equations, we find

$$S_{\text{G},i} = -\frac{1}{2} \langle S_{\text{G},ijk} \phi^j \phi^k \rangle - \frac{1}{3!} \langle S_{\text{G},ijkl} \phi^j \phi^k \phi^l \rangle + \cdots.$$

The first term on the right-hand side of this equation is the average of the second variation. We shall see currently that the second variation contains terms of the form $h^{\mu}_{\sigma\tau;} h^{\sigma\tau\nu}_{;}$, $h_{\sigma\tau}\, h^{\sigma\tau\mu\nu}_{;}$, etc. Hence this term is of the order a^2/λ^2. The next term is of order a^3/λ^2, and so on.

Subtracting Einstein's equation from its average, we get

$$\begin{aligned} S_{\text{G},ij} \phi^j = & -\frac{1}{2} \left(S_{\text{G},ijk} \phi^j \phi^k - \langle S_{\text{G},ijk} \phi^j \phi^k \rangle \right) \\ & -\frac{1}{3!} \left(S_{\text{G},ijkl} \phi^j \phi^k \phi^l - \langle S_{\text{G},ijkl} \phi^j \phi^k \phi^l \rangle \right) - \cdots. \end{aligned} \tag{15.1}$$

The first term on the right of this equation is of order a^2/λ^2, the second is of order a^3/λ^2, and so on. These terms describe the interaction of the gravitational waves with each other. Our present approximation, based on $a \ll 1$, is to neglect these terms and write simply the homogeneous equation

$$S_{G,ij}\phi^j = 0.$$

This equation is derivable from an action functional of the form

$$S_{GW}[\varphi, \phi] = \frac{1}{2} S_{G,ij}[\varphi]\phi^i\phi^j,$$

and it describes the gravitational waves as freely propagating in the background geometry without acting on each other. The waves *do* act on the background geometry, however, through the equation for $S_{G,\,i}[\phi]$. In the same spirit of approximation, we shall keep only the first term on the right of this equation. It then may be written in the form

$$S_{G,i} = -\frac{1}{2}\langle T_i^{GW}\rangle, \tag{15.2}$$

where T_i^{GW} is the energy–momentum–stress density of the waves:

$$T_i^{GW} = 2\frac{\delta S_{GW}}{\delta\varphi^i} = S_{G,ijk}\phi^j\phi^k. \tag{15.3}$$

Notice that it is only the *average* of this density that serves as a source for the background field.

It may not be the only source. All we know is that $S_{G,i}$, or more explicitly

$$g^{1/2}\left(R^{\mu\nu} - \frac{1}{2}g^{\mu\nu}R\right),$$

is of order a^2/λ^2 in the region of interest. The full Riemann tensor, which is of order $1/\mathcal{R}^2$ in magnitude, will register the effect of any sources located in other regions. Hence we can only say that

$$\frac{a^2}{\lambda^2} \lesssim \frac{1}{\mathcal{R}^2}.$$

Written in the form

$$a \lesssim \frac{\lambda}{\mathcal{R}} \ll 1,$$

the inequality is, of course, compatible with our assumption about the smallness of a.

Let us now return to the expression for the first variation computed in present chapter. With $\delta g_{\alpha\beta} = h_{\alpha\beta}$ the terms involving the Ricci tensor in this expression are of order a^3/λ^2. These are a factor of a smaller than terms we have already thrown away on the right-hand side of (15.1). Therefore, they may be completely ignored, and we find for Eq. 8.3) satisfied by the gravitational waves (see Chap. 8)

$$h^{\sigma}_{\mu\nu;\sigma} + h_{;\mu\nu} - h^{\sigma}_{\mu;\nu\sigma} - h^{\sigma}_{\nu;\mu\sigma} + g_{\mu\nu}(h^{\sigma\tau}_{;\sigma\tau} - h^{\sigma}_{;\sigma}) = 0, \tag{15.4}$$

where

$$h = h^{\mu}_{\mu},$$

indices on the $h_{\mu\nu}$ being raised and lowered by means of the background metric. Multiplying this equation by $g^{\mu\nu}$, we find

$$h^{\mu\nu}_{;\mu\nu} - h^{\mu}_{;\mu} = 0, \tag{15.5}$$

which allows the simplification

$$h^{\sigma}_{\mu\nu;\sigma} + h_{;\mu\nu} - h^{\sigma}_{\mu;\nu\sigma} - h^{\sigma}_{\nu;\mu\sigma} = 0.$$

When we vary the Ricci tensor terms appearing in the first variation, in the process of computing the second variation, we obtain terms that, by virtue of the equations for φ^i and ϕ^i, are of order a^3/λ^2 and a^4/λ^2 in magnitude. But the second variation itself is of order a^2/λ^2, and hence these terms may be thrown away. Most of the terms that must be retained come from variations like $\delta(\delta g_{\alpha\beta;\gamma\delta})$. This is readily computed to be

$$\begin{aligned}\delta(\delta g_{\alpha\beta;\gamma\delta}) &= \delta[(\delta g_{\alpha\beta,\gamma} - \Gamma^{\varepsilon}_{\alpha\gamma}\delta g_{\varepsilon\beta} - \Gamma^{\varepsilon}_{\beta\gamma}\delta g_{\alpha\varepsilon})_{,\delta} - \Gamma^{\xi}_{\alpha\delta}(\delta g_{\xi\beta,\gamma} - \Gamma^{\varepsilon}_{\xi\gamma}\delta g_{\varepsilon\beta} - \Gamma^{\varepsilon}_{\beta\gamma}\delta g_{\xi\varepsilon}) \\ &\quad - \Gamma^{\xi}_{\beta\delta}(\delta g_{\alpha\xi,\gamma} - \Gamma^{\varepsilon}_{\alpha\gamma}\delta g_{\varepsilon\xi} - \Gamma^{\varepsilon}_{\xi\gamma}\delta g_{\alpha\varepsilon}) - \Gamma^{\xi}_{\gamma\delta}(\delta g_{\alpha\beta,\xi} - \Gamma^{\varepsilon}_{\alpha\xi}\delta g_{\varepsilon\beta} \\ &\quad - \Gamma^{\varepsilon}_{\beta\xi}\delta g_{\alpha\varepsilon})] \\ &= (-\delta\Gamma^{\varepsilon}_{\alpha\gamma}\delta g_{\varepsilon\beta} - \delta\Gamma^{\varepsilon}_{\beta\gamma}\delta g_{\alpha\varepsilon})_{;\delta} - \delta\Gamma^{\xi}_{\alpha\delta}\delta g_{\xi\beta;\gamma} - \delta\Gamma^{\xi}_{\beta\delta}\delta g_{\alpha\xi;\gamma} - \delta\Gamma^{\xi}_{\gamma\delta}\delta g_{\alpha\beta;\xi}.\end{aligned}$$

Omitting terms that are of order a^3/λ^2 by virtue of the gravitational wave equation, we now find

$$\begin{aligned}T^{\mu\nu}_{\mathrm{GW}} = \left[\delta^2 \frac{\delta S_{\mathrm{G}}}{\delta g_{\mu\nu}}\right]_{\alpha\beta=h_{\alpha\beta}} &= \frac{1}{16\pi} g^{1/2}\Bigg\{\frac{1}{2}h^{\gamma\delta}(h^{\mu\nu}_{\gamma;\nu\delta} + h^{\nu\mu}_{\gamma;\delta} - h^{\mu\nu}_{;\gamma\delta} - h^{\mu\nu}_{\gamma\delta;}) \\ &+ \frac{1}{4}g^{\mu\alpha}g^{\nu\beta}g^{\gamma\delta}\Big[(h^{\varepsilon}_{\alpha;\beta} + h^{\varepsilon}_{\beta;\alpha} - h^{\varepsilon}_{\alpha\beta;})h_{\varepsilon\gamma} + (h^{\varepsilon}_{\gamma;\beta} + h^{\varepsilon}_{\beta;\gamma} - h^{\varepsilon}_{\gamma\beta;})h_{\varepsilon\alpha} \\ &+ (h^{\varepsilon}_{\beta;\alpha} + h^{\varepsilon}_{\alpha;\beta} - h^{\varepsilon}_{\beta\alpha;})h_{\varepsilon\gamma} + (h^{\varepsilon}_{\gamma;\alpha} + h^{\varepsilon}_{\alpha;\gamma} - h^{\varepsilon}_{\gamma\alpha;})h_{\varepsilon\beta} \\ &-(h^{\varepsilon}_{\alpha;\gamma} + h^{\varepsilon}_{\gamma;\alpha} - h^{\varepsilon}_{\alpha\gamma;})h_{\varepsilon\beta} - (h^{\varepsilon}_{\varepsilon\beta;\gamma} + h^{\varepsilon}_{\gamma;\beta} - h^{\varepsilon}_{\beta\gamma;})h_{\varepsilon\alpha}\Big]_{;\delta} \\ &+ \frac{1}{4}g^{\mu\alpha}g^{\nu\beta}g^{\gamma\delta}\Big[-(h^{\varepsilon}_{\gamma;\alpha} + h^{\varepsilon}_{\alpha;\gamma} - h^{\varepsilon}_{\gamma\alpha;})h_{\varepsilon\delta} - (h^{\varepsilon}_{\delta;\alpha} + h^{\varepsilon}_{\alpha;\delta} - h^{\varepsilon}_{\delta\alpha;})h_{\varepsilon\gamma}\Big]_{;\beta} \\ &- \frac{1}{4}g^{\mu\nu}g^{\alpha\beta}g^{\gamma\delta}\Big[(h^{\varepsilon}_{\alpha;\beta} + h^{\varepsilon}_{\beta;\alpha} - h^{\varepsilon}_{\alpha\beta;})h_{\varepsilon\gamma} + (h^{\varepsilon}_{\gamma;\beta} + h^{\varepsilon}_{\beta;\gamma} - h^{\varepsilon}_{\gamma\beta;})h_{\varepsilon\alpha}\end{aligned}$$

$$
\begin{aligned}
&-(h^{\varepsilon}_{\alpha;\gamma} + h^{\varepsilon}_{\gamma;\alpha} - h^{\varepsilon}_{\alpha\gamma;})h_{\varepsilon\beta} - (h^{\varepsilon}_{\beta;\gamma} + h^{\varepsilon}_{\gamma;\beta} - h^{\varepsilon}_{\beta\gamma;})h_{\varepsilon\alpha}\Big]_{;\delta} \\
&+\frac{1}{4}g^{\mu\alpha}g^{\nu\beta}g^{\gamma\delta}\Big[(h^{\varepsilon}_{\alpha;\delta} + h^{\varepsilon}_{\delta;\alpha} - h^{\varepsilon}_{\alpha\delta;})h_{\varepsilon\gamma;\beta} + (h^{\varepsilon}_{\gamma;\delta} + h^{\varepsilon}_{\delta;\gamma} - h^{\varepsilon}_{\gamma\delta;})h_{\alpha\varepsilon;\beta} \\
&+ (h^{\varepsilon}_{\beta;\delta} + h^{\varepsilon}_{\delta;\beta} - h^{\varepsilon}_{\beta\delta;})h_{\alpha\gamma;\varepsilon} + (h^{\varepsilon}_{\beta;\delta} + h^{\varepsilon}_{\delta;\beta} - h^{\varepsilon}_{\beta\delta;})h_{\varepsilon\gamma;\alpha} + (h^{\varepsilon}_{\gamma;\delta} + h^{\varepsilon}_{\delta;\gamma} - h^{\varepsilon}_{\gamma\delta;})h_{\beta\varepsilon;\alpha} \\
&+ (h^{\varepsilon}_{\alpha;\delta} + h^{\varepsilon}_{\delta;\alpha} - h^{\varepsilon}_{\alpha\delta;})h_{\beta\gamma;\varepsilon} - (h^{\varepsilon}_{\alpha;\delta} + h^{\varepsilon}_{\delta;\alpha} - h^{\varepsilon}_{\alpha\delta;})h_{\varepsilon\beta;\gamma} - (h^{\varepsilon}_{\beta;\delta} + h^{\varepsilon}_{\delta;\beta} - h^{\varepsilon}_{\beta\delta;})h_{\alpha\varepsilon;\gamma} \\
&- (h^{\varepsilon}_{\gamma;\delta} + h^{\varepsilon}_{\delta;\gamma} - h^{\varepsilon}_{\gamma\delta;})h_{\alpha\beta;\varepsilon} - (h^{\varepsilon}_{\gamma;\beta} + h^{\varepsilon}_{\beta;\gamma} - h^{\varepsilon}_{\gamma\beta;})h_{\varepsilon\delta;\alpha} - (h^{\varepsilon}_{\delta;\beta} + h^{\varepsilon}_{\beta;\delta} - h^{\varepsilon}_{\delta\beta;})h_{\gamma\varepsilon;\alpha} \\
&- (h^{\varepsilon}_{\alpha;\beta} + h^{\varepsilon}_{\beta;\alpha} - h^{\varepsilon}_{\alpha\beta;})h_{\gamma\delta;\varepsilon}\Big] - \frac{1}{4}g^{\mu\nu}g^{\alpha\beta}g^{\gamma\delta}\Big[(h^{\varepsilon}_{\alpha;\delta} + h^{\varepsilon}_{\delta;\alpha} - h^{\varepsilon}_{\alpha\delta;})h_{\varepsilon\gamma;\beta} \\
&+ (h^{\varepsilon}_{\gamma;\delta} + h^{\varepsilon}_{\delta;\gamma} - h^{\varepsilon}_{\gamma\delta;})h_{\alpha\varepsilon;\beta} + (h^{\varepsilon}_{\beta;\delta} + h^{\varepsilon}_{\delta;\beta} - h^{\varepsilon}_{\beta\delta;})h_{\alpha\gamma;\varepsilon} - (h^{\varepsilon}_{\alpha;\delta} + h^{\varepsilon}_{\delta;\alpha} - h^{\varepsilon}_{\alpha\delta;})h_{\varepsilon\beta;\gamma} \\
&- (h^{\varepsilon}_{\beta;\delta} + h^{\varepsilon}_{\delta;\beta} - h^{\varepsilon}_{\beta\delta;})h_{\alpha\varepsilon;\gamma} - (h^{\varepsilon}_{\gamma;\delta} + h^{\varepsilon}_{\delta;\gamma} - h^{\varepsilon}_{\gamma\delta;})h_{\alpha\beta;\varepsilon}\Big]\Big\} \\
=&\frac{1}{16\pi}g^{1/2}\Big[h^{\sigma\tau}h^{\mu\nu}_{\sigma;\tau} + h^{\sigma\tau}h^{\nu\mu}_{\sigma;\tau} - h^{\sigma\tau}h^{\mu\nu}_{;\sigma\tau} - h^{\sigma\tau}h^{\mu\nu}_{\sigma\tau;} \\
&- g^{\mu\nu}\left(\frac{1}{2}h^{\sigma\tau}h^{\rho}_{\sigma\rho;\tau} - \frac{1}{4}h^{\sigma\tau}h_{;\sigma\tau} - \frac{1}{4}h^{\sigma\tau}h^{\rho}_{\sigma\tau;\rho}\right) + h^{\mu\nu}_{\sigma;}h^{\sigma\tau}_{;\tau} + h^{\nu\mu}_{\sigma;}h^{\sigma\tau}_{;\tau} - h^{\mu\nu}_{;\sigma}h^{\sigma\tau}_{;\tau} \\
&- \frac{1}{2}h^{\sigma\tau\mu}_{;}h^{\nu}_{\sigma\tau;} - \frac{1}{2}h^{\mu\sigma\nu}_{;}h_{;\sigma} - \frac{1}{2}h^{\nu\sigma\mu}_{;}h_{;\sigma} + h^{\mu\sigma\tau}_{;}h^{\nu}_{\tau;\sigma} - h^{\mu\sigma\tau}_{;}h^{\nu}_{\sigma;\tau} + \frac{1}{2}h^{\mu\nu\sigma}_{;}h_{;\sigma} \\
&- g^{\mu\nu}\left(h^{\tau}_{\sigma\tau;}h^{\sigma\rho}_{;\rho} - h^{\sigma\tau}_{;\tau}h_{;\sigma} + \frac{1}{2}h^{\sigma\tau\rho}_{;}h_{\sigma\rho;\tau} - \frac{1}{4}h^{\sigma\tau\rho}_{;}h_{\sigma\tau;\rho} + \frac{1}{4}h_{;\sigma}h^{\sigma}_{;}\right)\Big].
\end{aligned}
$$

In the present approximation (which regard the gravitational waves as propagating without interacting with each other), this energy–momentum–stress density satisfies the usual divergence law

$$T_{\mathrm{GW}}{}^{\mu\nu}_{;\nu} = 0.$$

The proof of this law is very tedious if one attempts to work directly with the tensor components $h_{\mu\nu}$, but it becomes very easy if one uses the compact notation. From the wave equation $S_{\mathrm{G},ij}\phi^j = 0$ and the identity

$$\Phi^i_a S_{\mathrm{G},ijk} \equiv -S_{\mathrm{G},ik}\Phi^i_{a,j} - S_{\mathrm{G},ji}\Phi^i_{a,k},$$

which follows from functionally differentiating the contracted Bianchi identity ($S_{\mathrm{G},i}\Phi^i{}_a \equiv 0$) twice, we get

$$
\begin{aligned}
T^{\mathrm{GW}}_i\Phi^i_a &\equiv \Phi^i_a S_{\mathrm{G},ijk}\phi^j\phi^k \\
&\equiv -\left(S_{\mathrm{G},ik}\Phi^i_{a,j} + S_{\mathrm{G},ji}\Phi^i_{a,k}\right)\phi^j\phi^k = 0.
\end{aligned}
$$

Because Φ^i_a is basically just a differential operator, we have also

$$\langle T^{\mathrm{GW}}_i\rangle\Phi^i_a = \langle T^{\mathrm{GW}}_i\Phi^i_a\rangle = 0,$$

or, in tensor notation,

$$\langle T^{\mu\nu}_{\mathrm{GW}} \rangle_{;\nu} = 0,$$

which establishes the consistency of the equation

$$S_{\mathrm{G},i} = -\frac{1}{2} \langle T^{\mathrm{GW}}_i \rangle$$

with the contracted Bianchi identity.

Although $T^{\mu\nu}_{\mathrm{GW}}$ satisfies the divergence law, it is not gauge invariant, i.e., it is not invariant under the class of coordinate transformations discussed, in which the whole burden of transformation is placed on $h_{\mu\nu}$. Its average, however, is *nearly* gauge invariant, and it is of interest to examine semi-quantitatively just *how* nearly. Let the coordinate transformation be expressed in the form

$$\bar{x}^\mu = x^\mu + \xi^\mu,$$

the coordinates x^μ belonging, of course, to the special class. Then, in order that the coordinates $\bar{x}^\mu$ also be members of the special class, the ξ^μ must be no bigger than $a\lambda$ and their first derivatives must be no bigger than a. The gauge transformation law for $h_{\mu\nu}$ then takes the formthe coordinates x^μ belonging, of course, to the special class. Then, in order that the coordinates $\bar{x}^\mu$ also be members of the special class, the ξ^μ must be no bigger than $a\lambda$ and their first derivatives must be no bigger than a. The gauge transformation law for $h_{\mu\nu}$ then takes the form

$$\bar{h}_{\mu\nu} = h_{\mu\nu} + \Delta h_{\mu\nu}, \quad \Delta h_{\mu\nu} = -\xi_{\mu;\nu} - \xi_{\nu;\mu} + O(a^2),$$

or, in the compact notation,

$$\Delta\phi^i = \Phi^i_a[\varphi]\xi^a + O(a^2).$$

Under this transformation, T^{GW}_i suffers the change

$$\begin{aligned} \Delta T^{\mathrm{GW}}_i &= S_{\mathrm{G},ijk}(2\phi^j \Delta\phi^k + \Delta\phi^j \Delta\phi^k) \\ &= -(S_{\mathrm{G},kj}\Phi^k_{a,i} + S_{\mathrm{G},ik}\Phi^k_{a,j})\xi^a(2\phi^j + \Delta\phi^j) + 0(a^3/\lambda^2) \\ &= -S_{\mathrm{G},ik}\Phi^k_{a,j}\xi^a(2\phi^j + \Delta\phi^j) + S_{\mathrm{G},j}\Phi^j_{b,k}\xi^b\Phi^k_{a,i}\xi^a + 0(a^3/\lambda^2). \end{aligned}$$

The second term in the final expression is of order a^4/λ^2 and may be completely neglected. The first term, on the other hand, is of the same order as T^{GW}_i itself, namely a^2/λ^2. Therefore, dropping all higher order terms we have effectively

$$\Delta T^{\mathrm{GW}}_i = -S_{\mathrm{G},ik}\Phi^k_{a,j}\xi^a(2\phi^j + \Delta\phi^j),$$

which shows that T^{GW}_i is not at all gauge invariant. The expression on the right-hand side of this last equation, however, involves the second functional derivative $S_{\mathrm{G},ij}$, which is effectively a second-order diffferential operator. Therefore, making use of the averaging rules obtained, we find for the *average* of ΔT^{GW}_i,

$$\begin{aligned} \langle \Delta T^{\mathrm{GW}}_i \rangle &= -S_{\mathrm{G},ik}\langle \Phi^k_{a,j}\xi^a(2\phi^j + \Delta\phi^j)\rangle \sim \frac{1}{d^2}\langle \Phi^k_{a,j}\xi^a(2\phi^j + \Delta\phi^j)\rangle \\ &= O\left(\frac{a^2}{d^2}\right) \sim \frac{\lambda^2}{d^2}\langle T^{\mathrm{GW}}_i \rangle. \end{aligned}$$

We see that the change in $\langle T_i^{\rm GW}\rangle$ caused by a gauge transformation is smaller than $\langle T_i^{\rm GW}\rangle$ itself by the factor $(\lambda/d)^2$. It is this result that permits us to regard the energy and momentum of a gravitational wave as localized within a region of size d (by the averaging process). The error we make in doing this decreases inversely as d^2 with increasing d.

The approximate gauge invariance of $\langle T_i^{\rm GW}\rangle$ (or $\langle T_{\rm GW}^{\mu\nu}\rangle$) allows us to make a special choice of gauge.We begin by carrying out a gauge transformation for which the gauge parameter xi^μ satisfies the differential equation

$$(\xi_;^{\mu\nu} - \xi_;^{\nu\mu})_{;\nu} = h_{;\nu}^{\mu\nu} - h_;^\mu.$$

Because of the wave equation $h_{;\mu\nu}^{\mu\nu} - h_{;\mu}^{\mu\mu} = 0$ [see (15.5)], both the left and right-hand sides of this equation have vanishing covariant divergence, and hence the equation is consistent. From its analogy to Maxwell's equation, however, we know that its solution is not unique. If ξ^μ is a solution, then so is

$$\overline{\xi}^\mu = \xi^\mu + \Lambda_;^\mu,$$

for any Λ. We may get a particular solution by imposing the Lorentz condition $\xi_{;\mu}^\mu = 0$ on ξ^μ. Actually, this is not the most convenient condition for present purposes. We shall choose instead

$$\xi_{;\mu}^\mu = \frac{1}{2}h.$$

If this condition does not already hold, we can impose it by carrying out the transformation $\overline{\xi}^\mu = \xi^\mu + \Lambda_;^\mu$ with Λ chosen to be a solution of the inhomogeneous wave equation

$$\Lambda_{;\mu}^\mu = \frac{1}{2}h - \xi_{;\mu}^\mu.$$

With ξ^μ thus chosen,[2] we find

$$\overline{h} = h - 2\xi_{;\mu}^\mu = 0$$

and

$$\begin{aligned}\overline{h}_{;\nu}^{\mu\nu} &= h_{;\nu}^{\mu\nu} - \xi_{;\nu}^{\mu\nu} - \xi_{;\nu}^{\nu\mu}\\ &= h_{;\nu}^{\mu\nu} - \xi_{;\nu}^{\mu\nu} - \xi_{;\nu}^{\nu\mu} - h_;^\mu + 2\xi_{;\nu}^{\nu\mu}\\ &= h_{;\nu}^{\mu\nu} - h_;^\mu - (\xi_;^{\mu\nu} - \xi_;^{\nu\mu})_{;\nu} + 2\xi^\sigma R_{\sigma\nu}^{\nu\mu}\\ &= 0,\end{aligned}$$

[2] The resulting ξ^μ will be of order $a\lambda$, as required, if $h_{\mu\nu}$ is of order a.

to order a/λ. From now on we assume this gauge transformation already to have been carried out and drop the bars. Both $T^{\mu\nu}_{\mathrm{GW}}$ and the wave equation then take on considerably simpler forms. We find

$$0 = h^{\sigma}_{\mu\nu;\sigma} - h^{\sigma}_{\tau} R^{\tau}_{\mu\nu\sigma} - h^{\tau}_{\mu} R^{\sigma}_{\tau\nu\sigma} - h^{\sigma}_{\tau} R^{\tau}_{\nu\mu\sigma} - h^{\tau}_{\nu} R^{\sigma}_{\tau\mu\sigma} = h^{\sigma}_{\mu\nu;\sigma} + 2R^{\sigma\tau}_{\mu\nu} h_{\sigma\tau},$$

to order a/λ^2, and

$$\begin{aligned} T^{\mu\nu}_{\mathrm{GW}} = \frac{1}{16\pi} g^{1/2} \Big[& h^{\sigma\tau}\left(h^{\mu\nu}_{\sigma;\tau} + h^{\nu\mu}_{\sigma;\tau} - h^{\mu\nu}_{;\sigma\tau} - h^{\mu\nu}_{\sigma\tau;}\right) \\ & - \frac{1}{2} h^{\sigma\tau\mu}_{;} h^{\nu}_{\sigma\tau;} + h^{\mu\sigma\tau}_{;} h^{\nu}_{\tau;\sigma} - h^{\mu\sigma\tau}_{;} h^{\nu}_{\sigma;\tau} \\ & + g^{\mu\nu}\left(\frac{1}{4} h^{\sigma\tau\rho}_{;} h_{\sigma\tau;\rho} - \frac{1}{2} h^{\sigma\tau\rho}_{;} h_{\sigma\rho;\tau}\right)\Big], \end{aligned} \tag{15.6}$$

to order a^2/λ^2. In obtaining the latter expression we have also dropped terms that are of order a^3/λ^2 by virtue of the wave equation. In the wave equation itself, we have dropped the terms in the Ricci tensor (of order a^3/λ^2) but have retained the term in the Riemann tensor (of order $a/\mathcal{R}^2$) even though this term is small compared to the remaining term (of order a/λ^2). The reason for doing this is that the resulting equation is rigorous in the limit of a becoming infinitesimal (and $h_{\mu\nu}$ satisfying $h^{\mu\nu}_{;\nu} = 0$, $h = 0$) independently of the relative magnitudes of λ and $\mathcal{R}$.

15.1 Eikonal Approximation

The gravitational wave equation, like the electromagnetic wave equation, may be solved in the eikonal approximation. To obtain a locally plane monochromatic wave we make the Ansatz

$$h_{\mu\nu} = (\overset{0}{h}_{\mu\nu} + \overset{2}{h}_{\mu\nu} + \cdots)\cos\phi + (\overset{1}{h}_{\mu\nu} + \overset{3}{h}_{\mu\nu} + \cdots)\sin\phi,$$

and then insert it into the supplementary conditions $h^{\mu\nu}_{;\nu} = 0$, $h = 0$, and into the wave equation, making assumptions about the coefficients $\overset{n}{h}_{\mu\nu}$ analogous to those made about the coefficients $\overset{n}{A}_{\mu}$ in Chap. 14. We find

$$-\overset{0}{h}_{\mu\nu} k^{\nu} = 0, \quad \overset{n}{h}{}^{\mu}_{\mu} = 0, \quad \forall n,$$

$$\overset{1}{h}_{\mu\nu} k^{\nu} + \overset{0}{h}{}^{\nu}_{\mu\nu;} = 0,$$

$$-\overset{2}{h}_{\mu\nu} k^{\nu} + \overset{1}{h}{}^{\nu}_{\mu\nu;} = 0,$$

$$\overset{3}{h}_{\mu\nu}k^\nu + \overset{2}{h}{}^{\nu}_{\mu\nu;} = 0,$$

and so on, and

$$-\overset{0}{h}_{\mu\nu}k^2 = 0,$$

$$-\overset{1}{h}_{\mu\nu}k^2 - 2\overset{0}{h}_{\mu\nu;\sigma}k^\sigma - \overset{0}{h}_{\mu\nu}k^\sigma_{;\sigma} = 0,$$

$$-\overset{2}{h}_{\mu\nu}k^2 + 2\overset{1}{h}_{\mu\nu;\sigma}k^\sigma + \overset{1}{h}_{\mu\nu}k^\sigma_{;\sigma} + \overset{0}{h}{}^{\sigma}_{\mu\nu;\sigma} + 2R^{\sigma\tau}_{\mu\nu}\overset{0}{h}_{\sigma\tau} = 0,$$

$$-\overset{3}{h}_{\mu\nu}k^2 - 2\overset{2}{h}_{\mu\nu;\sigma}k^\sigma - \overset{2}{h}_{\mu\nu}k^\sigma_{;\sigma} + \overset{1}{h}{}^{\sigma}_{\mu\nu;\sigma} + 2R^{\sigma\tau}_{\mu\nu}\overset{1}{h}_{\sigma\tau} = 0,$$

and so on, where, as before,

$$k_\mu = \phi_{;\mu}.$$

Again we have

$$k^2 = 0, \quad k^\nu k_{\nu;\mu} = 0, \quad k_{\mu;\nu} = k_{\nu;\mu}, \quad k_{\mu;\nu}k^\nu = 0. \tag{15.7}$$

Defining

$$a^2 \equiv \frac{1}{2}\overset{0}{h}_{\mu\nu}\overset{0}{h}{}^{\mu\nu}, \quad e_{\mu\nu} \equiv \overset{0}{f}_{\mu\nu}, \quad \overset{n}{f}_{\mu\nu} \equiv a^{-1}\overset{n}{h}_{\mu\nu},$$

we have

$$\frac{1}{2}e_{\mu\nu}e^{\mu\nu} = 1, \quad e^\mu_\mu = 0, \quad \overset{n}{f}{}^\mu_\mu = 0, \quad \forall n,$$

$$e_{\mu\nu}k^\nu = 0, \quad \overset{n}{f}_{\mu\nu}k^\nu = (-1)^n a^{-1}\left(a\overset{n-1}{f}_{\mu\nu}\right)^{\nu}_{;}, \quad n \geq 1,$$

$$(a^2k^\mu)_{;\mu} = 0, \quad 2a_{;\mu}k^\mu + ak^\mu_{;\mu} = 0, \quad e_{\mu\nu;\sigma}k^\sigma = 0, \tag{15.8}$$

$$\overset{n}{f}_{\mu\nu;\sigma}k^\sigma = \frac{1}{2}(-1)^n\left[a^{-1}\left(a\overset{n-1}{f}_{\mu\nu}\right)^{\sigma}_{;\sigma} + 2R^{\sigma\tau}_{\mu\nu}\overset{n-1}{f}_{\sigma\tau}\right], \quad n \geq 1.$$

$e_{\mu\nu}$ is the *polarization tensor* of the wave. It is seen to be propagated in a parallel fashion along the null geodesics that constitute the rays orthogonal to the wave fronts. Consistency of the two equations involving k^μ satisfied by the coefficients $\overset{n}{f}_{\mu\nu}$ is verified by the following computation in which the Ricci tensor is, as usual, treated as vanishing:

$$
\begin{aligned}
0 &= \left(\overset{n}{h}_{\mu\nu}k^{\nu}\right)_{;\sigma}k^{\sigma} - \left(\overset{n}{h}_{\mu\nu}k^{\nu}\right)_{;\sigma}k^{\sigma} \\
&= (-1)^n \overset{n-1}{h}{}^{\nu}_{\mu\nu;\sigma}k^{\sigma} + \frac{1}{2}\overset{n}{h}_{\mu\nu}k^{\nu}k^{\sigma}_{;\sigma} - \frac{1}{2}(-1)^n k^{\nu}\left(\overset{n-1}{h}{}^{\sigma}_{\mu\nu;\sigma} + 2R^{\sigma\tau}_{\mu\nu}\overset{n-1}{h}_{\sigma\tau}\right) \\
&= (-1)^n \overset{n-1}{h}{}^{\nu}_{\mu\nu;\sigma}k^{\sigma} + (-1)^n \overset{n-1}{h}_{\tau\nu}R^{\tau\nu}_{\mu\sigma}k^{\sigma} + (-1)^n \overset{n-1}{h}_{\mu\tau}R^{\tau\nu}_{\nu\sigma}k^{\sigma} + \frac{1}{2}(-1)^n \overset{n-1}{h}{}^{\nu}_{\mu\nu;}k^{\sigma}_{;\sigma} \\
&\quad - \frac{1}{2}(-1)^n\left(k^{\nu}\overset{n-1}{h}_{\mu\nu;\sigma}\right)^{\sigma}_{;} + \frac{1}{2}(-1)^n k^{\nu\sigma}_{;}\overset{n-1}{h}_{\mu\nu;\sigma} - (-1)^n k^{\nu}R^{\sigma\tau}_{\mu\nu}\overset{n-1}{h}_{\sigma\tau} \\
&= (-1)^n\left(\overset{n-1}{h}_{\mu\nu;\sigma}k^{\sigma}\right)^{\nu}_{;} - \frac{1}{2}(-1)^n \overset{n-1}{h}_{\mu\nu;\sigma}k^{\sigma\nu}_{;} + \frac{1}{2}(-1)^n \overset{n-1}{h}{}^{\nu}_{\mu\nu;}k^{\sigma}_{;\sigma} \\
&\quad - \frac{1}{2}(-1)^n\left(k^{\nu}\overset{n-1}{h}_{\mu\nu}\right)^{\sigma}_{;\sigma} + \frac{1}{2}(-1)^n\left(k^{\nu}_{;\sigma}\overset{n-1}{h}_{\mu\nu}\right)^{\sigma}_{;} \\
&= -\frac{1}{2}(-1)^n\left(\overset{n-1}{h}_{\mu\nu}k^{\sigma}_{;\sigma}\right)^{\nu}_{;} - \frac{1}{2}\left(\overset{n-2}{h}{}^{\sigma}_{\mu\nu;\sigma} + 2R^{\sigma\tau}_{\mu\nu}\overset{n-2}{h}_{\sigma\tau}\right)^{\nu}_{;} + \frac{1}{2}(-1)^n \overset{n-1}{h}{}^{\nu}_{\mu\nu;}k^{\sigma}_{;\sigma} \\
&\quad + \frac{1}{2}\overset{n-2}{h}{}^{\nu\sigma\sigma}_{\mu\nu;\sigma} + \frac{1}{2}(-1)^n k^{\nu\sigma}_{\sigma;}\overset{n-1}{h}_{\mu\nu} \\
&= -\frac{1}{2}(-1)^n k_{\tau}R^{\tau\nu\sigma}_{\sigma}\overset{n-1}{h}_{\mu\nu} - \frac{1}{2}\overset{n-2}{h}{}^{\nu\sigma}_{\mu\nu;\sigma} - \frac{1}{2}\overset{n-2}{h}_{\tau\nu;\sigma}R^{\tau\sigma\nu}_{\mu} - \frac{1}{2}\overset{n-2}{h}_{\mu\tau;\sigma}R^{\tau\sigma\nu}_{\nu} - \frac{1}{2}\overset{n-2}{h}_{\mu\nu;\tau}R^{\tau\sigma\nu}_{\sigma} \\
&\quad - R^{\sigma\tau\nu}_{\mu\nu;}\overset{n-2}{h}_{\sigma\tau} - R^{\sigma\tau}_{\mu\nu}\overset{n-2}{h}{}^{\nu}_{\sigma\tau;} + \frac{1}{2}\overset{n-2}{h}{}^{\nu\sigma}_{\mu\nu;\sigma} + \frac{1}{2}\left(\overset{n-2}{h}_{\tau\nu}R^{\tau\nu}_{\mu\sigma} + \overset{n-2}{h}_{\mu\tau}R^{\tau\nu}_{\nu\sigma}\right)^{\sigma}_{;} \\
&= 0.
\end{aligned}
$$

The coefficients $\overset{n}{f}_{\mu\nu}$ in the eikonal expansion for a given wave are not uniquely determined, for we may always carry out a gauge transformation

$$\overline{h}_{\mu\nu} = h_{\mu\nu} - \xi_{\mu;\nu} - \xi_{\nu;\mu},$$

where the gauge parameter ξ_{μ} satisfies the homogeneous wave equation

$$\xi^{\nu}_{\mu;\nu} = 0,$$

as well as the supplementary condition

$$\xi^{\mu}_{;\mu} = 0,$$

for such gauge transformations leave the conditions $h^{\mu\nu}_{;\nu} = 0$ and $h = 0$ intact. If we write an eikonal expansion for ξ_{μ},

$$\xi_{\mu} = a\left(\overset{0}{a}_{\mu} + \overset{2}{a}_{\mu} + \cdots\right)\sin\phi - a\left(\overset{1}{a}_{\mu} + \overset{3}{a}_{\mu} + \cdots\right)\cos\phi,$$

we find that the coefficients $\overset{n}{a}_{\mu}$ must satisfy

$$k \cdot \overset{0}{a} = 0, \quad k \cdot \overset{n}{a} = (-1)^n a^{-1}\left(a\overset{n-1}{a}{}^{\mu}\right)_{;\mu}, \quad n \geq 1,$$

$$\overset{0}{a}_{\mu;\nu}k^\nu = 0, \quad \overset{n}{a}_{\mu;\nu}k^\nu = \frac{1}{2}(-1)^n a^{-1}\left(a\overset{n-1}{a}_\mu\right)_{;\nu}^{\ \nu}, \quad n \geq 1,$$

and that the coefficients $\overset{n}{f}_{\mu\nu}$ suffer the gauge transformations

$$\overline{e}_{\mu\nu} = e_{\mu\nu} - \overset{0}{a}_\mu k_\nu - \overset{0}{a}_\nu k_\mu,$$

$$\overline{\overset{n}{f}}_{\mu\nu} = \overset{n}{f}_{\mu\nu} - \overset{n}{a}_\mu k_\nu - \overset{n}{a}_\nu k_\mu + (-1)^n a^{-1}\left[\left(a\overset{n-1}{a}_\mu\right)_{;\nu} + \left(a\overset{n-1}{a}_\nu\right)_{;\mu}\right].$$

These transformations of course leave the relations satisfied by the $\overset{n}{f}_{\mu\nu}$ unchanged:

$$\frac{1}{2}\overline{e}_{\mu\nu}\overline{e}^{\mu\nu} = \frac{1}{2}e_{\mu\nu}e^{\mu\nu} - 2e_{\mu\nu}\overset{0}{a}{}^\mu k^\nu + \overset{0}{a}{}^2 k^2 + (\overset{0}{a}\cdot k)^2 = 1,$$

$$\overline{e}^\mu_\mu = e^\mu_\mu - 2\overset{0}{a}\cdot k = 0,$$

$$\overline{\overset{n}{f}}{}^\mu_\mu = \overset{n}{f}{}^\mu_\mu - 2\overset{n}{a}\cdot k + 2(-1)^n a^{-1}\left(a\overset{n-1}{a}_\mu\right)_;^{\ \mu} = 0,$$

$$\overline{e}_{\mu\nu}k^\nu = e_{\mu\nu}k^\nu - \overset{0}{a}_\mu k^2 - k_\mu\left(\overset{0}{a}\cdot k\right) = 0,$$

$$\begin{aligned}
&\overline{\overset{n}{f}}_{\mu\nu}k^\nu - (-1)^n a^{-1}\left(a\overline{\overset{n-1}{f}}_{\mu\nu}\right)_;^{\ \nu} = \overset{n}{f}_{\mu\nu}k^\nu - \overset{n}{a}_\mu k^2 - k_\mu\left(\overset{n}{a}\cdot k\right) \\
&\quad + (-1)^n a^{-1}\left[\left(a\overset{n-1}{a}_\mu\right)_{;\nu} + \left(a\overset{n-1}{a}_\nu\right)_{;\mu}\right]k^\nu \\
&\quad - (-1)^n a^{-1}\left\{a\overset{n-1}{f}_{\mu\nu} - a\overset{n-1}{a}_\mu k_\nu - a\overset{n-1}{a}_\nu k_\mu + (-1)^{n-1}\left[\left(a\overset{n-2}{a}_\mu\right)_{;\nu} + \left(a\overset{n-2}{a}_\nu\right)_{;\mu}\right]\right\}_;^{\ \nu} \\
&\quad = \overset{n-2}{a}{}_\sigma R^{\sigma\nu}_{\ \ \nu\mu} = 0,
\end{aligned}$$

$$\overline{e}_{\mu\nu;\sigma}k^\sigma = (e_{\mu\nu} - \overset{0}{a}_\mu k_\nu - \overset{0}{a}_\nu k_\mu)_{;\sigma}k^\sigma = 0,$$

$$\begin{aligned}
&\overline{\overset{n}{f}}_{\mu\nu;\sigma}k^\sigma - \frac{1}{2}(-1)^n\left[a^{-1}\left(a\overline{\overset{n-1}{f}}_{\mu\nu}\right)_{;\sigma}^{\ \sigma} + 2R^{\sigma\tau}_{\mu\nu}\overline{\overset{n-1}{f}}_{\sigma\tau}\right] \\
&\quad = \overset{n}{f}_{\mu\nu;\sigma}k^\sigma - \left(\overset{n}{a}_\mu k_\nu + \overset{n}{a}_\nu k_\mu\right)_{;\sigma}k^\sigma + (-1)^n\left\{a^{-1}\left[\left(a\overset{n-1}{a}_\mu\right)_{;\nu} + \left(a\overset{n-1}{a}_\nu\right)_{;\mu}\right]\right\}_{;\sigma}k^\sigma \\
&\quad - \frac{1}{2}(-1)^n a^{-1}\left(a\overset{n-1}{f}_{\mu\nu}\right)_{;\sigma}^{\ \sigma} + \frac{1}{2}(-1)^n a^{-1}\left[a\left(\overset{n-1}{a}_\nu + \overset{n-1}{a}_\nu k_\mu\right)\right]_{;\sigma}^{\ \sigma}
\end{aligned}$$

$$
\begin{aligned}
&+\frac{1}{2}a^{-1}\left[\left(a\overset{n-2}{a}{}_{\mu}\right)_{;\nu}+\left(a\overset{n-2}{a}{}_{\nu}\right)_{;\mu}\right]^{\sigma}_{;\sigma}-(-1)^{n}R^{\sigma\tau}_{\mu\nu}\overset{n-1}{f}{}_{\sigma\tau}\\
&+(-1)^{n}R^{\sigma\tau}_{\mu\nu}\left(\overset{n-1}{a}{}_{\tau}+\overset{n-1}{a}{}_{\tau}k_{\sigma}\right)+R^{\sigma\tau}_{\mu\nu}a^{-1}\left[\left(a\overset{n-2}{a}{}_{\sigma}\right)_{;\tau}+\left(a\overset{n-2}{a}{}_{\tau}\right)_{;\sigma}\right]\\
=&\frac{1}{2}(-1)^{n}a^{-1}\left[\left(a\overset{n-1}{a}{}_{\mu}\right)_{;\nu}+\left(a\overset{n-1}{a}{}_{\nu}\right)_{;\mu}\right]k^{\sigma}_{;\sigma}+(-1)^{n}a^{-1}\left(a\overset{n-1}{a}{}_{\mu}\right)_{;\sigma\nu}k^{\sigma}\\
&+(-1)^{n}\overset{n-1}{a}{}_{\tau}R^{\tau}_{\mu\nu\sigma}k^{\sigma}+(-1)^{n}a^{-1}\left(a\overset{n-1}{a}{}_{\nu}\right)_{;\sigma\mu}k^{\sigma}+(-1)^{n}a^{-1}\overset{n-1}{a}{}_{\tau}R^{\tau}_{\nu\mu\sigma}k^{\sigma}\\
&+(-1)^{n}a^{-1}\left(a\overset{n-1}{a}{}_{\mu}\right)_{;\sigma}k^{\sigma}_{\nu;}+(-1)^{n}a^{-1}\left(a\overset{n-1}{a}{}_{\nu}\right)_{;\sigma}k^{\sigma}_{\mu;}+\frac{1}{2}(-1)^{n}\overset{n-1}{a}{}_{\mu}k^{\sigma}_{\nu;\sigma}\\
&+\frac{1}{2}(-1)^{n}\overset{n-1}{a}{}_{\nu}k^{\sigma}_{\mu;\sigma}+\frac{1}{2}a^{-1}\left[\left(a\overset{n-2}{a}{}_{\mu}\right)_{;\nu}+\left(a\overset{n-2}{a}{}_{\nu}\right)_{;\mu}\right]^{\sigma}_{;\sigma}+(-1)^{n}R^{\sigma\tau}_{\mu\nu}\left(\overset{n-1}{a}{}_{\tau}+\overset{n-1}{a}{}_{\tau}k_{\sigma}\right)\\
&+R^{\sigma\tau}_{\mu\nu}a^{-1}\left[\left(a\overset{n-2}{a}{}_{\sigma}\right)_{;\tau}+\left(a\overset{n-2}{a}{}_{\tau}\right)_{;\sigma}\right]\\
=&\frac{1}{2}(-1)^{n}a^{-1}\left[\left(a\overset{n-1}{a}{}_{\mu}\right)_{;\nu}+\left(a\overset{n-1}{a}{}_{\nu}\right)_{;\mu}\right]k^{\sigma}_{;\sigma}\\
&-\frac{1}{2}(-1)^{n}a^{-1}\left(a\overset{n-1}{a}{}_{\mu}k^{\sigma}_{;\sigma}\right)_{;\nu}-\frac{1}{2}a^{-1}\left(a\overset{n-2}{a}{}_{\mu}\right)^{\sigma}_{;\sigma\nu}-\frac{1}{2}(-1)^{n}a^{-1}\left(a\overset{n-1}{a}{}_{\nu}k^{\sigma}_{;\sigma}\right)_{;\mu}\\
&-\frac{1}{2}a^{-1}\left(a\overset{n-2}{a}{}_{\nu}\right)^{\sigma}_{;\sigma\mu}+\frac{1}{2}(-1)^{n}\overset{n-1}{a}{}_{\mu}k^{\sigma}_{\nu;\sigma}+\frac{1}{2}(-1)^{n}\overset{n-1}{a}{}_{\nu}k^{\sigma}_{\mu;\sigma}\\
&+\frac{1}{2}a^{-1}\left[\left(a\overset{n-2}{a}{}_{\mu}\right)_{;\nu}+\left(a\overset{n-2}{a}{}_{\nu}\right)_{;\mu}\right]^{\sigma}_{;\sigma}+R^{\sigma\tau}_{\mu\nu}a^{-1}\left[\left(a\overset{n-2}{a}{}_{\sigma}\right)_{;\tau}+\left(a\overset{n-2}{a}{}_{\tau}\right)_{;\sigma}\right]\\
=&-\frac{1}{2}(-1)^{n}\overset{n-1}{a}{}_{\mu}k^{\tau}R^{\sigma}_{\tau\sigma\nu}-\frac{1}{2}(-1)^{n}\overset{n-1}{a}{}_{\nu}k^{\tau\sigma\mu}\sigma-\frac{1}{2}a^{-1}\left(a\overset{n-2}{a}{}_{\tau}\right)_{;\sigma}R^{\tau\sigma}_{\mu\nu}\\
&-\frac{1}{2}a^{-1}\left(a\overset{n-1}{a}{}_{\mu}\right)_{;\tau}R^{\tau\sigma}_{\sigma\nu}+\frac{1}{2}a^{-1}\left(a\overset{n-2}{a}{}_{\tau\mu\nu\sigma}\right)^{\sigma}_{;}-\frac{1}{2}a^{-1}\left(a\overset{n-2}{a}{}_{\tau}\right)_{;\sigma}R^{\tau\sigma}_{\nu\mu}\\
&-\frac{1}{2}a^{-1}\left(a\overset{n-1}{a}{}_{\nu}\right)_{;\tau}R^{\tau\sigma}_{\sigma\mu}+\frac{1}{2}a^{-1}\left(a\overset{n-2}{a}{}_{\tau\nu\mu\sigma}\right)^{\sigma}_{;}+R^{\sigma\tau}_{\mu\nu}a^{-1}\left[\left(a\overset{n-2}{a}{}_{\sigma}\right)_{;\tau}+\left(a\overset{n-2}{a}{}_{\tau}\right)_{;\sigma}\right]=0.
\end{aligned}
$$

Now suppose the wave sweeps over a pair of test particles initially at rest relative to one another and separated by a small interval η^{α} ($\ll\lambda$) satisfying $\eta\cdot u=0$, where u is the 4-velocity of either of the particles. Let us also suppose that $1/\mathcal{R}^2\ll a/\lambda^2$, so that the contribution of the wave to the total Riemannian tensor is much greater than that of the background geometry. [Note that this implies $a^2\lesssim\lambda^2/\mathcal{R}^2\ll a$ (see present chapter), which is consistent with $a\ll 1$.] Then the wave dominates the equation of geodesic deviation [see (5.3) in Chap. 5] which governs the behavior of η^{α}. Using the expression (8.2) for $\delta R^{\tau}_{\sigma\mu\nu}$ given in Chap. 8, we find[3]

---

[3] The quantity in the parentheses in this equation is gauge invariant to order a/λ^2. The u^{α} are 4-velocities normalized to unity relative to $g^{\text{tot}}_{\mu\nu}$.

$$\frac{D^2\eta^\alpha}{D\tau_{\text{tot}}^2} = -\frac{1}{2}(h^\alpha_{\gamma;\beta\delta} + h^\alpha_{\beta\delta;\gamma} - h^\alpha_{\delta;\beta\gamma} - h^\alpha_{\beta\gamma;\delta} + R^{\varepsilon\alpha}_{\gamma\delta}h_{\varepsilon\beta} - R^\varepsilon_{\beta\gamma\delta}h^\alpha_\varepsilon)u^\beta u^\gamma \eta^\delta,$$

where $D/D\tau_{\text{tot}}$ denotes covariant proper time differentiation based on $g^{\text{tot}}_{\mu\nu}$. Keeping only the dominant term of the eikonal expansion, i.e.,

$$h_{\alpha\beta} = ae_{\alpha\beta}\cos\phi, \quad h_{\alpha\beta;\gamma\delta} = -ae_{\alpha\beta}k_\gamma k_\delta \cos\phi,$$

and dropping the terms in the background Riemann tensor (of order $a/\mathcal{R}^2$), we convert this equation to

$$\frac{D^2\eta^\alpha}{D\tau_{\text{tot}}^2} = \frac{1}{2}a\Big(e^\alpha_\gamma k_\beta k_\delta + e_{\beta\delta}k^\alpha k_\gamma - e^\alpha_\delta k_\beta k_\gamma - e_{\beta\gamma}k^\alpha k_\delta\Big)u^\beta u^\gamma \eta^\delta \cos\phi.$$

A further simplification is achieved by carrying out a gauge transformation with a gauge parameter of the form described on p. 189, whose coefficient $\overset{0}{a}_\mu$ in its eikonal expression is given, along the world line of the test particle pair,[4] by

$$\overset{0}{a}_\mu = (k\cdot u)^{-1}e_{\mu\nu}u^\nu - \frac{1}{2}(k\cdot u)^{-2}k_\mu u^\nu u^\sigma e_{\nu\sigma}.$$

The new polarization tensor then becomes

$$\bar{e}_{\mu\nu} = e_{\mu\nu} - (k\cdot u)^{-1}k_\mu e_{\nu\sigma}u^\sigma - (k\cdot u)^{-1}k_\nu e_{\mu\sigma}u^\sigma + (k\cdot u)^{-2}k_\mu k_\nu u^\sigma u^\tau e_{\sigma\tau},$$

and satisfies

$$\bar{e}^\mu_\mu = 0, \quad \frac{1}{2}\bar{e}_{\mu\nu}\bar{e}^{\mu\nu} = 1,$$

$$\bar{e}_{\mu\nu}k^\nu = 0, \quad \bar{e}_{\mu\nu}u^\nu = 0.$$

We shall assume this transformation already to have been carried out and drop the bar. The equation of geodesic deviation then reduces to

$$\frac{D^2\eta^\alpha}{D\tau_{\text{tot}}^2} = -\frac{1}{2}ae^\alpha_\beta\eta^\beta(k\cdot u)^2\cos\phi.$$

In order to integrate this equation we introduce a local rest frame for the test particle pair, which is Fermi–Walker transported, relative to the *total* metric $g^{\text{tot}}_{\mu\nu}$ along the world line of the pair. Using parentheses around indices to denote components relative to the local frame, we have

---

[4] The coefficient $\overset{0}{a}_\mu$ generally cannot be chosen in this way throughout all of spacetime unless the background geometry is flat.

$$\frac{d^2\eta^{(\alpha)}}{d\tau_{tot}^2} = -\frac{1}{2}ae^{(\alpha)}_{(\beta)}\eta^{(\beta)}k_{(0)}^2 \cos\phi,$$

which may be rewritten effectively in the form

$$\frac{d^2\eta^{(\alpha)}}{d\tau_{tot}^2} = \frac{1}{2}\frac{d^2}{d\tau_{tot}^2}\left[ae^{(\alpha)}_{(\beta)}\eta^{(\beta)} \cos\phi\right],$$

because the proper time derivatives of $a, e^{(\alpha)}_{(\beta)}, \eta^{(\beta)}$, and $k_{(0)}$ are small compared to $k_{(0)}\, a$, $k_{(0)}\, e^{(\alpha)}_{(\beta)}$, $k_{(0)}\eta^{(\beta)}$, and $k_{(0)}^2$, respectively. The condition $a \ll 1$ now allows us to write the following solution:

$$\eta^{(\alpha)}(\tau_{tot}) = \left[\delta^{(\alpha)}_{(\beta)} + \frac{1}{2}ae^{(\alpha)}_{(\beta)} \cos\phi\right]\eta^{(\beta)}(0).$$

The motion of the test particle pair provides a simple characterization of the polarization of a gravitational wave. Let us for simplicity drop the parentheses on the indices above and replace τ_{tot} by t, the local time. Then the motion in the local frame is given by the formula

$$\eta_i(t) = \left(\delta_{ij} + \frac{1}{2}ae_{ij}\cos\omega t\right)\eta_j(0), \quad \omega = -k_{(0)}.$$

The temporal components of the polarization tensor $e_{\alpha\beta}$ vanish in this frame so that it becomes effectively a 3-tensor e_{ij}. In fact, if the 3-axis is chosen in the direction of $\hat{\boldsymbol{k}}(=\boldsymbol{k}/\omega)$, then it becomes effectively a 2-tensor because of the condition

$$e_{ij}\hat{k}_j = 0.$$

The remaining conditions

$$e_{ii} = 0, \quad \frac{1}{2}e_{ij}e_{ij} = 1$$

tell us, furthermore, that the 1-axis and 2-axis may be rotated into a position for which e_{ij} takes the form

$$(e_{ij}) = \begin{pmatrix} 1 & 0 & 0 \\ 0 & -1 & 0 \\ 0 & 0 & 0 \end{pmatrix}.$$

The relative motion of the two particles is seen to be at right angles to the direction of propagation of the wave. Thus gravitational waves, like electromagnetic waves, are *transverse waves*, i.e., their dominant action on test bodies is transverse to the propagation direction. The relative motion of the test particles also depends on the orientation of the vector η_i. The effect of the relative orientation is best displayed by replacing the pair of particles by a ring of particles forming, in the absence of

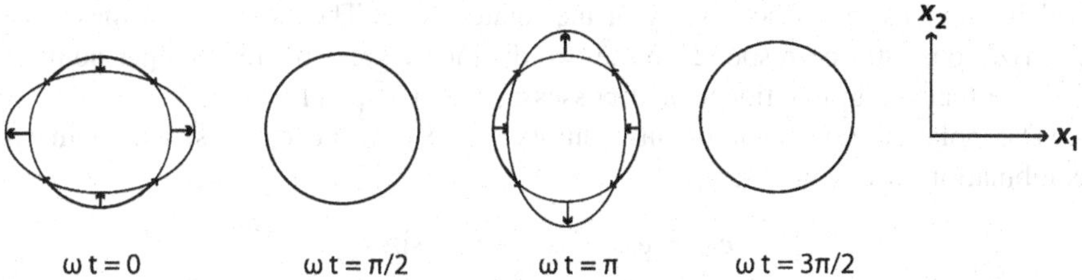

Fig. 15.1 Effect of a gravitational wave on a ring of test particles

the wave, a circle in the (1,2) plane. In the presence of the wave, the circle suffers the sequence of distortions in time depicted in the Fig. 15.1.

We shall use a special symbol, namely, $e_{+\mu\nu}$, for the polarization tensor characterizing nearly monochromatic gravitational waves giving rise to test particle motions having the above transverse orientation. This tensor may be decomposed into products of unit vectors as follows:

$$e_{+\mu\nu} = e_{\mathrm{I}\mu} e_{\mathrm{I}\nu} - e_{\mathrm{II}\mu} e_{\mathrm{II}\nu}, \tag{15.9}$$

where, in the special local frame,

$$(e_{\mathrm{I}\mu}) = (0, 1, 0, 0), \quad (e_{\mathrm{II}\mu}) = (0, 0, 1, 0).$$

We note that the state of the wave remains invariant under a rotation about $\hat{\boldsymbol{k}}$ through 180°, whereas in the electromagnetic case a rotation through 360° is required to return to the same state. Moreover, the negative field is obtained by a rotation through only 90°, whereas in the electromagnetic case a rotation through 180° is required. How do we get to an orthogonal state of polarization? In the electromagnetic case, the orthogonal state is obtained by rotation through 90°. In analogy with the above results we expect that the orthogonal state is attained in the gravitational case by rotating through only 45°. We introduce the new unit vectors

$$e'_{\mathrm{I}\mu} = \frac{1}{\sqrt{2}}(e_{\mathrm{I}\mu} + e_{\mathrm{II}\mu}), \quad e'_{\mathrm{II}\mu} = \frac{1}{\sqrt{2}}(-e_{\mathrm{I}\mu} + e_{\mathrm{II}\mu}), \quad e'_{\mathrm{I}} \cdot e'_{\mathrm{II}} = 0,$$

and define

$$\begin{aligned} e_{\times\mu\nu} &= e'_{\mathrm{I}\mu} e'_{\mathrm{I}\nu} - e'_{\mathrm{II}\mu} e'_{\mathrm{II}\nu} = \frac{1}{2}(e_{\mathrm{I}\mu} + e_{\mathrm{II}\mu})(e_{\mathrm{I}\nu} + e_{\mathrm{II}\nu}) - \frac{1}{2}(-e_{\mathrm{I}\mu} + e_{\mathrm{II}\mu})(-e_{\mathrm{I}\nu} + e_{\mathrm{II}\nu}) \\ &= e_{\mathrm{I}\mu} e_{\mathrm{II}\nu} + e_{\mathrm{I}\nu} e_{\mathrm{II}\mu}. \end{aligned} \tag{15.10}$$

We have

$$e^{\mu}_{\times\mu} = 0, \quad \frac{1}{2} e_{\times\mu\nu} e^{\mu\nu}_{\times} = 1, \quad e_{\times\mu\nu} e^{\mu\nu}_{+} = 0,$$

which confirms the orthogonality of the rotated state. The existence of these two polarization states corresponds to the two distinct degrees of freedom per point of 3-space that the gravitational field possesses (see Chap. 11).

The polarization tensor of an arbitrary wave may be expressed as a linear combination of $e_{+\mu\nu}$ and $e_{\times\mu\nu}$:

$$e_{\mu\nu} = e_{+\mu\nu} \cos\alpha + e_{\times\mu\nu} \sin\alpha,$$

$$(e_{ij}) = \begin{pmatrix} \cos\alpha & \sin\alpha & 0 \\ \sin\alpha & -\cos\alpha & 0 \\ 0 & 0 & 0 \end{pmatrix}.$$

Actually this canonical form can be employed not merely along the world line of the test particle pair, but throughout spacetime, with $e_{\mathrm{I}\mu}$ and $e_{\mathrm{II}\mu}$ chosen to be the orthonormal unit vectors introduced in Chap. 14 to describe the polarization of electromagnetic waves. The proof is as follows. First choose a spacelike[5] hypersurface Σ and introduce a set of tangent orthonormal triads throughout it, with the third member of each triad pointing parallel to the projection of k^μ on Σ at that point. $e_{\mathrm{I}\mu}$ and $e_{\mathrm{II}\mu}$ may be chosen as the other two members of each triad. Next, introduce a set of timelike unit vectors orthogonal to Σ and adjoin to them the triads, to form a set of orthonormal tetrads throughout Σ. Finally, carry the tetrads by parallel transport along the null geodesics generated by k^μ, thus defining a set of local canonical frames throughout spacetime.

In each of the local frames $e_{\mathrm{I}\mu}$ and $e_{\mathrm{II}\mu}$ will point in the direction of two of the axes, and the propagation vector k^μ will take the form

$$(k^\mu) = (\omega, 0, 0, \omega).$$

The six conditions $e^\mu_\mu = 0$, $e_{\mu\nu}k^\nu = 0$, and $e_{\mu\nu}\, e^{\mu\nu} = 2$ then tell us that the most general polarization tensor $e_{\mu\nu}$ takes in each frame the form

$$(e_{\mu\nu}) = \begin{pmatrix} a & b & c & -a \\ b & \cos\alpha & \sin\alpha & -b \\ c & \sin\alpha & -\cos\alpha & -c \\ -a & -b & -c & a \end{pmatrix},$$

$$(e^{\mu\nu}) = \begin{pmatrix} a & -b & -c & a \\ -b & \cos\alpha & \sin\alpha & -b \\ -c & \sin\alpha & -\cos\alpha & -c \\ a & -b & -c & a \end{pmatrix},$$

so that there are 10 components, 6 conditions, and 4 independent parameters a, b, c, and α. Moreover, by virtue of the conditions $k_{\mu;\nu}k^\nu = 0$ and $e_{\mu\nu;\sigma}k^\sigma = 0$ [see (15.7) and (15.8) in Sect. 15.1], the parameters a, b, c, α, and ω will be

[5] The words 'spacelike', 'orthonormal', 'parallel transport', etc., are all to be understood as relative to the background geometry.

constant along each geodesic generated by k^{μ}. Now carry out the gauge transformation described in Sect. 15.1, with the coefficients $\overset{0}{a}_{\mu}$ chosen to have, in each local frame, the form

$$\left(\overset{0}{a}_{\mu}\right) = -\frac{1}{\omega}\left(\frac{1}{2}a, b, c, -\frac{1}{2}a\right).$$

Then $\overset{0}{a} \cdot k = 0, \overset{0}{a}_{\mu;\nu}k^{\nu} = 0$, and

$$\begin{aligned}
\left(\bar{e}_{\mu\nu}\right) &= \left(e_{\mu\nu} - \overset{0}{a}_{\mu}k_{\nu} - \overset{0}{a}_{\nu}k_{\mu}\right) \\
&= (e_{\mu\nu}) + \frac{1}{\omega}\begin{pmatrix} a/2 \\ b \\ c \\ -a/2 \end{pmatrix}(-\omega, 0, 0, \omega) + \frac{1}{\omega}\begin{pmatrix} -\omega \\ 0 \\ 0 \\ \omega \end{pmatrix}\left(\frac{1}{2}a, b, c, -\frac{1}{2}a\right) \\
&= (e_{\mu\nu}) + \begin{pmatrix} -a/2 & 0 & 0 & a/2 \\ -b & 0 & 0 & b \\ -c & 0 & 0 & c \\ a/2 & 0 & 0 & -a/2 \end{pmatrix} + \begin{pmatrix} -a/2 & -b & -c & a/2 \\ 0 & 0 & 0 & 0 \\ 0 & 0 & 0 & 0 \\ a/2 & b & c & -a/2 \end{pmatrix} \\
&= \begin{pmatrix} 0 & 0 & 0 & 0 \\ 0 & \cos\alpha & \sin\alpha & 0 \\ 0 & \sin\alpha & -\cos\alpha & 0 \\ 0 & 0 & 0 & 0 \end{pmatrix},
\end{aligned}$$

as required.

From now on we shall assume that the polarization tensor has this canonical form and drop the bar. We then write

$$e_{\mu\nu} = e_{+\mu\nu}\cos\alpha + e_{\times\mu\nu}\sin\alpha,$$

where $e_{+\mu\nu}$ and $e_{\times\mu\nu}$ are given in terms of $e_{\mathrm{I}\mu}$ and $e_{\mathrm{II}\mu}$ by (15.9) and (15.10) in Sect. 15.1, respectively. The conditions

$$e_{\mathrm{I}\mu;\nu}k^{\nu} = 0, \quad e_{\mathrm{II}\mu;\nu}k^{\nu} = 0,$$

insure that

$$e_{+\mu\nu;\sigma}k^{\sigma} = 0, \quad e_{\times\mu\nu;\sigma}k^{\sigma} = 0,$$

and the condition $e_{\mu\nu;\sigma}k^{\sigma} = 0$ requires that α be constant along each geodesic generated by k^{μ}, i.e.,

$$\alpha_{;\mu}k^{\mu} = 0.$$

15.2 Lines-of-Force Representation and Circularly Polarized Waves

Consider the ring of test particles introduced in Sect. 15.1. If this ring is centered on the origin then the particle at position x_i moves according to the law

$$\ddot{x}_i = F_i,$$

where the force per unit mass F_i has the form

$$F_i = -\frac{1}{2}\omega^2 a e_{ij} x_j \cos \omega t.$$

The magnitude of this force increases as the magnitude of x increases. Moreover, it satisfies the divergence law

$$F_{i,i} = -\frac{1}{2}\omega^2 a e_{ii} \cos \omega t = 0,$$

and hence the force *field* may be represented by a lines-of-force diagram, with the density of lines proportional to the magnitude of F. The lines-of-force diagram appropriate to the instant $t = 0$ for the state of polarization depicted in Fig. 15.1 is shown in Fig. 15.2.

The density of lines increases as distance from the origin increases. The direction of the field lines reverses after half a period ($\omega t = \pi$), and the field vanishes at $\omega t = \pi/2$, $3\pi/2$, etc. The lines-of-force diagram appropriate to the orthogonal state of polarization ($e_{\times \mu\nu}$) is obtained from the above diagram by rotation through 45°.

Waves of the type we have been considering up to now are all said to be linearly polarized. On the other hand, elliptically polarized waves are also

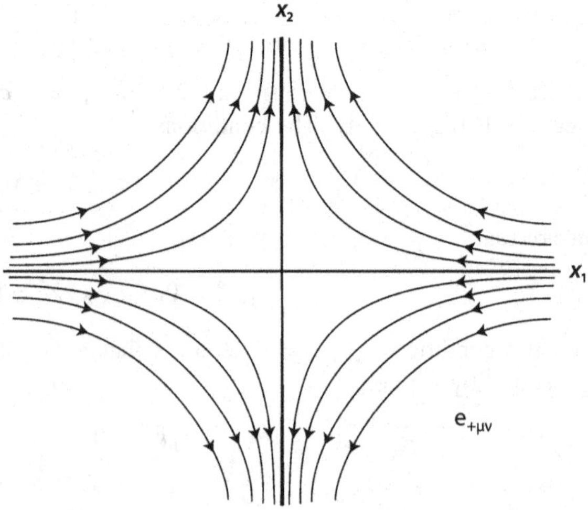

Fig. 15.2 Lines-of-force diagram at the instant $t = 0$ for the state of polarization depicted in Fig. 15.1

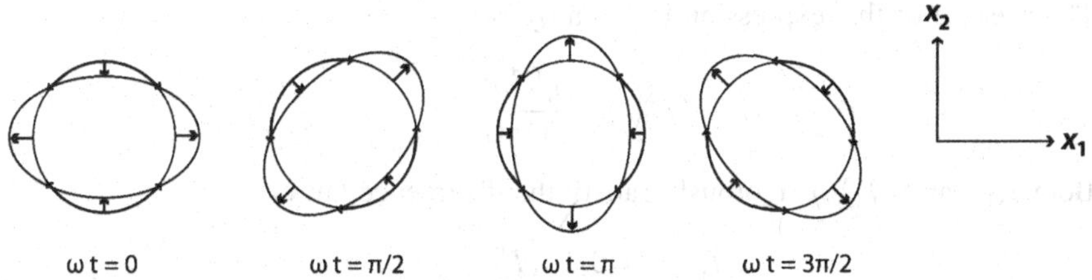

Fig. 15.3 Effect of a circularly polarized gravitational wave with right-handed helicity on a ring of test particles. The wave is propagating out of the paper

possible, just as they are in the electromagnetic case. These are obtained by superposing two linearly polarized waves 90° out of phase:

$$h_{\mu\nu} = a_{\rm I}(e_{+\mu\nu} + \cdots)\cos\phi - a_{\rm II}(e_{\times\mu\nu} + \cdots)\sin\phi.$$

If $a_{\rm I} = a_{\rm II}$, the wave is circularly polarized with right-handed helicity. If $a_{\rm I} = -a_{\rm II}$, the wave is circularly polarized with left-handed helicity. Figure 15.3 shows the successive distortions of the test particle ring in the case of circular polarization with right-handed helicity. It will be noted that each particle in the ring executes a small circle once each period, and the ring pattern as a whole rotates through 180° in the right-handed sense each period. If one uses a lines-of-force diagram to represent a circularly polarized wave then the diagram must be rotated through 180° each period.

15.3 Energy, Momentum, and Angular Momentum in Gravitational Waves

The energy–momentum–stress density $T^{\mu\nu}_{\rm GW}$ is easily computed in the case of nearly monochromatic gravitational waves. Keeping only the dominant terms in the eikonal approximation, we have for a linearly polarized gravitational wave

$$h_{\mu\nu} = ae_{\mu\nu}\cos\phi, \quad h_{\mu\nu;\sigma} = -ae_{\mu\nu}k_\sigma \sin\phi, \quad h_{\mu\nu;\sigma\tau} = -ae_{\mu\nu}k_\sigma k_\tau \cos\phi.$$

Inserting these expressions into the expression (15.6) for $T^{\mu\nu}_{\rm GW}$ in present chapter, and making use of the relations satisfied by the polarization tensor $e_{\mu\nu}$ (see Sect. 15.1), we find that the only non-vanishing terms are those in $h^{\sigma\tau}\, h^{\mu\nu}_{\sigma\tau;}$ and $h^{\sigma\tau\mu}_{;}\, h^{\nu}_{\sigma\tau;}$. Reintroducing the gravitation constant, we, therefore, get

$$\begin{aligned} T^{\mu\nu}_{\rm GW} &= \frac{1}{16\pi G} g^{1/2}\left(-h^{\sigma\tau}h^{\mu\nu}_{\sigma\tau;} - \frac{1}{2}h^{\sigma\tau\mu}_{;}h^{\nu}_{\sigma\tau;}\right) \\ &= \frac{g^{1/2}a^2}{16\pi G} k^\mu k^\nu (2\cos^2\phi - \sin^2\phi) \\ &= \frac{g^{1/2}a^2}{16\pi G} k^\mu k^\nu (\cos^2\phi + \cos 2\phi). \end{aligned}$$

The average of this expression is given by

$$\langle T^{\mu\nu}_{\mathrm{GW}}\rangle = \frac{g^{1/2}a^2}{32\pi G}k^\mu k^\nu.$$

Both $T^{\mu\nu}_{\mathrm{GW}}$ and $\langle T^{\mu\nu}_{\mathrm{GW}}\rangle$ rigorously satisfy the divergence laws

$$T^{\mu\nu}_{\mathrm{GW};\nu} = 0, \quad \langle T^{\mu\nu}_{\mathrm{GW}}\rangle_{;\nu} = 0,$$

in this approximation, by virtue of the relations $k^2 = 0$, $k^\mu_{;\nu}k^\nu = 0$, and $(a^2k^\nu)_{;\nu} = 0$.

In order to discuss the angular momentum in the wave, we shall confine our attention, as in the electromagnetic case, to a nearly monochromatic plane circularly polarized wave propagating in a flat spacetime. Again it is necessary to include the next higher terms in the eikonal expansion. For right-handed helicity we have

$$h_{\mu\nu} = a(e_{+\mu\nu}\cos\phi - e_{\times\mu\nu}\sin\phi) + \left(\overset{1}{f}_{+\mu\nu}\sin\phi + \overset{1}{f}_{\times\mu\nu}\cos\phi\right),$$

$$\begin{aligned} h_{\mu\nu,\sigma} = &-ak_\sigma(e_{+\mu\nu}\sin\phi + e_{\times\mu\nu}\cos\phi) + \left(a_{,\sigma}e_{+\mu\nu} + aa\overset{1}{f}_{+\mu\nu}k_\sigma\right)\cos\phi \\ &- \left(a_{,\sigma}e_{\times\mu\nu} + a\overset{1}{f}_{\times\mu\nu}k_\sigma\right)\sin\phi, \end{aligned}$$

$$\begin{aligned} h_{\mu\nu,\sigma\tau} = &-ak_\sigma k_\tau(e_{+\mu\nu}\cos\phi - e_{\times\mu\nu}\sin\phi) \\ &- \left[(a_{,\sigma}k_\tau + a_{,\tau}k_\sigma)e_{+\mu\nu} + a\overset{1}{f}_{+\mu\nu}k_\sigma k_\tau\right]\sin\phi \\ &- \left[(a_{,\sigma}k_\tau + a_{,\tau}k_\sigma)e_{\times\mu\nu} + a\overset{1}{f}_{\times\mu\nu}k_\sigma k_\tau\right]\cos\phi. \end{aligned}$$

Here k_μ and the polarization tensors are constants, and a and the f's satisfy (see Sect. 15.1)

$$a_{,\mu}k^\mu = 0, \quad \overset{1}{f}_{+\mu\nu}k^\mu k^\nu = 0, \quad \overset{1}{f}_{\times\mu\nu}k^\mu k^\nu = 0,$$

$$\overset{1}{f}_{+\mu\nu}k^\nu = -a^{-1}e_{+\mu\nu}a_{,}^{\ \nu}, \quad \overset{1}{f}_{\times\mu\nu}k^\nu = -a^{-1}e_{\times\mu\nu}a_{,}^{\ \nu},$$

$$\overset{1}{f}_{+\mu\nu,\sigma}k^\sigma = -\frac{1}{2}a^{-1}a^{\ \sigma}_{,\sigma}e_{+\mu\nu}, \quad \overset{1}{f}_{\times\mu\nu,\sigma}k^\sigma = -\frac{1}{2}a^{-1}a^{\ \sigma}_{,\sigma}e_{\times\mu\nu}.$$

Evidently we may choose the f's so as to satisfy

$$e_{+\mu\nu}\overset{1}{f}{}^{\mu\nu}_{+} = e_{\times\mu\nu}\overset{1}{f}{}^{\mu\nu}_{\times}, \quad e_{+\mu\nu}\overset{1}{f}{}^{\mu\nu}_{\times} = e_{\times\mu\nu}\overset{1}{f}{}^{\mu\nu}_{+} = 0.$$

We shall also assume the gauge to be chosen so that the polarization tensors have their canonical forms in terms of the orthonormal (constant) vectors $e_{\mathrm{I}\mu}$ and $e_{\mathrm{II}\mu}$ [see (15.9) and (15.10)]. We then have the additional relations

$$e^{\mu}_{+\mu\sigma}e^{\sigma\nu}_{+} = (e^{\mu}_{\mathrm{I}}e_{\mathrm{I}\sigma} - e^{\mu}_{\mathrm{II}}e_{I\sigma})(e^{\sigma}_{\mathrm{I}}e^{\nu}_{\mathrm{I}} - e^{\sigma}_{\mathrm{II}}e^{\nu}_{\mathrm{II}}) = e^{\mu}_{\mathrm{I}}e^{\nu}_{\mathrm{I}} + e^{\mu}_{\mathrm{II}}e^{\nu}_{\mathrm{II}},$$

$$e^{\mu}_{\times\mu\sigma}e^{\sigma\nu}_{\times} = (e^{\mu}e_{\mathrm{II}\sigma} + e^{\mu}_{\mathrm{II}}e_{\mathrm{II}\sigma})(e^{\sigma}_{\mathrm{I}}e^{\nu}_{\mathrm{II}} + e^{\sigma}_{\mathrm{II}}e^{\nu}_{\mathrm{I}}) = e^{\mu}_{\mathrm{I}}e^{\nu}_{\mathrm{I}} + e^{\mu}_{\mathrm{II}}e^{\nu}_{\mathrm{II}},$$

$$e^{\mu}_{+\mu\sigma}e^{\sigma\nu}_{\times} = (e^{\mu}_{\mathrm{I}}e_{\mathrm{I}\sigma} - e^{\mu}_{\mathrm{II}}e_{\mathrm{II}\sigma})(e^{\sigma}_{\mathrm{I}}e^{\nu}_{\mathrm{II}} + e^{\sigma}_{\mathrm{II}}e^{\nu}_{\mathrm{I}}) = e^{\mu}_{\mathrm{I}}e^{\nu}_{\mathrm{II}} - e^{\mu}_{\mathrm{II}}e^{\nu}_{\mathrm{I}},$$

$$e^{\mu}_{\times\mu\sigma}e^{+\sigma\nu}_{+} = (e^{\mu}_{\mathrm{I}}e_{\mathrm{II}\sigma} + e^{\mu}_{\mathrm{II}}e_{\mathrm{I}\sigma})(e^{\sigma}_{\mathrm{I}}e^{\nu}_{\mathrm{I}} - e^{\sigma}_{\mathrm{II}}e^{\nu}_{\mathrm{II}}) = -e^{\mu}_{\mathrm{I}}e^{\nu}_{\mathrm{II}} + e^{\mu}_{\mathrm{II}}e^{\nu}_{\mathrm{I}}.$$

We are now ready to compute the following quantities:

$$\begin{aligned}
h^{\sigma\tau}h^{\mu\nu}_{\sigma,\tau} &= -a(e^{\sigma\tau}_{+}\cos\phi - e^{\sigma\tau}_{\times}\sin\phi)\left[\left(a^{\nu}_{,}k_{\tau} + a_{,\tau}k^{\nu}\right)e^{\mu}_{+\sigma} + a\overset{1}{f}{}^{\mu}_{+\sigma}k^{\nu}k_{\tau}\right]\sin\phi \\
&\quad - a(e^{\sigma\tau}_{+}\cos\phi - e^{\sigma\tau}_{\times}\sin\phi)\left[\left(a^{\nu}_{,}k_{\tau} + a_{,\tau}k^{\nu}\right)e^{\mu}_{\times\sigma} + a\overset{1}{f}{}^{\mu}_{\times\sigma}k^{\tau}\right]\cos\phi \\
&\quad - a^{2}\left(\overset{1}{f}{}^{\sigma\tau}_{+}\sin\phi + \overset{1}{f}{}^{\sigma\tau}_{\times}\cos\phi\right)k^{\nu}k_{\tau}(e^{\mu}_{+\mu\sigma}\cos\phi - e^{\mu}_{\times\sigma}\sin\phi) \\
&= -aa_{,\tau}k^{\nu}(e^{\mu}_{+\sigma}e^{\sigma\tau}_{+}\sin\phi\cos\phi - e^{\mu}_{+\sigma}e^{\sigma\tau}_{\times}\sin^{2}\phi + e^{\mu}_{\times\sigma}e^{\sigma\tau}_{+}\cos^{2}\phi \\
&\quad - e^{\mu}_{\times\sigma}e^{\sigma\tau}_{\times}\sin\phi\cos\phi - e^{\mu}_{+\sigma}e^{\sigma\tau}_{+}\sin\phi\cos\phi - e^{\mu}_{+\sigma}e^{\sigma\tau}_{\times}\cos^{2}\phi + e^{\mu}_{\times\sigma}e^{\sigma\tau}_{+}\sin^{2}\phi \\
&\quad + e^{\mu}_{\times\sigma}e^{\sigma\tau}_{\times}\sin\phi\cos\phi) = e^{\mu}_{+\sigma}e^{\sigma\tau}_{\times}k^{\nu}(a^{2})_{,\tau},
\end{aligned}$$

$$h^{\sigma\tau}h^{\mu\nu}_{,\sigma\tau} = 0,$$

$$\begin{aligned}
h^{\sigma\tau}h^{\mu\nu}_{\sigma\tau,} &= -a^{2}k^{\mu}k^{\nu}(e^{\sigma\tau}_{+}\cos\phi - e^{\sigma\tau}_{\times}\sin\phi)(e_{+\sigma\tau}\cos\phi - e_{\times\sigma\tau}\sin\phi) \\
&\quad - a(e^{\sigma\tau}_{+}\cos\phi - e^{\sigma\tau}_{\times}\sin\phi)\left[(a^{\mu}_{,}k^{\nu} + a^{\nu}_{,}k^{\mu})e_{+\sigma\tau} + a\overset{1}{f}{}_{+\sigma\tau}k^{\mu}k^{\nu}\right]\sin\phi \\
&\quad - a(e^{\sigma\tau}_{+}\cos\phi - e^{\sigma\tau}_{\times}\sin\phi)\left[(a^{\mu}_{,}k^{\nu} + a^{\nu}_{,}k^{\mu})e_{\times\sigma\tau} + a\overset{1}{f}{}_{\times\sigma\tau}k^{\nu}\right]\cos\phi \\
&\quad - a^{2}\left(\overset{1}{f}{}^{\sigma\tau}_{+}\sin\phi + \overset{1}{f}{}^{\sigma\tau}_{\times}\cos\phi\right)k^{\mu}k^{\nu}(e_{+\sigma\tau}\cos\phi - e_{\times\sigma\tau}\sin\phi) \\
&= -2a^{2}k^{\mu}k^{\nu} - 2a(a^{\mu}_{,}k^{\nu} + a^{\nu}_{,}k^{\mu})\sin\phi\cos\phi - 2a^{2}e^{\sigma\tau}_{+}\overset{1}{f}{}_{+\sigma\tau}k^{\mu}k^{\nu}\sin\phi\cos\phi \\
&\quad + 2a(a^{\mu}_{,}k^{\nu} + a^{\nu}_{,}k^{\mu})\sin\phi\cos\phi + 2a^{2}e^{\sigma\tau}_{\times}\overset{1}{f}{}_{\times\sigma\tau}k^{\mu}k^{\nu}\sin\phi\cos\phi = -2a^{2}k^{\mu}k^{\nu},
\end{aligned}$$

$$
\begin{aligned}
&h_{,}^{\sigma\tau,\mu}h_{\sigma\tau,}^{\nu}\\
&=a^2k^\mu k^\nu(e_{+}^{+\sigma\tau}\sin\phi+e_{\times}^{\sigma\tau}\cos\phi)(e_{+\sigma\tau}\sin\phi+e_{\times\sigma\tau}\cos\phi)\\
&\quad -ak^\mu(e_{+}^{\sigma\tau}\sin\phi+e_{\times}^{\sigma\tau}\cos\phi)\left[\left(a_{,}^{\nu}e_{+\sigma\tau}+a\overset{1}{f}_{+\sigma\tau}k^\nu\right)\cos\phi-\left(a_{,}^{\nu}e_{\times\sigma\tau}+a\overset{1}{f}_{\times\sigma\tau}k^\nu\right)\sin\phi\right]\\
&\quad -ak^\nu\left[\left(a_{,}^{\mu}e_{+}^{\sigma\tau}+a\overset{1}{f}{}_{+}^{+\sigma\tau}k^\mu\right)\cos\phi-\left(a_{,}^{\mu}e_{\times}^{\sigma\tau}+a\overset{1}{f}{}_{\times}^{\sigma\tau}k^\mu\right)\sin\phi\right]\times(e_{+\sigma\tau}\sin\phi+e_{\times\sigma\tau}\cos\phi)\\
&=2a^2k^\mu k^\nu,
\end{aligned}
$$

$$
\begin{aligned}
h_{,}^{\mu\sigma,\tau}h_{\tau,\sigma}^{\nu} &= -ak^\tau(e_{+}^{\mu\sigma}\sin\phi\\
&\quad + e_{\times}^{\mu\sigma}\cos\phi)\left[\left(a_{,\sigma}e_{+\tau}^{\nu}+a\overset{1}{f}{}_{+\tau}^{\nu}k_\sigma\right)\cos\phi-\left(a_{,\sigma}e_{\times\tau}^{\nu}+a\overset{1}{f}{}_{\times\tau}^{\nu}k_\sigma\right)\sin\phi\right]\\
&\quad - ak_\sigma\left[\left(a_{,}^{\tau}e_{+}^{\mu\sigma}+\overset{1}{f}{}_{+}^{\mu\sigma}k^\tau\right)\cos\phi-\left(a_{,}^{\tau}e_{\times}^{\mu\sigma}+\overset{1}{f}{}_{\times}^{\mu\sigma}k^\tau\right)\sin\phi\right]\times(e_{+\tau}^{\nu}\sin\phi\\
&\quad + e_{\times\tau}^{\nu}\cos\phi)\\
&= 0,
\end{aligned}
$$

$$
\begin{aligned}
h_{,}^{\mu\sigma,\tau}h_{\sigma,\tau}^{\nu} &= -ak^\tau(e_{+}^{\mu\sigma}\sin\phi\\
&\quad + e_{\times}^{\mu\sigma}\cos\phi)\left[\left(a_{,\tau}e_{+\sigma}^{\nu}+a\overset{1}{f}{}_{+\sigma}^{\nu}k_\tau\right)\cos\phi-\left(a_{,\tau}e_{\times\sigma}^{\nu}+a\overset{1}{f}{}_{\times\sigma}^{\nu}k_\tau\right)\sin\phi\right]\\
&\quad - ak_\tau\left[\left(a_{,}^{\tau}e_{+}^{\mu\sigma}+\overset{1}{f}{}_{+}^{\mu\sigma}k^\tau\right)\cos\phi-\left(a_{,}^{\tau}e_{\times}^{\mu\sigma}+\overset{1}{f}{}_{\times}^{\mu\sigma}k^\tau\right)\sin\phi\right]\times(e_{+\sigma}^{\nu}\sin\phi\\
&\quad + e_{\times\sigma}^{\nu}\cos\phi)\\
&= 0,
\end{aligned}
$$

$$
h_{,}^{\sigma\tau\rho}h_{\sigma\rho,\tau}=0,\quad h_{,}^{\sigma\tau\rho}h_{\sigma\tau,\rho}=0.
$$

Inserting these results into the expression (15.6) for $T_{\mathrm{GW}}^{\mu\nu}$ in current chapter, we find

$$
\begin{aligned}
T_{\mathrm{GW}}^{\mu\nu} &= \frac{1}{16\pi G}\left[e_{+\sigma}^{\mu}e_{\times}^{\sigma\tau}k^\nu(a^2)_{,\tau}+e_{+\sigma}^{\nu}e_{\times}^{\sigma\tau}k^\mu(a^2)_{,\tau}+2a^2k^\mu k^\nu-a^2k^\mu k^\nu\right]\\
&= \frac{a^2}{16\pi G}k^\mu k^\nu+\frac{1}{16\pi G}\left[k^\mu(e_{\mathrm{I}}^{\nu}e_{\mathrm{II}}^{\sigma}-e_{\mathrm{II}}^{\nu}e_{\mathrm{I}}^{\sigma})+k^\nu(e_{\mathrm{I}}^{\mu}e_{\mathrm{II}}^{\sigma}-e_{\mathrm{II}}^{\mu}e_{\mathrm{I}}^{\sigma})\right](a^2)_{,\sigma},
\end{aligned}
$$

which may be compared with the analogous expression for a circularly polarized electromagnetic wave given in (14.1) in Chap. 14. Proceeding exactly as in that case, we find for the total energy–momentum vector and angular momentum tensor

$$
P^\mu=\int_\Sigma T_{\mathrm{GW}}^{\mu\nu}\mathrm{d}\Sigma_\nu=N\hbar k^\mu,
$$

$$
J^{\mu\nu}=X_{\mathrm{E}}^{\mu}P^\nu-X_{\mathrm{E}}^{\nu}P^\mu+2N\hbar(e_{\mathrm{I}}^{\mu}e_{\mathrm{II}}^{\nu}-e_{\mathrm{II}}^{\mu}e_{\mathrm{I}}^{\nu}).
$$

Here N is the number of coherent *gravitons* out of which, in the quantum theory, the gravitational wave is built. It is given by

$$N = \frac{1}{16\pi G\hbar}\int_{\Sigma} a^2 k^\mu \mathrm{d}\Sigma_\mu = \frac{1}{16\pi G\hbar}\int \omega a^2 \mathrm{d}^3\boldsymbol{x},$$

where $\omega = k^0$. In the expression for $J^{\mu\nu}$, it is assumed that the gauge is chosen so that $e_{\mathrm{I}}^0 = 0 = e_{\mathrm{II}}^0$, and X_{E}^μ is the center of energy:

$$X_{\mathrm{E}}^\mu = \frac{1}{16\pi GN\hbar}\int x^\mu \omega a^2 \mathrm{d}^3\boldsymbol{x}.$$

When the origin of coordinates is shifted so that the world line of the center of energy passes through it, we find $J_i = S\hat{k}_i$, with

$$S = 2N\hbar.$$

Gravitons evidently have spin angular momentum $2\hbar$, twice the value for photons. Like photons, their spins can be oriented only parallel or antiparallel to the propagation vector $\hat{\boldsymbol{k}}$.

15.4 Weak Radiation in Flat Spacetime

We consider the situation where

$$\lim_{x\to\infty} h_{\mu\nu}(x) = 0,$$

and define

$$\left(\frac{\partial}{\partial x^0}\right)^{-1} h_{\mu\nu}(x) \equiv \frac{1}{2}\int_{-\infty}^{\infty} \frac{x^0 - x'^0}{|x^0 - x'^0|} h_{\mu\nu}(x'^0)\mathrm{d}x'^0.$$

Then

$$\frac{\partial}{\partial x^0}\left(\frac{\partial}{\partial x^0}\right)^{-1} h_{\mu\nu}(x) = \int_{-\infty}^{\infty} \delta(x^0 - x'^0) h_{\mu\nu}(x'^0)\mathrm{d}x'^0 = h_{\mu\nu}(x^0),$$

$$\begin{aligned}\left(\frac{\partial}{\partial x^0}\right)^{-1}\frac{\partial}{\partial x^0} h_{\mu\nu}(x) &= \frac{1}{2}\int_{-\infty}^{\infty} \frac{x^0 - x'^0}{|x^0 - x'^0|}\frac{\partial}{\partial x'^0} h_{\mu\nu}(x'^0)\mathrm{d}x'^0 \\ &= -\frac{1}{2}\int_{-\infty}^{\infty} h_{\mu\nu}(x'^0)\frac{\partial}{\partial x'^0}\frac{x^0 - x'^0}{|x^0 - x'^0|}\mathrm{d}x'^0\end{aligned}$$

$$= \frac{1}{2} \int_{-\infty}^{\infty} h_{\mu\nu}(x'^0) \frac{\partial}{\partial x^0} \frac{x^0 - x'^0}{|x^0 - x'^0|} \mathrm{d}x'^0$$

$$= \frac{\partial}{\partial x^0} \left(\frac{\partial}{\partial x^0} \right)^{-1} h_{\mu\nu}(x) = h_{\mu\nu}(x).$$

Consequently,

$$\frac{\partial}{\partial x^0} \left(\frac{\partial}{\partial x^0} \right)^{-1} = \left(\frac{\partial}{\partial x^0} \right)^{-1} \frac{\partial}{\partial x^0} = 1.$$

Similarly,

$$\frac{\partial}{\partial x^\mu} \left(\frac{\partial}{\partial x^0} \right)^{-1} = \left(\frac{\partial}{\partial x^0} \right)^{-1} \frac{\partial}{\partial x^\mu}.$$

Now define

$$\nabla^{-2} h_{\mu\nu}(x) \equiv -\frac{1}{4\pi} \int \frac{1}{|\boldsymbol{x} - \boldsymbol{x}'|} h_{\mu\nu}(x') \mathrm{d}^3 \boldsymbol{x}'.$$

Then

$$\nabla^2 \nabla^{-2} h_{\mu\nu}(x) = \nabla^{-2} \nabla^2 h_{\mu\nu}(x) = h_{\mu\nu}(x),$$

so that

$$\nabla^2 \nabla^{-2} = \nabla^{-2} \nabla^2 = 1.$$

Also

$$\left[\nabla^{-2}, \frac{\partial}{\partial x^\mu} \right] = 0, \quad \left[\nabla^{-2}, \left(\frac{\partial}{\partial x^0} \right)^{-1} \right] = 0,$$

$$\left(\frac{\partial}{\partial x^0} \right)^{-2} \left(\frac{\partial}{\partial x^0} \right)^{2} h_{\mu\nu} = \left(\frac{\partial}{\partial x^0} \right)^{-1} \left(\frac{\partial}{\partial x^0} \right)^{-1} \frac{\partial}{\partial x^0} \frac{\partial}{\partial x^0} h_{\mu\nu}$$

$$= \left(\frac{\partial}{\partial x^0} \right)^{-1} \frac{\partial}{\partial x^0} h_{\mu\nu} = h_{\mu\nu},$$

noting that $\partial h_{\mu\nu}/\partial x^0 \to 0$ as $x \to \infty$ because the operator affects only the dependence on x^0. Suppose the gauge has been chosen so that

$$h = 0, \quad h^{\mu\nu}_{\ ,\nu} = 0, \quad \Box^2 h_{\mu\nu} = 0,$$

as described in current chapter. Then

$$\begin{aligned}\nabla^{-2}h_{\mu\nu} &= \nabla^{-2}\left(\frac{\partial}{\partial x^0}\right)^{-2}\left(\frac{\partial}{\partial x^0}\right)^{2}h_{\mu\nu}\\ &= \nabla^{-2}\left(\frac{\partial}{\partial x^0}\right)^{-2}\nabla^2 h_{\mu\nu}\\ &= \nabla^{-2}\nabla^2\left(\frac{\partial}{\partial x^0}\right)^{-2}h_{\mu\nu}\\ &= \left(\frac{\partial}{\partial x^0}\right)^{-2}h_{\mu\nu},\end{aligned}$$

using the fact that

$$\lim_{x\to\infty}\left(\frac{\partial}{\partial x^0}\right)^{-2}h_{\mu\nu} = 0,$$

since the operator affects only the dependence on x^0, to justify the last step.

We now make a gauge transformation

$$\xi_\mu = \left(\frac{\partial}{\partial x^0}\right)^{-1}h_{0\mu} - \frac{1}{2}\left(\frac{\partial}{\partial x^0}\right)^{-2}h_{00,\mu},\quad \xi_0 = \frac{1}{2}\left(\frac{\partial}{\partial x^0}\right)^{-1}h_{00},$$

$$\xi^\mu_{\mu,} = 0,\quad \Box^2\xi_\mu = 0,$$

with

$$h^{\rm TT}_{\mu\nu} = h_{\mu\nu} - \xi_{\mu,\nu} - \xi_{\nu,\mu},$$

whence

$$h^{\rm TT} = 0,\quad \Box^2 h^{\rm TT}_{\mu\nu} = 0,\quad h^{{\rm TT}\nu}_{\mu\nu,} = 0,$$

$$\begin{aligned}h^{\rm TT}_{0\mu} &= h_{0\mu} - \xi_{0,\mu} - \xi_{\mu,0}\\ &= h_{0\mu} - \frac{1}{2}\left(\frac{\partial}{\partial x^0}\right)^{-1}h_{00,\mu} - h_{0\mu} + \frac{1}{2}\left(\frac{\partial}{\partial x^0}\right)^{-1}h_{00,\mu} = 0.\end{aligned}$$

In an arbitrary gauge,

$$h_{\mu\nu} = h^{\rm TT}_{\mu\nu} - \xi_{\mu,\nu} - \xi_{\nu,\mu},$$

with no restrictions on ξ_μ except that it should vanish at ∞.

Consider the transverse 3-dimensional projection operator

$$P_{ij} \equiv \delta_{ij} - \frac{\partial}{\partial x^i}\nabla^{-2}\frac{\partial}{\partial x^j}.$$

We have

$$\frac{\partial}{\partial x^j} P_{ij} = P_{ij} \frac{\partial}{\partial x^j} = 0, \quad P_{ii} = 2, \quad P_{ik}P_{kj} = P_{ij}.$$

and the obvious theorem

$$h_{ij}^{\mathrm{TT}} = P_{ik}P_{jl}h_{kl}.$$

Note that, in this form, the statement $h^{\mathrm{TT}} = 0$ is just one of the constraints on the Cauchy initial value data for the linearized Einstein equations:

$$h^{\mathrm{TT}} = P_{ij}h_{ij} = h_{ii} - \nabla^{-2}h_{ij,ij} = \nabla^{-2}(h_{ii,jj} - h_{ij,ij}).$$

and the 00 field equation is [see (15.4)]

$$\begin{aligned} 0 &= h^{\mu}_{00,\mu} + h_{,00} - 2h^{\mu}_{0,0\mu} - (h^{\mu\nu}_{,\mu\nu} - h^{\mu}_{,\mu}) \\ &= -h_{00,00} + h_{00,ii} - h_{00,00} + h_{ii,00} + 2h_{00,00} - 2h_{0i,0i} - h_{00,00} + 2h_{0i,0i} - h_{ij,ij} \\ &\quad - h^{\mu}_{00,\mu} + h^{\mu}_{ii,\mu} \\ &= h_{ii,jj} - h_{ij,ij}. \end{aligned}$$

15.5 Generation of Gravitational Waves

We have

$$g^{\mathrm{tot}}_{\mu\nu} = g_{\mu\nu} + h^{\mathrm{GW}}_{\mu\nu} = g^{\mathrm{B}}_{\mu\nu} + h_{\mu\nu} + h^{\mathrm{GW}}_{\mu\nu},$$

where $h_{\mu\nu}$ is smooth, or

$$\varphi^i_{\mathrm{tot}} = \varphi^i_{\mathrm{B}} + \phi^i + \phi^i_{\mathrm{GW}}.$$

From (15.2) and (15.3), we have

$$S_{\mathrm{G},i}[\varphi] = -\frac{1}{2}\langle T_i^{\mathrm{GW}}\rangle = S_{\mathrm{G},ijk}[\varphi]\langle \phi^j_{\mathrm{GW}}\phi^k_{\mathrm{GW}}\rangle.$$

We include $\langle T_i^{\mathrm{GW}}\rangle$ with the matter. From Chap. 12, we get

$$S_{\mathrm{G},ij}[\varphi_{\mathrm{B}}]\phi^j = -\frac{1}{2}\left(T_i + \langle T_i^{\mathrm{GW}}\rangle + \mathcal{T}_i\right),$$

where

$$\begin{aligned} \mathcal{T}_i &= 2\{S_{\mathrm{G},i}[\varphi_{\mathrm{B}} + \phi] - S_{\mathrm{G},i}[\varphi_{\mathrm{B}}] - S_{\mathrm{G},ij}[\varphi_{\mathrm{B}}]\phi^j\} \\ &= S_{\mathrm{G},ijk}[\varphi_{\mathrm{B}}]\phi^j\phi^k + \cdots \end{aligned}$$

is smooth. If the source $\mathcal{T}_i$ is non-stationary, it will produce an outward flux $\langle T_i^{\mathrm{GW}} \rangle$ and cause secular changes in ϕ^i and $\mathcal{T}_i$. We shall neglect these in a first approximation. Then in a flat background with gauge $l^{\mu\nu}_{,\nu} = 0$, we get (see Chap. 12)

$$\Box^2 l^{\mu\nu} = -16\pi(T^{\mu\nu} + \mathcal{T}^{\mu\nu}), \quad G = 1.$$

This equation is rigorous if $h^{\mathrm{GW}}_{\mu\nu}$ is included in $\mathcal{T}^{\mu\nu}$. We include $h^{\mathrm{GW}}_{\mu\nu}$ with $h_{\mu\nu}$ for the present and consider the secular changes later.

In the slow motion approximation, the velocity of moving matter is $v \ll 1$. In any case,

$$v \lesssim \left(\frac{M}{\mathcal{R}}\right)^{1/2}.$$

The frequency of the waves satisfies

$$\omega \sim \frac{v}{\mathcal{R}} \ll \frac{1}{\mathcal{R}},$$

and the wavelength

$$\lambda \sim \frac{1}{\omega} \sim \frac{\mathcal{R}}{v} \gg \mathcal{R}.$$

Now

$$l^{\mu\nu}_{\pm}(x) = 4 \int \frac{\mathcal{T}^{\mu\nu}_{\mathrm{tot}}(t \pm |\boldsymbol{x} - \boldsymbol{x}'|, \boldsymbol{x}')}{|\boldsymbol{x} - \boldsymbol{x}'|} \mathrm{d}^3\boldsymbol{x}',$$

where

$$t = x^0, \quad \mathcal{T}^{\mu\nu}_{\mathrm{tot}} = T^{\mu\nu} + \mathcal{T}^{\mu\nu}, \quad \mathcal{T}^{\mu\nu}_{\mathrm{tot},\nu} = 0,$$

the last of these following from the dynamical equations. Hence,

$$\begin{aligned} l^{\mu\nu}_{\pm}(x) &= 4 \int \mathrm{e}^{-x' \cdot \nabla} \frac{\mathcal{T}^{\mu\nu}_{\mathrm{tot}}(t \pm r, \boldsymbol{x}')}{r} \mathrm{d}^3\boldsymbol{x}' \\ (r \equiv |\boldsymbol{x}|) &= \frac{4A^{\mu\nu}(t \pm r)}{r} - \left[\frac{4B^{i\mu\nu}(t \pm r)}{r}\right]_{,i} + \frac{1}{2}\left[\frac{4C^{ij\mu\nu}(t \pm r)}{r}\right]_{,ij} + \cdots, \end{aligned}$$

where

$$A^{\mu\nu}(t) \equiv \int \mathcal{T}^{\mu\nu}_{\mathrm{tot}}(t, \boldsymbol{x}')\mathrm{d}^3\boldsymbol{x}', \quad B^{i\mu\nu}(t) \equiv \int x'^i \mathcal{T}^{\mu\nu}_{\mathrm{tot}}(t, \boldsymbol{x}')\mathrm{d}^3\boldsymbol{x}',$$

$$C^{ij\mu\nu}(t) \equiv \int x'^i x'^j \mathcal{T}^{\mu\nu}_{\mathrm{tot}}(t, \boldsymbol{x}')\mathrm{d}^3\boldsymbol{x}',$$

and so on. This expansion is valid only asymptotically. We should check that the gauge condition is still satisfied:

$$\begin{aligned} l^{\mu\nu}_{\pm,\nu}(x) &= l^{\mu 0}_{\pm,0}(t,\boldsymbol{x}) + l^{\mu i}_{\pm,i}(t,\boldsymbol{x}) \\ &= \frac{4\dot{A}^{\mu 0}}{r} + 4\left(\frac{A^{\mu i} - \dot{B}^{i\mu 0}}{r}\right)_{,i} - 4\left(\frac{B^{i\mu j} - \dot{C}^{ij\mu 0}/2}{r}\right)_{,ij} + \cdots, \end{aligned}$$

where everything is evaluated at $t \pm r$. But $A^{\mu 0} = P^{\mu}$. Hence, neglecting secular changes inside the asymptotic region, we have $\dot{A}^{\mu 0} = \dot{P}^{\mu} = 0$. Moreover,

$$\begin{aligned} A^{\mu i} - \dot{B}^{i\mu 0} &= \int \left(T^{\mu i}_{\text{tot}} - x^i T^{\mu 0}_{\text{tot},0}\right)\mathrm{d}^3\boldsymbol{x} = \int \left(T^{\mu i}_{\text{tot}} + x^i T^{\mu j}_{\text{tot},j}\right)\mathrm{d}^3\boldsymbol{x} \\ &= \int \left(T^{\mu i}_{\text{tot}} - \delta^i_{ij} T^{\mu j}_{\text{tot}}\right)\mathrm{d}^3\boldsymbol{x} = 0. \end{aligned}$$

Discarding the surface integral corresponds to neglecting secular changes. Furthermore,

$$\begin{aligned} B^{i\mu j} + B^{j\mu i} - \dot{C}^{ij\mu 0} &= \int \left(x^i T^{\mu j}_{\text{tot}} + x^j T^{\mu i}_{\text{tot}} - x^i x^j T^{\mu 0}_{\text{tot},0}\right)\mathrm{d}^3\boldsymbol{x} \\ &= \int \left(x^i T^{\mu j}_{\text{tot}} + x^j T^{\mu i}_{\text{tot}} + x^i x^j T^{\mu k}_{\text{tot},k}\right)\mathrm{d}^3\boldsymbol{x} = 0, \end{aligned}$$

and so on. We have the useful identities

$$x^i T^{00}_{\text{tot},0} + \left(x^i T^{j0}_{\text{tot}}\right)_{,j} = T^{i0}_{\text{tot}},$$

$$\begin{aligned} &\frac{1}{2} x^i x^j T^{00}_{\text{tot},00} + \left(x^i T^{jk}_{\text{tot}} + x^j T^{ik}_{\text{tot}}\right)_{,k} - \frac{1}{2}\left(x^i x^j T^{kl}_{\text{tot}}\right)_{,kl} \\ &= \frac{1}{2} x^i x^j T^{00}_{\text{tot},00} + \frac{1}{2}\left(x^i T^{jk}_{\text{tot}} + x^j T^{ik}_{\text{tot}}\right)_{,k} + \frac{1}{2}\left(x^i x^j T^{k0}_{\text{tot},0}\right)_{,k} \\ &= \frac{1}{2} x^i x^j T^{00}_{\text{tot},00} + T^{ij}_{\text{tot}} + \frac{1}{2} x^i x^j T^{k0}_{\text{tot},0k} \\ &= T^{ij}_{\text{tot}}. \end{aligned}$$

Now choose coordinates so that $P^0 = M$, $P^i = 0$, $X^i = 0$, and define

$$I_{ij} = \int x^i x^j T^{00}_{\text{tot}} \mathrm{d}^3\boldsymbol{x}.$$

Then

$$A^{00} = M, \quad A^{0i} = 0,$$

$$A^{ij} = \int T^{ij}_{\text{tot}} \mathrm{d}^3\boldsymbol{x} = \frac{1}{2}\int x^i x^j T^{00}_{\text{tot},00} \mathrm{d}^3\boldsymbol{x} = \frac{1}{2}\ddot{I}_{ij},$$

$$B^{i00} = \int x^i T^{00}_{\text{tot}} \mathrm{d}^3\boldsymbol{x} = 0,$$

$$\begin{aligned} B^{j0i} &= \int x^j T^{0i}_{\text{tot}} \mathrm{d}^3\boldsymbol{x} = \int x^j \left[x^i T^{00}_{\text{tot},0} + \left(x^i T^{k0}_{\text{tot}} \right)_{,k} \right] \mathrm{d}^3\boldsymbol{x} \\ &= \int \left(-x^i T^{0j}_{\text{tot}} + x^i x^j T^{00}_{\text{tot},0} \right) \mathrm{d}^3\boldsymbol{x} \\ &= \frac{1}{2} \int \left(x^j T^{0i}_{\text{tot}} - x^i T^{0j}_{\text{tot}} + x^i x^j T^{00}_{\text{tot},0} \right) \mathrm{d}^3\boldsymbol{x} \\ &= -\frac{1}{2} S_{ij} + \frac{1}{2} \dot{I}_{ij}, \end{aligned}$$

$$C^{ij00} = I_{ij}.$$

Therefore,

$$l^{00}_{\pm}(x) = \frac{4M}{r} + 2 \left[\frac{I_{ij}(t \pm r)}{r} \right]_{,ij} + \cdots,$$

$$l^{0i}_{\pm}(x) = \left(\frac{2S_{ij}}{r} \right)_{,j} - 2 \left[\frac{\dot{I}_{ij}(t \pm r)}{r} \right]_{,j} + \cdots,$$

where the first term on the right-hand side of each of these, familiar from the quasi-stationary case (see Chap. 12), changes only secularly, and

$$l^{ij}_{\pm}(x) = 2 \frac{\ddot{I}_{ij}(t \pm r)}{r} + \cdots.$$

Terminating the series here corresponds to neglecting retardation across the source and working only to the quadrupole approximation. Furthermore,

$$l_{\pm} = -l^{00}_{\pm} + l^{ii}_{\pm} = -\frac{4M}{r} - 2 \left[\frac{I_{ij}(t \pm r)}{r} \right]_{,ij} + 2 \frac{\ddot{I}_{ij}(t \pm r)}{r} + \cdots,$$

$$h^{\pm}_{\mu\nu} = l^{\pm}_{\mu\nu} - \frac{1}{2} \eta_{\mu\nu} l_{\pm},$$

$$h^{\pm}_{00} = \frac{2M}{r} + \left[\frac{I_{ij}(t \pm r)}{r} \right]_{,ij} + \frac{\ddot{I}_{ii}(t \pm r)}{r} + \cdots,$$

$$h^{\pm}_{0i} = -\left(\frac{2S_{ij}}{r} \right)_{,j} + 2 \left[\frac{\dot{I}_{ij}(t \pm r)}{r} \right]_{,j} + \cdots,$$

$$h^{\pm}_{ij} = \frac{2M}{r} \delta_{ij} + \delta_{ij} \left[\frac{I_{kl}(t \pm r)}{r} \right]_{,kl} + \frac{2\ddot{I}_{ij}(t \pm r) - \delta_{ij} \ddot{I}_{kk}(t \pm r)}{r} + \cdots.$$

Now

$$\left[\frac{I_{ij}(t\pm r)}{r}\right]_{,j} = -\frac{I_{ij}(t\pm r)x_j}{r^3} \pm \frac{\dot{I}_{ij}(t\pm r)x_j}{r^2},$$

$$\begin{aligned}\left[\frac{I_{ij}(t\pm r)}{r}\right]_{,jk} &= -\frac{I_{ik}(t\pm r)}{r^3} + 3\frac{I_{ij}(t\pm r)x_jx_k}{r^5} \pm \frac{\dot{I}_{ik}(t\pm r)}{r^2} \\ &\quad \mp 3\frac{\dot{I}_{ij}(t\pm r)x_jx_k}{r^4} + \frac{\ddot{I}_{ij}(t\pm r)x_jx_k}{r^3},\end{aligned}$$

$$\left[\frac{I_{ij}(t\pm r)}{r}\right]_{,ij} = \frac{3x_ix_jQ_{ij}(t\pm r)}{r^5} \mp \frac{3x_ix_j\dot{Q}_{ij}(t\pm r)}{r^4} + \frac{x_ix_j\ddot{I}_{ij}(t\pm r)}{r^3},$$

where Q_{ij} is the energy quadrupole moment tensor

$$Q_{ij} \equiv I_{ij} - \frac{1}{3}\delta_{ij}I_{kk}, \quad Q_{ii} = 0.$$

We make the gauge transformation

$$\xi_0^{\pm} = \frac{2}{3}\frac{\dot{I}_{ii}(t\pm r)}{r}, \quad \xi_i = 0.$$

Then

$$\begin{aligned}\bar{h}_{00}^{\pm} &= h_{00}^{\pm} - 2\xi_{0,0}^{\pm} \\ &= \frac{2M}{r} + \frac{3x_ix_jQ_{ij}^{\pm}}{r^5} \mp \frac{3x_ix_j\dot{Q}_{ij}^{\pm}}{r^4} + \frac{x_ix_j\ddot{I}_{ij}^{\pm}}{r^3} + \frac{\ddot{I}_{ii}^{\pm}}{r} - \frac{4}{3}\frac{\ddot{I}_{ii}^{\pm}}{r} + \cdots \\ &= \frac{2M}{r} + \frac{3x_ix_jQ_{ij}^{\pm}}{r^5} \mp \frac{3x_ix_j\dot{Q}_{ij}^{\pm}}{r^4} + \frac{x_ix_j\ddot{Q}_{ij}^{\pm}}{r^3} + \cdots \\ &= \frac{2M}{r} + \left(\frac{Q_{ij}^{\pm}}{r}\right)_{,ij} + \cdots,\end{aligned}$$

$$\begin{aligned}\bar{h}_{0i}^{\pm} &= h_{0i}^{\pm} - \xi_{0,i}^{\pm} \\ &= -\left(\frac{2S_{ij}}{r}\right)_{,j} - \frac{2\dot{I}_{ij}^{\pm}x_j}{r^3} \pm \frac{2\ddot{I}_{ij}^{\pm}x_j}{r^2} + \frac{2}{3}\frac{\dot{I}_{jj}^{\pm}x_i}{r^3} \mp \frac{2}{3}\frac{\ddot{I}_{jj}^{\pm}x_i}{r^2} + \cdots \\ &= -\left(\frac{2S_{ij}}{r}\right)_{,j} - \frac{2\dot{Q}_{ij}^{\pm}x_j}{r^3} \pm \frac{2\ddot{Q}_{ij}^{\pm}x_j}{r^2} + \cdots \\ &= -\left(\frac{2S_{ij}}{r}\right)_{,j} + 2\left(\frac{\dot{Q}_{ij}^{\pm}}{r}\right)_{,j} + \cdots,\end{aligned}$$

$$\begin{aligned}\overline{h}_{ij}^{\pm} &= \frac{2M}{r}\delta_{ij} + \delta_{ij}\left(\frac{3x_kx_lQ_{kl}^{\pm}}{r^5} \mp \frac{3x_kx_l\,\dot{Q}_{kl}^{\pm}}{r^4} + \frac{x_kx_l\ddot{I}_{kl}^{\pm}}{r^3}\right) - \frac{1}{3}\delta_{ij}\frac{\dot{I}_{kk}^{\pm}}{r} + \frac{2\ddot{I}_{ij}^{\pm} - 2\delta_{ij}\ddot{I}_{kk}^{\pm}/3}{r}\\ &\quad + \cdots\\ &= \frac{2M}{r}\delta_{ij} + \delta_{ij}\left(\frac{3x_kx_lQ_{kl}^{\pm}}{r^5} \mp \frac{3x_kx_l\,\dot{Q}_{kl}^{\pm}}{r^4} + \frac{x_kx_l\,\ddot{Q}_{kl}^{\pm}}{r^3}\right) + \frac{2\ddot{Q}_{ij}^{\pm}}{r} + \cdots\\ &= \frac{2M}{r}\delta_{ij} + \delta_{ij}\left(\frac{Q_{kl}^{\pm}}{r}\right)_{,kl} + \frac{2\ddot{Q}_{ij}^{\pm}}{r} + \cdots,\end{aligned}$$

where

$$Q_{ij}^{\pm} \equiv Q_{ij}(t \pm r), \quad I_{ij}^{\pm} \equiv I_{ij}(t \pm r).$$

Now

$$\begin{aligned}\Box^2\frac{f(t\pm r)}{r} &= \left(\nabla^2 - \frac{\partial^2}{\partial t^2}\right)\frac{f(t\pm r)}{r}\\ &= \nabla\cdot\left[-\frac{\boldsymbol{x}}{r^3}f(t\pm r) \pm \frac{\boldsymbol{x}}{r^2}\dot{f}(t\pm r)\right] - \frac{1}{r}\ddot{f}(t\pm r)\\ &= 4\pi\delta(\boldsymbol{x})f(t\pm r) + \frac{1}{r^2}(\mp 1 \pm 3 \mp 2)\dot{f}(t\pm r) + \frac{1}{r}\ddot{f}(t\pm r) - \frac{1}{r}\ddot{f}(t\pm r)\\ &= 4\pi\delta(\boldsymbol{x})f(t).\end{aligned}$$

Hence,

$$\Box^2\left[\frac{f(t-r)}{r} - \frac{f(t+r)}{r}\right] = 0, \quad \Box^2(\xi_\mu^- - \xi_\mu^+) = 0, \quad \Box^2 h_{\mu\nu}^{\text{rad}} = 0,$$

where $h_{\mu\nu}^{\text{rad}}$ is the free radiation part of the asymptotic field defined by

$$h_{\mu\nu}^{\text{rad}} \equiv \overline{h}_{\mu\nu}^{-} - \overline{h}_{\mu\nu}^{+}.$$

Note that $h^{\text{rad}} \neq 0$. If the source of gravitational waves remains non-stationary for only a finite amount of time and if we assume retarded boundary conditions, then at large distances from the source and at times $t \sim r$, we have

$$h_{\mu\nu}^{-} = h_{\mu\nu}^{\text{rad}} + O(1/r),$$

with the $O(1/r)$ terms being stationary. More generally,

$$\overline{h}_{\mu\nu}^{-} = h_{\mu\nu}^{\text{SW}} + \frac{1}{2}h_{\mu\nu}^{\text{rad}},$$

where $h_{\mu\nu}^{\text{SW}}$ is the standing wave 'potential' given by

$$h_{\mu\nu}^{\text{SW}} \equiv \frac{1}{2}\left(\overline{h}_{\mu\nu}^{-} + \overline{h}_{\mu\nu}^{+}\right).$$

Now

$$
\begin{aligned}
h_{ij}^{\mathrm{TT}} &= P_{ik}P_{jl}h_{kl}^{\mathrm{rad}} = h_{ij}^{\mathrm{rad}} - \nabla^{-2}h_{ik,kj}^{\mathrm{rad}} - \nabla^{-2}h_{jk,ki}^{\mathrm{rad}} + \nabla^{-4}h_{kl,klij}^{\mathrm{rad}} \\
&= h_{ij}^{\mathrm{rad}} - \left(\frac{\partial}{\partial t}\right)^{-2} h_{ik,kj}^{\mathrm{rad}} - \left(\frac{\partial}{\partial t}\right)^{-2} h_{jk,ki}^{\mathrm{rad}} + \left(\frac{\partial}{\partial t}\right)^{-4} h_{kl,klij}^{\mathrm{rad}} = \delta_{ij}\left(\frac{Q_{kl}^{-} - Q_{kl}^{+}}{r}\right)_{,kl} \\
&\quad + 2\frac{\ddot{Q}_{ij}^{-} - \ddot{Q}_{ij}^{+}}{r} - \left(\frac{\partial}{\partial t}\right)^{-2}\left(\frac{Q_{kl}^{-} - Q_{kl}^{+}}{r}\right)_{,klij} - 2\left(\frac{Q_{ik}^{-} - Q_{ik}^{+}}{r}\right)_{,kj} \\
&\quad - \left(\frac{\partial}{\partial t}\right)^{-2}\left(\frac{Q_{kl}^{-} - Q_{kl}^{+}}{r}\right)_{,klij} - 2\left(\frac{Q_{jk}^{-} - Q_{jk}^{+}}{r}\right)_{,ki} + \left(\frac{\partial}{\partial t}\right)^{-4}\left(\frac{Q_{kl}^{-} - Q_{kl}^{+}}{r}\right)_{,klmmij} \\
&\quad + 2\left(\frac{\partial}{\partial t}\right)^{-2}\left(\frac{Q_{kl}^{-} - Q_{kl}^{+}}{r}\right)_{,klij} + \cdots \\
&= \delta_{ij}\left[\frac{x_k x_l\,(\ddot{Q}_{kl}^{-} - \ddot{Q}_{kl}^{+})}{r^3} + \frac{3x_k x_l(\dot{Q}_{kl}^{-} + \dot{Q}_{kl}^{+})}{r^4} + \cdots\right] \\
&\quad + 2\frac{\ddot{Q}_{ij}^{-} - \ddot{Q}_{ij}^{+}}{r} + \left(\frac{\partial}{\partial t}\right)^{-2}\left[\frac{x_k x_l(\ddot{Q}_{kl}^{-} - \ddot{Q}_{kl}^{+})}{r^3} + \frac{3x_k x_l(\dot{Q}_{kl}^{-} + \dot{Q}_{kl}^{+})}{r^4} + \cdots\right]_{,ij} \\
&\quad - 2\left[\frac{(\ddot{Q}_{ik}^{-} - \ddot{Q}_{ik}^{+})x_k x_j}{r^3} - \frac{\dot{Q}_{ij}^{-} + \dot{Q}_{ij}^{+}}{r^2} + \frac{3(\dot{Q}_{ik}^{-} + \dot{Q}_{ik}^{+})x_k x_j}{r^4} + \cdots\right] \\
&\quad - 2\left[\frac{(\ddot{Q}_{jk}^{-} - \ddot{Q}_{jk}^{+})x_k x_i}{r^3} - \frac{\dot{Q}_{ij}^{-} + \dot{Q}_{ij}^{+}}{r^2} + \frac{3(\dot{Q}_{jk}^{-} + \dot{Q}_{jk}^{+})x_k x_i}{r^4} + \cdots\right] + \cdots \\
&= \frac{1}{r}[2(\ddot{Q}_{ij}^{-} - \ddot{Q}_{ij}^{+}) - 2\hat{x}_i\hat{x}_k(\ddot{Q}_{kj}^{-} - \ddot{Q}_{kj}^{+}) - 2\hat{x}_j\hat{x}_k(\ddot{Q}_{ki}^{-} - \ddot{Q}_{ki}^{+}) + \delta_{ij}\hat{x}_k\hat{x}_l(\ddot{Q}_{kl}^{-} - \ddot{Q}_{kl}^{+}) \\
&\quad + \hat{x}_i\hat{x}_j\hat{x}_k\hat{x}_l(\ddot{Q}_{kl}^{-} - \ddot{Q}_{kl}^{+})] + \frac{1}{r^2}[4(\dot{Q}_{ij}^{-} + \dot{Q}_{ij}^{+}) - 8\hat{x}_i\hat{x}_k(\dot{Q}_{kj}^{-} + \dot{Q}_{kj}^{+}) \\
&\quad - 8\hat{x}_j\hat{x}_k(\dot{Q}_{ki}^{-} + \dot{Q}_{ki}^{+}) + 2\delta_{ij}\hat{x}_k\hat{x}_l(\dot{Q}_{kl}^{-} + \dot{Q}_{kl}^{+}) + 10\hat{x}_i\hat{x}_j\hat{x}_k\hat{x}_l(\dot{Q}_{kl}^{-} + \dot{Q}_{kl}^{+})] + \cdots,
\end{aligned}
$$

in which we keep terms to order $1/r^2$, and $\hat{\boldsymbol{x}} \equiv \boldsymbol{x}/r$. We also have

$$
\begin{aligned}
h_{ij,a}^{\mathrm{TT}} &= \frac{\hat{x}_a}{r}[-2(\dddot{Q}_{ij}^{-} + \dddot{Q}_{ij}^{+}) + 2\hat{x}_i\hat{x}_k(\dddot{Q}_{kj}^{-} + \dddot{Q}_{kj}^{+}) + 2\hat{x}_j\hat{x}_k(\dddot{Q}_{ki}^{-} + \dddot{Q}_{ki}^{+}) - \delta_{ij}\hat{x}_k\hat{x}_l(\dddot{Q}_{kl}^{-} \\
&\quad + \dddot{Q}_{kl}^{+}) - \hat{x}_i\hat{x}_j\hat{x}_k\hat{x}_l(\dddot{Q}_{kl}^{-} + \dddot{Q}_{kl}^{+})] + \frac{\hat{x}_a}{r^2}[-6(\ddot{Q}_{ij}^{-} - \ddot{Q}_{ij}^{+}) + 14\hat{x}_i\hat{x}_k(\ddot{Q}_{kj}^{-} - \ddot{Q}_{kj}^{+}) \\
&\quad + 14\hat{x}_j\hat{x}_k(\ddot{Q}_{ki}^{-} - \ddot{Q}_{ki}^{+}) - 5\delta_{ij}\hat{x}_k\hat{x}_l(\ddot{Q}_{kl}^{-} - \ddot{Q}_{kl}^{+}) - 15\hat{x}_i\hat{x}_j\hat{x}_k\hat{x}_l(\ddot{Q}_{kl}^{-} - \ddot{Q}_{kl}^{+})] \\
&\quad + \frac{1}{r^2}[-2\delta_{ia}x_k(\ddot{Q}_{kj}^{-} - \ddot{Q}_{kj}^{+}) - 2\delta_{ja}x_k(\ddot{Q}_{ki}^{-} - \ddot{Q}_{ki}^{+}) + 2\delta_{ij}x_k(\ddot{Q}_{ka}^{-} - \ddot{Q}_{ka}^{+}) \\
&\quad - 2\hat{x}_i(\ddot{Q}_{ja}^{-} - \ddot{Q}_{ja}^{+}) - 2\hat{x}_j(\ddot{Q}_{ia}^{-} - \ddot{Q}_{ia}^{+}) + \delta_{ia}\hat{x}_j\hat{x}_k\hat{x}_l(\ddot{Q}_{kl}^{-} - \ddot{Q}_{kl}^{+}) + \delta_{ja}\hat{x}_i\hat{x}_k\hat{x}_l(\ddot{Q}_{kl}^{-} \\
&\quad - \ddot{Q}_{kl}^{+}) + 2x_i x_j x_k(\ddot{Q}_{ka}^{-} - \ddot{Q}_{ka}^{+})] + \cdots,
\end{aligned}
$$

$$h_{ij}^{\mathrm{TT}}\hat{x}_j = O\left(\frac{1}{r^2}\right), \quad h_{ij,k}^{\mathrm{TT}}\hat{x}_j = O\left(\frac{1}{r^2}\right).$$

Asymptotically, $h_{\mu\nu}^{\mathrm{TT}}$ should be separated into a smooth part and a gravitational wave remainder that may be regarded as contributing to $T_{\mathrm{GW}}^{\mu\nu}$. The terms that we have retained above constitute effectively this wave part; the unwritten terms may be lumped with the smooth part. (They drop off more rapidly with r and are negligible at infinity.) Now (see current chapter)

$$T_{\mathrm{GW}}^{0a} = \frac{1}{16\pi}\left[-(h_{ij}^{\mathrm{TT}}h_{ai,0}^{\mathrm{TT}})_{,j} + h_{ij}^{\mathrm{TT}}h_{ij,a0}^{\mathrm{TT}} + \frac{1}{2}h_{ij,0}^{\mathrm{TT}}h_{ij,a}^{\mathrm{TT}}\right],$$

$$\begin{aligned}
T_{\mathrm{GW}}^{ab} = &\frac{1}{16\pi}\Big[h_{ij}^{\mathrm{TT}}(h_{ai,bj}^{\mathrm{TT}} + h_{bi,aj}^{\mathrm{TT}} - h_{ab,ij}^{\mathrm{TT}} - h_{ij,ab}^{\mathrm{TT}}) - \frac{1}{2}h_{ij,a}^{\mathrm{TT}}h_{ij,b} + h_{ai,j}^{\mathrm{TT}}h_{bj,i}^{\mathrm{TT}} - h_{ai,}^{\mathrm{TT}\mu}h_{bi,\mu}^{\mathrm{TT}} \\
&+ \delta_{ab}\left(\frac{1}{4}h_{ij,}^{\mu}h_{ij,\mu}^{\mathrm{TT}} - \frac{1}{2}h_{ij,k}^{\mathrm{TT}}h_{ik,j}^{\mathrm{TT}}\right)\Big] = \frac{1}{16\pi}\Big\{(h_{ij}^{\mathrm{TT}}h_{ai,b}^{\mathrm{TT}})_{,j} + (h_{ij}^{\mathrm{TT}}h_{bi,a}^{\mathrm{TT}})_{,j} - (h_{ij}^{\mathrm{TT}}h_{ab}^{\mathrm{TT}})_{,ij}. \\
&- \frac{1}{2}(h_{ij}^{\mathrm{TT}}h_{ij}^{\mathrm{TT}})_{,ab} + \frac{1}{2}h_{ij,a}^{\mathrm{TT}}h_{ij,b}^{\mathrm{TT}} + (h_{ai}^{\mathrm{TT}}h_{bj})_{,ij} + \frac{1}{2}(h_{ai}^{\mathrm{TT}}h_{bi}^{\mathrm{TT}})_{,00} - \frac{1}{2}(h_{ai}^{\mathrm{TT}}h_{bi})_{,jj}. \\
&+ \delta_{ab}\left[-\frac{1}{8}(h_{ij}^{\mathrm{TT}}h_{ij}^{\mathrm{TT}})_{,00} + \frac{1}{8}(h_{ij}^{\mathrm{TT}}h_{ij})_{,kk} - \frac{1}{2}(h_{ij}^{\mathrm{TT}}h_{ik})_{,jk}\right]\Big\}.
\end{aligned}$$

Now

$$(h_{ij}^{\mathrm{TT}}h_{ai,0}^{\mathrm{TT}})_{,j}\underset{t\sim r\to\infty}{\longrightarrow}\frac{\partial}{\partial t}O\left(\frac{1}{r^2}\right) + O\left(\frac{1}{r^3}\right),$$

and

$$\begin{aligned}
h_{ij}^{\mathrm{TT}}h_{ij,a0}^{\mathrm{TT}}\underset{t\sim r\to\infty}{\longrightarrow}&\frac{\hat{x}_a}{r^2}[-4\ddot{Q}^{-}:\ddddot{Q}^{-} + 8\hat{x}\cdot\ddot{Q}^{-}\cdot\ddddot{Q}^{-}\cdot\hat{x} + 2\hat{x}\cdot\ddot{Q}^{-}\cdot\hat{x}\hat{x}\cdot\ddddot{Q}^{-}\cdot\hat{x} + 4\hat{x}\cdot\ddot{Q}^{-}\cdot\ddddot{Q}^{-}\cdot\hat{x} \\
&- 4\hat{x}\cdot\ddot{Q}^{-}\cdot\ddddot{Q}^{-}\cdot\hat{x} - 4\hat{x}\cdot\ddot{Q}^{-}\cdot\hat{x}\hat{x}\cdot\ddddot{Q}^{-}\cdot\hat{x} + 2\hat{x}\cdot\ddot{Q}^{-}\cdot\hat{x}\hat{x}\cdot\ddddot{Q}^{-}\cdot\hat{x} - 2\hat{x}\cdot\ddot{Q}^{-}\cdot\hat{x}\hat{x}\cdot\ddddot{Q}^{-}\cdot\hat{x} \\
&+ 4\hat{x}\cdot\ddot{Q}^{-}\cdot\ddddot{Q}^{-}\cdot\hat{x} - 4\hat{x}\cdot\ddot{Q}^{-}\cdot\hat{x}\hat{x}\cdot\ddddot{Q}^{-}\cdot\hat{x} - 4\hat{x}\cdot\ddot{Q}^{-}\cdot\ddddot{Q}^{-}\cdot\hat{x} + 2\hat{x}\cdot\ddot{Q}^{-}\cdot\hat{x}\hat{x}\cdot\ddddot{Q}^{-}\cdot\hat{x} \\
&- 2\hat{x}\cdot\ddot{Q}^{-}\cdot\hat{x}\hat{x}\cdot\ddddot{Q}^{-}\cdot\hat{x} + 4\hat{x}\cdot\ddot{Q}^{-}\cdot\hat{x}\hat{x}\cdot\ddddot{Q}^{-}\cdot\hat{x} - 3\hat{x}\cdot\ddot{Q}^{-}\cdot\hat{x}\hat{x}\cdot\ddddot{Q}^{-}\cdot\hat{x} + \hat{x}\cdot\ddot{Q}^{-}\cdot\hat{x}\hat{x}\cdot\ddddot{Q}^{-}\cdot\hat{x} \\
&- 2\hat{x}\cdot\ddot{Q}^{-}\cdot\hat{x}\hat{x}\cdot\ddddot{Q}^{-}\cdot\hat{x} + 4\hat{x}\cdot\ddot{Q}^{-}\cdot\hat{x}\hat{x}\cdot\ddddot{Q}^{-}\cdot\hat{x} - \hat{x}\cdot\ddot{Q}^{-}\cdot\hat{x}\hat{x}\cdot\ddddot{Q}^{-}\cdot\hat{x} + \hat{x}\cdot\ddot{Q}^{-}\cdot\hat{x}\hat{x}\cdot\ddddot{Q}^{-}\cdot\hat{x}] \\
&+ O\left(\frac{1}{r^3}\right) \\
&= \frac{\hat{x}_a}{r^2}[-4\ddot{Q}^{-}:\ddddot{Q}^{-} + 8\hat{x}\cdot\ddot{Q}^{-}\cdot\ddddot{Q}^{-}\cdot\hat{x} - 2\hat{x}\cdot\ddot{Q}^{-}\cdot\hat{x}\hat{x}\cdot\ddddot{Q}^{-}\cdot\hat{x}] + O\left(\frac{1}{r^3}\right),
\end{aligned}$$

$$\frac{1}{2}h_{ij,0}^{\mathrm{TT}}h_{ij,a}^{\mathrm{TT}}\underset{t\sim r\to\infty}{\longrightarrow}\frac{\hat{x}_a}{r^2}\left[-2\dddot{Q}^{-}:\dddot{Q}^{-} + 4\hat{x}\cdot\dddot{Q}^{-}\cdot\dddot{Q}^{-}\cdot\hat{x} - (\hat{x}\cdot\dddot{Q}^{-}\cdot\hat{x})^2\right] + O\left(\frac{1}{r^3}\right).$$

Averaging $T_{\mathrm{GW}}^{\mu\nu}$ in spacetime over a few wavelengths is equivalent to averaging the products of differentiated Q's over a few periods if the source motion is quasi-periodic, or over a total orbit if the source is an unbounded system (collision). Moreover,

$$\langle \ddot{Q}^- \ddddot{Q}^- \rangle = -\langle \dddot{Q}^- \dddot{Q}^- \rangle,$$

and so on. Hence,

$$\langle T^{0a}_{\mathrm{GW}} \rangle \underset{t\sim r\to\infty}{\longrightarrow} \frac{\hat{x}_a}{16\pi r^2} \left\langle 2\dddot{\boldsymbol{Q}}^- \boldsymbol{:} \dddot{\boldsymbol{Q}}^- - 4\hat{\boldsymbol{x}}\cdot\dddot{\boldsymbol{Q}}^-\cdot\dddot{\boldsymbol{Q}}^-\cdot\hat{\boldsymbol{x}} + (\hat{\boldsymbol{x}}\cdot\dddot{\boldsymbol{Q}}^-\cdot\hat{\boldsymbol{x}})^2 \right\rangle + O\left(\frac{1}{r^3}\right).$$

We also have

$$\begin{aligned}
h^{\mathrm{TT}}_{ij} h^{\mathrm{TT}}_{ai,b} \underset{t\sim r\to\infty}{\longrightarrow} \frac{\hat{x}_b}{r^2} \Big[& -4\ddot{Q}^-_{ji}\dddot{Q}^-_{ia} + 4\ddot{Q}^-_{ji}\dddot{Q}^-_{ik}\hat{x}_k\hat{x}_a + 4\ddot{Q}^-_{ji}\hat{x}_i\dddot{Q}^-_{ak}\hat{x}_k - 2\ddot{Q}^-_{ja}\hat{x}_k\hat{x}_l\dddot{Q}^-_{kl} \\
& - 2\ddot{Q}^-_{ji}\hat{x}_i\hat{x}_a\hat{x}_k\hat{x}_l\dddot{Q}^-_{kl} + 4\hat{x}_j\hat{x}_k\ddot{Q}^-_{ki}\dddot{Q}^-_{ia} - 4\hat{x}_j\hat{x}_k\ddot{Q}^-_{ki}\dddot{Q}^-_{il}\hat{x}_l\hat{x}_a - 4\hat{x}_j\hat{x}_k\ddot{Q}^-_{ki}\hat{x}_i\hat{x}_l\dddot{Q}^-_{la} \\
& + 2\hat{x}_j\hat{x}_k\ddot{Q}^-_{ka}\hat{x}_i\hat{x}_l\dddot{Q}^-_{il} + 2\hat{x}_j\hat{x}_k\ddot{Q}^-_{ki}\hat{x}_i\hat{x}_a\hat{x}_l\hat{x}_m\dddot{Q}^-_{lm} - 2\hat{x}_k\hat{x}_l\ddot{Q}^-_{kl}\dddot{Q}^-_{ja} + 2\hat{x}_k\hat{x}_l\ddot{Q}^-_{kl}\dddot{Q}^-_{ji}\hat{x}_i\hat{x}_a \\
& + 2\hat{x}_k\hat{x}_l\ddot{Q}^-_{kl}\hat{x}_j\hat{x}_i\dddot{Q}^-_{ia} - \delta_{ja}\hat{x}_k\hat{x}_l\ddot{Q}^-_{kl}\hat{x}_m\hat{x}_n\dddot{Q}^-_{mn} - \hat{x}_j\hat{x}_a\hat{x}_k\hat{x}_l\ddot{Q}^-_{kl}\hat{x}_m\hat{x}_n\dddot{Q}^-_{mn}\Big] + O\left(\frac{1}{r^3}\right) \\
= \frac{\hat{x}_b}{r^2} \Big[& -4\ddot{Q}^-_{ji}\dddot{Q}^-_{ia} + 4\ddot{Q}^-_{ji}\dddot{Q}^-_{ik}\hat{x}_k\hat{x}_a + 4\ddot{Q}^-_{ji}\hat{x}_i\dddot{Q}^-_{ak}\hat{x}_k - 2\ddot{Q}^-_{ji}\hat{x}_i\hat{x}_a\hat{x}_k\hat{x}_l\dddot{Q}^-_{kl} + 4\hat{x}_j\hat{x}_k\ddot{Q}^-_{ki}\dddot{Q}^-_{ia} \\
& - 2\hat{x}_k\hat{x}_l\ddot{Q}^-_{kl}\hat{x}_j\hat{x}_i\dddot{Q}^-_{ia} + 2\hat{x}_j\hat{x}_i\ddot{Q}^-_{ia}\hat{x}_k\hat{x}_l\dddot{Q}^-_{kl} + 2\hat{x}_k\hat{x}_l\ddot{Q}^-_{kl}\dddot{Q}^-_{ji}\hat{x}_i\hat{x}_a\Big] + \frac{\partial}{\partial t}O\left(\frac{1}{r^2}\right) + O\left(\frac{1}{r^3}\right),
\end{aligned}$$

$$\begin{aligned}
\left(h^{\mathrm{TT}}_{ij} h^{\mathrm{TT}}_{ai,b}\right)_{,j} \underset{t\sim r\to\infty}{\longrightarrow} \frac{1}{r^3} [& -4\ddot{Q}^-_{bi}\dddot{Q}^-_{ia} + 4\ddot{Q}^-_{bi}\dddot{Q}^-_{ij}\hat{x}_j\hat{x}_a + 4\ddot{Q}^-_{bi}\hat{x}_i\hat{x}_j\dddot{Q}^-_{ja} - 2\ddot{Q}^-_{bi}\hat{x}_i\hat{x}_a\hat{x}_k\hat{x}_l\dddot{Q}^-_{kl} \\
& + 4\hat{x}_b\hat{x}_i\ddot{Q}^-_{ij}\dddot{Q}^-_{ja} - 2\hat{x}_k\hat{x}_l\ddot{Q}^-_{kl}\hat{x}_b\hat{x}_i\dddot{Q}^-_{ia} + 2\hat{x}_b\hat{x}_i\ddot{Q}^-_{ia}\hat{x}_k\hat{x}_l\dddot{Q}^-_{kl} + 2\hat{x}_k\hat{x}_l\ddot{Q}^-_{kl}\dddot{Q}^-_{bi}\hat{x}_i\hat{x}_a \\
& + 12\hat{x}_b\hat{x}_j\ddot{Q}^-_{ji}\dddot{Q}^-_{ia} - 20\hat{x}_b\hat{x}_j\ddot{Q}^-_{ji}\dddot{Q}^-_{ik}\hat{x}_k\hat{x}_a - 20\hat{x}_b\hat{x}_j\ddot{Q}^-_{ji}\hat{x}_i\dddot{Q}^-_{ak}\hat{x}_k + 14\hat{x}_b\hat{x}_j\ddot{Q}^-_{ji}\hat{x}_i\hat{x}_a\hat{x}_k\hat{x}_l\dddot{Q}^-_{kl} \\
& - 20\hat{x}_b\hat{x}_k\ddot{Q}^-_{ki}\dddot{Q}^-_{ia} + 14\hat{x}_k\hat{x}_l\ddot{Q}^-_{kl}\hat{x}_b\hat{x}_i\dddot{Q}^-_{ia} - 14\hat{x}_b\hat{x}_i\ddot{Q}^-_{ia}\hat{x}_k\hat{x}_l\dddot{Q}^-_{kl} - 14\hat{x}_b\hat{x}_k\hat{x}_l\ddot{Q}^-_{kl}\hat{x}_j\dddot{Q}^-_{ji}\hat{x}_i\hat{x}_a \\
& + 4\hat{x}_a\hat{x}_b\ddot{Q}^-_{ij}\dddot{Q}^-_{ij} + 4\hat{x}_b\ddot{Q}^-_{ai}\dddot{Q}^-_{ik}\hat{x}_k + 4\hat{x}_b\hat{x}_i\ddot{Q}^-_{ij}\dddot{Q}^-_{ja} - 2\hat{x}_b\ddot{Q}^-_{ai}\hat{x}_i\hat{x}_k\hat{x}_l\dddot{Q}^-_{kl} \\
& - 4\hat{x}_b\hat{x}_a\hat{x}_i\ddot{Q}^-_{ij}\dddot{Q}^-_{jk}\hat{x}_k + 12\hat{x}_b\hat{x}_k\ddot{Q}^-_{ki}\dddot{Q}^-_{ia} + 4\hat{x}_b\hat{x}_j\ddot{Q}^-_{ji}\dddot{Q}^-_{ia} - 4\hat{x}_b\hat{x}_k\ddot{Q}^-_{kj}\hat{x}_j\hat{x}_i\dddot{Q}^-_{ia} \\
& - 6\hat{x}_b\hat{x}_k\hat{x}_l\ddot{Q}^-_{kl}\hat{x}_i\dddot{Q}^-_{ia} - 2\hat{x}_b\hat{x}_k\hat{x}_l\ddot{Q}^-_{kl}\hat{x}_j\dddot{Q}^-_{ja} + 6\hat{x}_b\hat{x}_i\ddot{Q}^-_{ia}\hat{x}_k\hat{x}_l\dddot{Q}^-_{kl} + 2\hat{x}_b\hat{x}_j\ddot{Q}^-_{ja}\hat{x}_k\hat{x}_l\dddot{Q}^-_{kl} \\
& + 4\hat{x}_b\hat{x}_i\ddot{Q}^-_{ia}\hat{x}_k\hat{x}_l\dddot{Q}^-_{kl} + 4\hat{x}_b\hat{x}_k\ddot{Q}^-_{kj}\dddot{Q}^-_{ji}\hat{x}_i\hat{x}_a + 2\hat{x}_b\hat{x}_k\hat{x}_l\ddot{Q}^-_{kl}\hat{x}_i\dddot{Q}^-_{ia}] + \frac{\partial}{\partial t}O\left(\frac{1}{r^2}\right) \\
& + \frac{\partial^2}{\partial t^2}O\left(\frac{1}{r^2}\right) + \frac{\partial}{\partial t}O\left(\frac{1}{r^3}\right) + O\left(\frac{1}{r^4}\right),
\end{aligned}$$

$$\begin{aligned}
\left\langle \left(h^{\mathrm{TT}}_{ij} h^{\mathrm{TT}}_{ai,b}\right)_{,j} + \left(h^{\mathrm{TT}}_{ij} h^{\mathrm{TT}}_{bi,a}\right)_{,j} \right\rangle \underset{t\sim r\to\infty}{\longrightarrow} \frac{1}{r^3} \Big\langle & 8\hat{x}_a\hat{x}_i\ddot{Q}^-_{ij}\dddot{Q}^-_{jb} + 8\hat{x}_b\hat{x}_i\ddot{Q}^-_{ij}\dddot{Q}^-_{ja} \\
& + 12\hat{x}_a\ddot{Q}^-_{bi}\hat{x}_i\hat{x}_j\hat{x}_k\dddot{Q}^-_{jk} + 12\hat{x}_b\ddot{Q}^-_{ai}\hat{x}_i\hat{x}_j\hat{x}_k\dddot{Q}^-_{jk}\Big\rangle + O\left(\frac{1}{r^4}\right),
\end{aligned}$$

$$h^{\mathrm{TT}}_{ij} h^{\mathrm{TT}}_{ab} \underset{t\sim r\to\infty}{\longrightarrow} O\left(\frac{1}{r^2}\right),$$

$$\left(h_{ij}^{\rm TT}h_{ab}^{\rm TT}\right)_{,it}\underset{t\sim r\to\infty}{\longrightarrow}\frac{\partial}{\partial t}O\left(\frac{1}{r^2}\right)+O\left(\frac{1}{r^3}\right),$$

$$\left(h_{ij}^{\rm TT}h_{ab}^{\rm TT}\right)_{,ijt}\underset{t\sim r\to\infty}{\longrightarrow}\frac{\partial^2}{\partial t^2}O\left(\frac{1}{r^2}\right)+\frac{\partial}{\partial t}O\left(\frac{1}{r^3}\right)+O\left(\frac{1}{r^4}\right),$$

$$\left\langle\left(h_{ij}^{\rm TT}h_{ab}^{\rm TT}\right)_{,ij}\right\rangle\underset{t\sim r\to\infty}{\longrightarrow}O\left(\frac{1}{r^4}\right),$$

$$\left\langle\left(h_{ij}^{\rm TT}h_{ij}^{\rm TT}\right)_{,ab}\right\rangle\underset{t\sim r\to\infty}{\longrightarrow}O\left(\frac{1}{r^4}\right),$$

$$\begin{aligned}
\frac{1}{2}h_{ij,a}^{\rm TT}h_{ij,b}^{\rm TT}\underset{t\sim r\to\infty}{\longrightarrow}&\frac{\hat{x}_a\hat{x}_b}{2r^2}\left[4\dddot{\boldsymbol{Q}}^-:\dddot{\boldsymbol{Q}}^- - 8\hat{\boldsymbol{x}}\cdot\dddot{\boldsymbol{Q}}^-\cdot\dddot{\boldsymbol{Q}}^-\cdot\hat{\boldsymbol{x}}+2(\hat{\boldsymbol{x}}\cdot\dddot{\boldsymbol{Q}}^-\cdot\hat{\boldsymbol{x}})^2\right]\\
&+\frac{1}{2r^3}[8\hat{x}_a\dddot{Q}_{bi}^-\ddot{Q}_{ij}^-\hat{x}_j+8\hat{x}_a\hat{x}_i\dddot{Q}_{ij}^-\ddot{Q}_{jb}^- - 4\hat{x}_a\dddot{Q}_{bi}^-\hat{x}_i\,\hat{x}_j\hat{x}_k\,\ddot{Q}_{jk}^- - 4\hat{x}_a\hat{x}_i\hat{x}_j\dddot{Q}_{ij}^-\hat{x}_k\,\ddot{Q}_{ka}^-\\
&-4\hat{x}_a\hat{x}_b\hat{\boldsymbol{x}}\cdot\dddot{\boldsymbol{Q}}^-\cdot\ddot{\vec{Q}}^-\cdot\hat{\boldsymbol{x}}-4\hat{x}_a\dddot{Q}_{bi}^-\hat{x}_i\hat{x}_k\hat{x}_l\,\ddot{Q}_{kl}^-+4\hat{x}_a\hat{x}_k\hat{x}_l\dddot{Q}_{kl}^-\hat{x}_i\,\ddot{Q}_{ib}^- - 4\hat{x}_a\hat{x}_i\dddot{Q}_{ij}^-\ddot{Q}_{jb}^-\\
&-4\hat{x}_a\hat{x}_k\hat{x}_l\dddot{Q}_{kl}^-\hat{x}_i\,\ddot{Q}_{ib}^-+2\hat{x}_a\hat{x}_b\hat{x}_i\hat{x}_j\dddot{Q}_{ij}^-\hat{x}_k\hat{x}_l\,\ddot{Q}_{kl}^-+2\hat{x}_a\hat{x}_i\dddot{Q}_{ib}^-\hat{x}_k\hat{x}_l\,\ddot{Q}_{kl}^-+4\hat{x}_a\hat{x}_k\hat{x}_l\dddot{Q}_{kl}^-\hat{x}_i\,\ddot{Q}_{ib}^-\\
&-4\hat{x}_a\hat{x}_i\dddot{Q}_{ib}^-\hat{x}_k\hat{x}_l\,\ddot{Q}_{kl}^- - 4\hat{x}_a\hat{x}_b\hat{\boldsymbol{x}}\cdot\dddot{\boldsymbol{Q}}^-\cdot\dddot{\boldsymbol{Q}}^-\cdot\hat{\boldsymbol{x}}+4\hat{x}_a\hat{x}_k\hat{x}_l\dddot{Q}_{kl}^-\hat{x}_i\,\ddot{Q}_{ib}^- - 4\hat{x}_a\hat{x}_k\hat{x}_l\dddot{Q}_{kl}^-\hat{x}_i\,\ddot{Q}_{ib}^-\\
&-4\hat{x}_a\hat{x}_i\,\dddot{Q}_{ij}^-\ddot{Q}_{jb}^-+2\hat{x}_a\hat{x}_i\,\dddot{Q}_{ib}^-\hat{x}_k\hat{x}_l\,\ddot{Q}_{kl}^-+2\hat{x}_a\hat{x}_b\hat{x}_i\hat{x}_j\,\dddot{Q}_{ij}^-\hat{x}_k\hat{x}_l\,\ddot{Q}_{kl}^-+4\hat{x}_a\hat{x}_k\hat{x}_l\,\dddot{Q}_{kl}^-\hat{x}_i\,\ddot{Q}_{ib}^-\\
&+4\hat{x}_a\hat{x}_k\hat{x}_l\,\dddot{Q}_{kl}^-\hat{x}_i\,\ddot{Q}_{ib}^- - 6\hat{x}_a\hat{x}_k\hat{x}_l\,\dddot{Q}_{kl}^-\hat{x}_i\,\ddot{Q}_{ib}^-+4\hat{x}_a\hat{x}_k\hat{x}_l\,\dddot{Q}_{kl}^-\hat{x}_i\,\ddot{Q}_{ib}^-\\
&-2\hat{x}_a\hat{x}_b\hat{x}_i\hat{x}_j\,\dddot{Q}_{ij}^-\hat{x}_k\hat{x}_l\,\ddot{Q}_{kl}^- - 2\hat{x}_a\hat{x}_k\hat{x}_l\,\dddot{Q}_{kl}^-\hat{x}_i\,\ddot{Q}_{ib}^-+4\hat{x}_a\hat{x}_b\hat{x}_i\hat{x}_j\,\dddot{Q}_{ij}^-\hat{x}_k\hat{x}_l\,\ddot{Q}_{kl}^-\\
&-2\hat{x}_a\hat{x}_k\hat{x}_l\,\dddot{Q}_{kl}^-\hat{x}_i\,\ddot{Q}_{ib}^-+4\hat{x}_a\hat{x}_k\hat{x}_l\,\dddot{Q}_{kl}^-\hat{x}_i\,\ddot{Q}_{ib}^- - 2\hat{x}_a\hat{x}_b\hat{x}_i\hat{x}_j\dddot{Q}_{ij}^-\hat{x}_k\hat{x}_l\,\ddot{Q}_{kl}^-\\
&-2\hat{x}_a\hat{x}_k\hat{x}_l\dddot{Q}_{kl}^-\hat{x}_i\,\ddot{Q}_{ib}^- + \text{same terms with a and b interchanged}]+\frac{\partial}{\partial t}O\left(\frac{1}{r^3}\right)\\
&+O\left(\frac{1}{r^4}\right),
\end{aligned}$$

$$\begin{aligned}
\frac{1}{2}\left\langle h_{ij,a}^{\rm TT}h_{ij,b}^{\rm TT}\right\rangle\underset{t\sim r\to\infty}{\longrightarrow}&\frac{\hat{x}_a\hat{x}_b}{r^2}\left\langle 2\dddot{\boldsymbol{Q}}^-:\dddot{\boldsymbol{Q}}^- - 4\hat{\boldsymbol{x}}\cdot\dddot{\boldsymbol{Q}}^-\cdot\dddot{\boldsymbol{Q}}^-\cdot\hat{\boldsymbol{x}}+(\hat{\boldsymbol{x}}\cdot\dddot{\boldsymbol{Q}}^-\cdot\hat{\boldsymbol{x}})^2\right\rangle+\frac{1}{r^3}\left\langle 4\hat{x}_a\hat{x}_i\ddot{Q}_{ij}^-\dddot{Q}_{jb}^-\right.\\
&\left.+4\hat{x}_b\hat{x}_i\ddot{Q}_{ij}^-\dddot{Q}_{ja}^-+6\hat{x}_a\ddot{Q}_{bi}^-\hat{x}_i\hat{x}_j\hat{x}_k\dddot{Q}_{jk}^-+6\hat{x}_b\ddot{Q}_{ai}^-\hat{x}_i\hat{x}_j\hat{x}_k\dddot{Q}_{jk}^-\right\rangle+0\left(\frac{1}{r^4}\right),
\end{aligned}$$

$$\left\langle\left(h_{ai}^{\rm TT}h_{bj}^{\rm TT}\right)_{,ij}\right\rangle\underset{t\sim r\to\infty}{\longrightarrow}O\left(\frac{1}{r^4}\right),\quad\left\langle\left(h_{ai}^{\rm TT}h_{bi}^{\rm TT}\right)_{,00}\right\rangle\underset{t\sim r\to\infty}{\longrightarrow}0,$$

$$\left\langle\left(h_{ai}^{\rm TT}h_{bi}^{\rm TT}\right)_{,jj}\right\rangle\underset{t\sim r\to\infty}{\longrightarrow}O\left(\frac{1}{r^4}\right),$$

$$\left\langle T^{ab}_{\mathrm{GW}}\right\rangle_{t\sim r\to\infty}\frac{\hat{x}_a\hat{x}_b}{16\pi r^2}\left\langle 2\dddot{\boldsymbol{Q}}^{-}:\dddot{\boldsymbol{Q}}^{-}-4\hat{\boldsymbol{x}}\cdot\dddot{\boldsymbol{Q}}^{-}\cdot\dddot{\boldsymbol{Q}}^{-}\cdot\hat{\boldsymbol{x}}+(\hat{\boldsymbol{x}}\cdot\dddot{\boldsymbol{Q}}^{-}\cdot\hat{\boldsymbol{x}})^2\right\rangle+\frac{1}{16\pi r^3}\left\langle 12\hat{x}_a\hat{x}_i\ddot{Q}^{-}_{ij}\dddot{Q}^{-}_{jb}\right.$$
$$\left.+\,12\hat{x}_b\hat{x}_i\ddot{Q}^{-}_{ij}\dddot{Q}^{-}_{ja}+18\hat{x}_a\ddot{Q}^{-}_{bi}\hat{x}_i\hat{x}_j\hat{x}_k\dddot{Q}^{-}_{jk}+18\hat{x}_b\ddot{Q}^{-}_{ai}\hat{x}_i\hat{x}_j\hat{x}_k\dddot{Q}^{-}_{jk}\right\rangle+0\left(\frac{1}{r^4}\right).$$

Secular changes in P^μ and J^{ij} are given by

$$\frac{\mathrm{d}P^\mu}{\mathrm{d}t}=\int\limits_V T^{\mu 0}_{\mathrm{tot},0}\mathrm{d}^3\boldsymbol{x}=-\int\limits_V T^{\mu i}_{\mathrm{tot},i}\mathrm{d}^3\boldsymbol{x}=-\int\limits_S T^{\mu i}_{\mathrm{tot}}\mathrm{d}^2S_i,$$

$$\frac{\mathrm{d}S^{\mathrm{ij}}}{\mathrm{d}t}=\int\limits_V\left(x^iT^{j0}_{\mathrm{tot}}-x^jT^{i0}_{\mathrm{tot}}\right)_{,0}\mathrm{d}^3\boldsymbol{x}=-\int\limits_V\left(x^iT^{jk}_{\mathrm{tot},k}-x^jT^{ik}_{\mathrm{tot},k}\right)\mathrm{d}^3\boldsymbol{x}$$
$$=-\int\limits_V\left[\left(x^iT^{jk}_{\mathrm{tot}}\right)_{,k}-T^{ji}_{\mathrm{tot}}-\left(x^jT^{ik}_{\mathrm{tot}}\right)_{,k}+T^{ij}_{\mathrm{tot}}\right]\mathrm{d}^3\boldsymbol{x}=-\int\limits_S\left(x^iT^{jk}_{\mathrm{tot}}-x^jT^{ik}_{\mathrm{tot}}\right)\mathrm{d}^2S_k,$$

$$\frac{1}{4\pi}\int\limits_{4\pi}\hat{x}_i\hat{x}_j\mathrm{d}^2\Omega=\frac{1}{3}\delta_{ij},\quad\frac{1}{4\pi}\int\limits_{4\pi}\hat{x}_i\mathrm{d}^2\Omega=0,$$

$$\frac{1}{4\pi}\int\limits_{4\pi}\hat{x}_i\hat{x}_j\hat{x}_k\hat{x}_l\mathrm{d}^2\Omega=\frac{1}{15}(\delta_{ij}\delta_{kl}+\delta_{ik}\delta_{jl}+\delta_{il}\delta_{jk}),$$

$$\left\langle\frac{\mathrm{dM}}{\mathrm{d}t}\right\rangle=-\int\limits_S\langle T^{oi}_{\mathrm{GW}}\rangle\mathrm{d}^2S_i$$
$$=-\frac{1}{16\pi}\int\limits_{4\pi}\langle 2\dddot{\boldsymbol{Q}}^{-}:\dddot{\boldsymbol{Q}}^{-}-4\hat{\boldsymbol{x}}\cdot\dddot{\boldsymbol{Q}}^{-}\cdot\dddot{\boldsymbol{Q}}^{-}\cdot\hat{\boldsymbol{x}}+\hat{\boldsymbol{x}}\cdot\dddot{\boldsymbol{Q}}^{-}\cdot\hat{\boldsymbol{x}}\hat{\boldsymbol{x}}\cdot\dddot{\boldsymbol{Q}}^{-}\cdot\hat{\boldsymbol{x}}\rangle\mathrm{d}^2\Omega$$
$$=-\frac{1}{4}\left(2-\frac{4}{3}+\frac{2}{15}\right)\langle\dddot{\boldsymbol{Q}}^{-}:\dddot{\boldsymbol{Q}}^{-}\rangle=-\frac{1}{5}\langle\dddot{\boldsymbol{Q}}^{-}:\dddot{\boldsymbol{Q}}^{-}\rangle,$$

$$\left\langle\frac{\mathrm{dP^i}}{\mathrm{d}t}\right\rangle=-\int\limits_S\langle T^{ij}_{\mathrm{GW}}\rangle\mathrm{d}^2S_j=0,$$

which tells us that quadrupole gravitational radiation cannot be used as a propellant, i.e., we must go to octupole terms, and

$$
\begin{aligned}
\left\langle \frac{\mathrm{dS_{ij}}}{\mathrm{d}t} \right\rangle &= -\int\limits_S \left(x^i \langle T^{jk}_{\mathrm{GW}} \rangle - x^j \langle T^{ik}_{\mathrm{GW}} \rangle \right) \mathrm{d}^2 S_k \\
&= -\frac{1}{16\pi} \int\limits_{4\pi} \Big\langle 12\hat{x}_i\hat{x}_k\, \ddot{Q}^{-}_{kl}\dddot{Q}^{-}_{lj} - 12\hat{x}_j\hat{x}_k\, \ddot{Q}^{-}_{kl}\dddot{Q}^{-}_{li} + 18\hat{x}_i\ddot{Q}^{-}_{jk}\hat{x}_k\hat{x}_l\hat{x}_m\dddot{Q}^{-}_{lm} \\
&\quad - 18\hat{x}_j\ddot{Q}^{-}_{ik}\hat{x}_k\hat{x}_l\hat{x}_m\dddot{Q}^{-}_{lm} \Big\rangle \mathrm{d}^2\Omega \\
&= -\frac{1}{4}\left\langle 4\ddot{Q}^{-}_{ik}\dddot{Q}^{-}_{kj} - 4\ddot{Q}^{-}_{jk}\dddot{Q}^{-}_{ki} + \frac{36}{15}\ddot{Q}^{-}_{jk}\dddot{Q}^{-}_{ki} - \frac{36}{15}\ddot{Q}^{-}_{ik}\dddot{Q}^{-}_{kj} \right\rangle = -\frac{4}{5}\left\langle \ddot{Q}^{-}_{ik}\dddot{Q}^{-}_{jk} \right\rangle,
\end{aligned}
$$

$$
\left\langle \frac{\mathrm{dS_i}}{\mathrm{d}t} \right\rangle = \frac{1}{2}\varepsilon_{ijk}\left\langle \frac{\mathrm{dS_{jk}}}{\mathrm{d}t} \right\rangle = -\frac{2}{5}\varepsilon_{ijk}\left\langle \ddot{Q}^{-}_{jl}\dddot{Q}^{-}_{kl} \right\rangle.
$$

The expansions leading to the above results are essentially expansions in powers of $\mathcal{R}/r$ and λ/r, useful in the radiation zone ($r \gg \lambda \gg \mathcal{R}$). In the near zone ($\lambda \gg r \gg \mathcal{R}$), an expansion in r/λ is more useful (see current chapter):

$$
\begin{aligned}
\overline{h}^{\pm}_{00} &= \frac{2M}{r} + \frac{3x_ix_j}{r^5}\left(Q_{ij} \pm r\dot{Q}_{ij} + \frac{1}{2}r^2\ddot{Q}_{ij} \pm \frac{1}{6}r^3\dddot{Q}_{ij} + \frac{1}{24}r^4\ddddot{Q}_{ij} \pm \frac{1}{120}r^5\overset{(5)}{Q}_{ij} + \cdots \right) \\
&\quad + \frac{3x_ix_j}{r^5}\left(\mp r\dot{Q}_{ij} - r^2\ddot{Q}_{ij} \mp \frac{1}{2}r^3\dddot{Q}_{ij} - \frac{1}{6}r^4\ddddot{Q}_{ij} \mp \frac{1}{24}r^5\overset{(5)}{Q}_{ij} + \cdots \right) \\
&\quad + \frac{x_ix_j}{r^5}\left(r^2\ddot{Q}_{ij} \pm r^3\dddot{Q}_{ij} + \frac{1}{2}r^4\ddddot{Q}_{ij} \pm \frac{1}{6}r^5\overset{(5)}{Q}_{ij} + \cdots \right) + \cdots,
\end{aligned}
$$

$$
\begin{aligned}
\overline{h}^{\pm}_{0i} &= \frac{2S_{ij}\hat{x}_j}{r^2} - \frac{2x_j}{r^3}\left(\dot{Q}_{ij} \pm r\ddot{Q}_{ij} + \frac{1}{2}r^2\dddot{Q}_{ij} \pm \frac{1}{6}r^3\ddddot{Q}_{ij} + \frac{1}{24}r^4\overset{(5)}{Q}_{ij} + \cdots \right) \\
&\quad + \frac{2x_j}{r^3}\left(\pm r\ddot{Q}_{ij} + r^2\dddot{Q}_{ij} \pm \frac{1}{2}r^3\ddddot{Q}_{ij} + \frac{1}{6}r^4\overset{(5)}{Q}_{ij} + \cdots \right) + \cdots,
\end{aligned}
$$

$$
\begin{aligned}
\overline{h}^{\pm}_{ij} &= \frac{2M}{r}\delta_{ij} \\
&\quad + \delta_{ij}\frac{3x_kx_l}{r^5}\left(Q_{kl} \pm r\dot{Q}_{kl} + \frac{1}{2}r^2\ddot{Q}_{kl} \pm \frac{1}{6}r^3\dddot{Q}_{kl} + \frac{1}{24}r^4\ddddot{Q}_{kl} \pm \frac{1}{120}r^5\overset{(5)}{Q}_{kl} + \cdots \right) \\
&\quad + \delta_{ij}\frac{3x_kx_l}{r^5}\left(\mp r\dot{Q}_{kl} - r^2\ddot{Q}_{kl} \mp \frac{1}{2}r^3\dddot{Q}_{kl} - \frac{1}{6}r^4\ddddot{Q}_{kl} \mp \frac{1}{24}r^5\overset{(5)}{Q}_{kl} + \cdots \right) \\
&\quad + \delta_{ij}\frac{x_kx_l}{r^5}\left(r^2\ddot{Q}_{kl} \pm r^3\dddot{Q}_{kl} + \frac{1}{2}r^4\ddddot{Q}_{kl} \pm \frac{1}{6}r^5\overset{(5)}{Q}_{kl} + \cdots \right) \\
&\quad + \frac{2}{r}\left(\ddot{Q}_{ij} \pm r\dddot{Q}_{ij} + \frac{1}{2}r^2\ddddot{Q}_{ij} \pm \frac{1}{6}r^3\overset{(5)}{Q}_{ij} + \cdots \right) + \cdots,
\end{aligned}
$$

$$
h^{\mathrm{SW}}_{00} = \frac{2M}{r} + \frac{3\hat{x}_i\hat{x}_jQ_{ij}}{r^3} - \frac{1}{2}\frac{\hat{x}_i\hat{x}_j\ddot{Q}_{ij}}{r} + \frac{1}{8}r\hat{x}_i\hat{x}_j\ddddot{Q}_{ij} + \cdots,
$$

where only the first two terms are significant in the near zone,

$$h_{0i}^{\mathrm{SW}} = \frac{2S_{ij}\hat{x}_j}{r^2} - \frac{2\dot{Q}_{ij}\hat{x}_j}{r^2} + \dddot{Q}_{ij}\hat{x}_j + \frac{1}{4}r^2\overset{\cdot\cdot\cdot\cdot\cdot}{Q}_{ij}\hat{x}_j + \cdots,$$

where only the first term is significant in the near zone,

$$\begin{aligned} h_{ij}^{\mathrm{SW}} = \delta_{ij}&\left(\frac{2M}{r} + \frac{3\hat{x}_k\hat{x}_l Q_{kl}}{r^3} - \frac{1}{2}\frac{\hat{x}_k\hat{x}_l\ddot{Q}_{kl}}{r} + \frac{1}{8}r\hat{x}_k\hat{x}_l\ddddot{Q}_{kl} + \cdots\right) \\ &+ \frac{2}{r}\ddot{Q}_{ij} + r\ddddot{Q}_{ij} + \cdots, \end{aligned}$$

where only the first two terms are significant in the near zone, and finally,

$$h_{00}^{\mathrm{rad}} = -\frac{2}{15}x_i x_j\,\overset{\cdot\cdot\cdot\cdot\cdot}{Q}_{ij} + \cdots, \quad h_{0i}^{\mathrm{rad}} = -\frac{4}{3}\ddddot{Q}_{ij}x_j + \cdots,$$

$$h_{ij}^{\mathrm{rad}} = -4\,\dddot{Q}_{ij} - \frac{2}{3}r^2\overset{\cdot\cdot\cdot\cdot\cdot}{Q}_{ij} - \frac{2}{15}\delta_{ij}x_k x_l\,\overset{\cdot\cdot\cdot\cdot\cdot}{Q}_{kl} + \cdots.$$

The parameters M, S_{ij}, Q_{ij} may (in principle) be determined by observing orbits in the near zone but still in the Newtonian region. Quasi-stationarity makes the orbit equations reduce effectively to

$$\frac{\mathrm{d}^2 z^i}{\mathrm{d}t^2} = \frac{1}{2}h_{00,i}^{\mathrm{SW}},$$

as can be seen from Chap. 6. There is no need to know the details of $\mathcal{T}_{\mathrm{tot}}^{\mu\nu}$.

We make the gauge transformation

$$\xi_0 = \frac{1}{3}x_i x_j\ddddot{Q}_{ij} + \cdots, \quad \xi_i = -2\dddot{Q}_{ij}x_j - \frac{1}{5}r^2\overset{\cdot\cdot\cdot\cdot\cdot}{Q}_{ij}x_j + \cdots,$$

with $\Box^2\xi_\mu = 0$. Then

$$\overline{h}_{00}^{\mathrm{rad}} = h_{00}^{\mathrm{rad}} - 2\xi_{0,0} = -\frac{4}{5}x_i x_j\overset{\cdot\cdot\cdot\cdot\cdot}{Q}_{ij} + O(r^4),$$

$$\overline{h}_{0i}^{\mathrm{rad}} = h_{0i}^{\mathrm{rad}} - \xi_{0,i} - \xi_{i,0} = O(r^3),$$

$$\begin{aligned} \overline{h}_{ij}^{\mathrm{rad}} &= h_{ij}^{\mathrm{rad}} - \xi_{i,j} - \xi_{j,i} \\ &= -\frac{4}{15}r^2\overset{\cdot\cdot\cdot\cdot\cdot}{Q}_{ij} + \frac{2}{15}x_i x_k\overset{\cdot\cdot\cdot\cdot\cdot}{Q}_{kj} + \frac{2}{15}x_j x_k\overset{\cdot\cdot\cdot\cdot\cdot}{Q}_{ki} - \frac{2}{15}\delta_{ij}x_k x_l\,\overset{\cdot\cdot\cdot\cdot\cdot}{Q}_{kl} + O(r^4), \end{aligned}$$

$$\overline{\overline{h}}_{\mu\nu} = \overline{h}_{\mu\nu} - \xi_{\mu,\nu} - \xi_{\nu,\mu} = h_{\mu\nu}^{\mathrm{SW}} + \frac{1}{2}\overline{h}_{\mu\nu}^{\mathrm{rad}}.$$

Now suppose the source is Newtonian ($\mathcal{R} \gg M$). Then the above expression for $\overline{h}_{\mu\nu}^{\mathrm{rad}}$ may be used *even inside the source*. The appropriate expression for $h_{\mu\nu}^{\mathrm{SW}}$ in this region, of course, will no longer take the form of a multipole expansion, but will be that corresponding to the weak Newtonian field produced by the given mass distribution at any instant. In the absence of the component $\bar{h}^{\mathrm{rad}}_{\mu\nu}/2$, the dynamics of the source would be precisely that of Newtonian physics, and the energy and angular momentum of the source would be conserved. The component $\bar{h}^{\mathrm{rad}}_{\mu\nu}/2$ accounts for the actual loss of energy and angular momentum by radiation. It describes the *radiation reaction*. It introduces a slight additional force per unit mass given by (see Chap. 6)

$$F_i = -\frac{1}{4}\left(2\bar{h}^{\mathrm{rad}}_{i0,0} - \bar{h}^{\mathrm{rad}}_{00,i}\right) = -\frac{2}{5}\overset{:::}{Q}_{ij}x_j + O(r^3) \quad \text{(slow motion)}.$$

Let m_n and x_n be the mass and position of the nth particle of which the source is composed. Then the rates of change of energy, momentum, and angular momentum caused by this radiation reaction force are

$$\frac{\mathrm{dM}}{\mathrm{d}t} = \sum_n m_n F_{ni}\dot{x}_{ni} = -\frac{2}{5}\sum_n m_n \dot{x}_{ni}x_{nj}\overset{:::}{Q}_{ij} = -\frac{1}{5}\dot{Q}\colon \overset{:::}{Q},$$

$$\frac{\mathrm{dP^i}}{\mathrm{d}t} = \sum_n m_n F_{ni} = -\frac{2}{5}\sum_n m_n x_{nj}\overset{:::}{Q}_{ij} = 0,$$

because the origin is at the center of mass, and

$$\frac{\mathrm{dS_i}}{\mathrm{d}t} = \varepsilon_{ijk}\sum_n m_n x_{nj}F_{nk} = -\frac{2}{5}\varepsilon_{ijk}\sum_n m_n x_{nj}\overset{:::}{Q}_{kl}x_{nl}$$

$$= -\frac{2}{5}\varepsilon_{ijk}\sum_n m_n\left(x_{nj}x_{nl} - \frac{1}{3}\delta_{jl}r_n^2\right)\overset{:::}{Q}_{kl} = -\frac{2}{5}\varepsilon_{ijk}Q_{jl}\overset{:::}{Q}_{kl}.$$

Taking averages over periods or orbits, we get

$$\left\langle\frac{\mathrm{dM}}{\mathrm{d}t}\right\rangle = -\frac{1}{5}\left\langle\dot{Q}\colon\overset{:::}{Q}\right\rangle = -\frac{1}{5}\langle\dddot{Q}\colon\dddot{Q}\rangle, \quad \left\langle\frac{\mathrm{dP^i}}{\mathrm{d}t}\right\rangle = 0,$$

$$\left\langle\frac{\mathrm{dS_i}}{\mathrm{d}t}\right\rangle = -\frac{2}{5}\varepsilon_{ijk}\left\langle Q_{jl}\overset{:::}{Q}_{kl}\right\rangle = -\frac{2}{5}\varepsilon_{ijk}\langle\ddot{Q}_{jl}\dddot{Q}_{kl}\rangle,$$

in agreement with the results obtained by integrating fluxes in present chapter.

Appendix A
Spinning Bodies

A.1: Nonrelativistic Spinning Body in an Impressed Electromagnetic Field

Mass of nth componentparticle	m_n
Charge of nth component particle	e_n
Position of center of mass	$\boldsymbol{x}$
Position of nth componentparticle	$\boldsymbol{x} + \boldsymbol{x}_n$

$$\sum_n m_n \boldsymbol{x}_n = 0$$

Total mass $\quad m = \sum_n m_n$

Total charge $\quad e = \sum_n e_n$

Moment of inertia tensor $\quad \underline{I} = \sum_n m_n \left(x_n^2 \underline{1} - \boldsymbol{x}_n \boldsymbol{x}_n \right)$

Charge moment tensor $\quad \underline{D} = \sum_n e_n \left(x_n^2 \underline{1} - \boldsymbol{x}_n \boldsymbol{x}_n \right)$

Electric dipole moment $\quad \zeta = \sum_n e_n \boldsymbol{x}_n$

Magnetic dipole moment $\quad \mu = \frac{1}{2} \sum_n e_n \boldsymbol{x}_n \times \dot{\boldsymbol{x}}_n$

Electric quadrupole moment tensor

$$\underline{Q} = \sum_n e_n \left(\boldsymbol{x}_n \boldsymbol{x}_n - \frac{1}{3} x_n^2 \underline{1} \right) = -\underline{D} + \frac{1}{3} \underline{1} \operatorname{tr} \underline{D}$$

Electromagnetic potential vector $(A_\mu) = (-\phi, \boldsymbol{A})$

Electromagnetic field tensor $_{\mu\nu} = A_{\nu,\mu} - A_{\mu,\nu}$

Electric field $\boldsymbol{E} = (F.0) = -\nabla\phi - \boldsymbol{A}_{,0}$

Magnetic field $H_a = \frac{1}{2}\varepsilon_{abc}F_{bc} = \varepsilon_{abc}A_{c,b}$

$$\boldsymbol{H} = \nabla \times \boldsymbol{A}_{,ab} = \varepsilon_{abc}H_c$$

where dots denote the time derivative and commas denote differentiation with respect to spacetime coordinates $(x^\mu) = (t, \boldsymbol{x})$.

The impressed electromagnetic field will be assumed to satisfy Maxwell's empty-space field equations in the region occupied by the spinning body:

$$F^{\mu\nu}_{\ \ ,\nu} = 0, \quad F_{\mu\nu,\sigma} + F_{\nu\sigma,\mu} + F_{\sigma\mu,\nu} = 0,$$

where

$$F^{\mu\nu} = \eta^{\mu\sigma}\eta^{\nu\tau}F_{\sigma\tau}, \quad (\eta_{\mu\nu}) = (\eta^{\mu\nu}) = \mathrm{diag}(-1,1,1,1),$$

so that

$$\begin{aligned} \nabla\cdot\boldsymbol{E} = 0, \quad \nabla\times\boldsymbol{H} + \dot{\boldsymbol{E}} = 0, \\ \nabla\times\boldsymbol{E} - \dot{\boldsymbol{H}} = 0, \quad \nabla\cdot\boldsymbol{H} = 0. \end{aligned}$$

The Lagrangian is

$$\begin{aligned} L' &= \frac{1}{2}\sum_n m_n(\dot{\boldsymbol{x}} + \dot{\boldsymbol{x}}_n)^2 - \sum_n e_n\phi(t, \boldsymbol{x} + \boldsymbol{x}_n) + \sum_n e_n A(t, \boldsymbol{x} + \boldsymbol{x}_n)\cdot(\dot{\boldsymbol{x}} + \dot{\boldsymbol{x}}_n) \\ &= \frac{1}{2}m\dot{\boldsymbol{x}}^2 + \frac{1}{2}\sum_n m_n\dot{\boldsymbol{x}}_n^2 - e\phi - \sum_n e_n\boldsymbol{x}_n\cdot\nabla\phi - \frac{1}{2}\sum_n e_n\boldsymbol{x}_n\cdot\nabla\nabla\phi\cdot\boldsymbol{x}_n - \cdots \\ &\quad + e\boldsymbol{A}\cdot\dot{\boldsymbol{x}} + \sum_n e_n\dot{\boldsymbol{x}}_n\cdot\boldsymbol{A} + \sum_n e_n\boldsymbol{x}_n\cdot\nabla\boldsymbol{A}\cdot\dot{\boldsymbol{x}} + \sum_n e_n\boldsymbol{x}_n\cdot\nabla\boldsymbol{A}\cdot\dot{\boldsymbol{x}}_n + \cdots \\ &= L + \frac{\mathrm{d}}{\mathrm{d}t}\left(\sum_n e_n\boldsymbol{x}_n\cdot\boldsymbol{A} + \frac{1}{2}e_n\boldsymbol{x}_n\cdot\nabla\boldsymbol{A}\cdot\boldsymbol{x}_n + \cdots\right), \end{aligned}$$

where all fields are evaluated at the center of mass and time t, and

$$\begin{aligned} L &= \frac{1}{2}m\dot{\boldsymbol{x}}^2 - e\phi + e\boldsymbol{A}\cdot\dot{\boldsymbol{x}} + \frac{1}{2}\sum_n m_n\dot{\boldsymbol{x}}_n^2 - \sum_n e_n\boldsymbol{x}_n\cdot(\nabla\phi + \boldsymbol{A}_{,0}) \\ &\quad - \frac{1}{2}\sum_n e_n\boldsymbol{x}_n\cdot\nabla(\nabla\phi + \boldsymbol{A}_{,0})\cdot\boldsymbol{x}_n + \sum_n e_n(\boldsymbol{x}_n\cdot\nabla\boldsymbol{A}\cdot\dot{\boldsymbol{x}} - \dot{\boldsymbol{x}}\cdot\nabla\boldsymbol{A}\cdot\boldsymbol{x}_n) \\ &\quad + \frac{1}{2}\sum_n e_n(\boldsymbol{x}_n\cdot\nabla\boldsymbol{A}\cdot\dot{\boldsymbol{x}}_n - \dot{\boldsymbol{x}}_n\cdot\nabla\boldsymbol{A}\cdot\boldsymbol{x}_n) + \cdots \\ &= \frac{1}{2}m\dot{\boldsymbol{x}}^2 - e\phi + e\boldsymbol{A}\cdot\dot{\boldsymbol{x}} + \frac{1}{2}\sum_n m_n\dot{\boldsymbol{x}}_n^2 + \boldsymbol{\zeta}\cdot\boldsymbol{E} + \frac{1}{2}\underline{Q} : \nabla\boldsymbol{E} \\ &\quad + \boldsymbol{\zeta}\cdot(\dot{\boldsymbol{x}}\times\boldsymbol{H}) + \boldsymbol{\mu}\cdot\boldsymbol{H} + \cdots. \end{aligned}$$

Consider now a rigid body. Its orientation is specified by three parameters q^i, e.g., the Euler angles. The spin angular velocity vector $\boldsymbol{\omega}$ is given by

$$\dot{q}^i = R^i_a \omega_a,$$

where

$$R^i_{a,j} R^j_b - R^i_{b,j} R^j_a = \varepsilon_{abc} R^i_c, \quad \dot{\boldsymbol{x}}_n = \boldsymbol{\omega} \times \boldsymbol{x}_n,$$

with the comma denoting differentiation with respect to the q^i. Also

$$\varepsilon_{abc}\omega_b x_{nc} = \dot{x}_{na} = x_{na,i}\dot{q}^i = x_{na,i} R^i_b \omega_b,$$

whence

$$x_{na,i} R^i_b = \varepsilon_{abc} x_{nc}.$$

The conjugate momenta are

$$\begin{aligned} \boldsymbol{p} &= \frac{\partial L}{\partial \dot{\boldsymbol{x}}} = m\dot{\boldsymbol{x}} + e\boldsymbol{A} + \boldsymbol{H} \times \boldsymbol{\zeta} + \cdots, \\ p_i &= \frac{\partial L}{\partial \dot{q}^i} = \sum_n m_n \dot{x}_{na} x_{na,i} + \frac{1}{2}\sum_n e_n \varepsilon_{abc} x_{na} x_{nb,i} H_c + \cdots. \end{aligned}$$

The spin angular momentum is

$$\begin{aligned} S_a &= p_i R^i_a \\ &= \sum_n m_n \dot{x}_{nb} x_{nb,i} R^i_a + \frac{1}{2}\sum_n e_n \varepsilon_{bcd} x_{nb} x_{nc,i} R^i_a H_d + \cdots \\ &= \sum_n m_n \varepsilon_{bcd}\omega_c x_{nd} \varepsilon_{bae} x_{ne} + \frac{1}{2}\sum_n e_n \varepsilon_{bcd} x_{nb} \varepsilon_{cae} x_{ne} H_d + \cdots \\ &= \sum_n m_n (\delta_{ca}\delta_{de} - \delta_{ce}\delta_{da})\omega_c x_{nd} x_{ne} \\ &\qquad + \frac{1}{2}\sum_n e_n (\delta_{da}\delta_{be} - \delta_{de}\delta_{ba}) x_{nb} x_{ne} H_d + \cdots \\ &= I_{ac}\omega_c + \frac{1}{2} D_{ad} H_d + \cdots, \end{aligned}$$

or

$$\boldsymbol{S} = \underline{I} \cdot \boldsymbol{\omega} + \frac{1}{2}\underline{D} \cdot \boldsymbol{H} + \cdots.$$

The Poisson brackets for the spin angular momentum are

$$\begin{aligned} (S_a, S_b) &= (p_i R^i_a, p_j R^j_b) = p_i R^i_{a,j} R^j_b - p_j R^j_{b,i} R^i_a \\ &= \varepsilon_{abc} p_i R^i_c = \varepsilon_{abc} S_c. \end{aligned}$$

Assuming that $\underline{I}$ has an inverse, so that

$$\omega = \underline{I}^{-1} \cdot \left(S - \frac{1}{2} \underline{D} \cdot H \right) + \cdots,$$

the Hamiltonian is

$$\begin{aligned} \mathcal{H} &= p \cdot \dot{x} + p_i \dot{q}^i - L \\ &= m\dot{x}^2 + eA \cdot \dot{x} + \dot{x} \cdot (H \times \zeta) + \sum_n m_n \dot{x}_n^2 + \mu \cdot H + \cdots - L \\ &= \frac{1}{2} m\dot{x}^2 + e\phi + \frac{1}{2} \sum_n m_n \dot{x}_n^2 - \zeta \cdot E - \frac{1}{2} \underline{Q} : \nabla E + \cdots, \end{aligned}$$

with

$$\begin{aligned} \dot{x} &= \frac{1}{m}(p - eA - H \times \zeta) + \cdots, \\ \dot{x}_n^2 &= (\omega \times x_n)^2 = \omega \cdot [x_n \times (\omega \times x_n)] = \omega \cdot \left[\omega x_n^2 - x_n(\omega \cdot x_n) \right] \\ &= \omega \cdot (x_n^2 \underline{1} - x_n x_n) \cdot \omega, \end{aligned}$$

$$\frac{1}{2} \sum_n m_n \dot{x}_n^2 = \frac{1}{2} \omega \cdot \underline{I} \cdot \omega,$$

so that

$$\begin{aligned} \mathcal{H} = \frac{1}{2m}(p - eA - H \times \zeta)^2 + e\phi + \frac{1}{2} \left(S - \frac{1}{2} H \cdot \underline{D} \right) \cdot \underline{I}^{-1} \cdot \left(S - \frac{1}{2} \underline{D} \cdot H \right) \\ - \zeta \cdot E - \frac{1}{2} \underline{Q} : \nabla E + \cdots. \end{aligned}$$

Now

$$(S_a, x_{nb}) = (p_i R^i_a, x_{nb}) = -x_{nb,i} R^i_a = \varepsilon_{abc} x_{nc},$$

and therefore

$$(S_a, \zeta_b) = \varepsilon_{abc} \zeta_c, \quad (S_a, D_{bc}) = \varepsilon_{abd} D_{dc} + \varepsilon_{acd} D_{bd},$$

$$(S_a, I^{-1}_{bc}) = \varepsilon_{abd} I^{-1}_{dc} + \varepsilon_{acd} I^{-1}_{bd}, \qquad (S_a, Q_{bc}) = \varepsilon_{abd} Q_{dc} + \varepsilon_{acd} Q_{bd},$$

and so on. Hence

$$\begin{aligned} \dot{S}_a &= (S_a, \mathcal{H}) \\ &= -\dot{x}_b \varepsilon_{bcd} H_c \varepsilon_{ade} \zeta_e + \frac{1}{2} \omega_b D_{bc} \varepsilon_{acd} H_d - \varepsilon_{abc} \zeta_c E_b \\ &\quad - \frac{1}{2} (\varepsilon_{abd} Q_{dc} + \varepsilon_{acd} Q_{bd}) E_{c,b} + \cdots \\ &= (\delta_{ba} \delta_{ce} - \delta_{be} \delta_{ca}) \dot{x}_b H_c \zeta_e + \varepsilon_{acd} \mu_c H_d + \varepsilon_{acb} \zeta_c E_b - \cdots \\ &= \dot{x}_a H_b \zeta_b - H_a \dot{x}_b \zeta_b + \varepsilon_{abc} (\mu_b H_c + \zeta_b E_c) - \frac{1}{2} (\varepsilon_{abd} Q_{dc} + \varepsilon_{acd} Q_{bd}) E_{c,b} + \cdots, \end{aligned}$$

or

$$\dot{\boldsymbol{S}} = \boldsymbol{\mu} \times \boldsymbol{H} + \boldsymbol{\zeta} \times (\boldsymbol{E} + \dot{\boldsymbol{x}} \times \boldsymbol{H}) - \frac{1}{2}\nabla \times (\boldsymbol{E} \cdot \underline{Q}) + \frac{1}{2}(\underline{Q} \cdot \nabla) \times \boldsymbol{E} + \cdots,$$

in which we have used

$$\boldsymbol{\mu} = \frac{1}{2}\sum_n e_n \boldsymbol{x}_n \times \dot{\boldsymbol{x}}_n = \frac{1}{2}\sum_n e_n \boldsymbol{x}_n \times (\boldsymbol{\omega} \times \boldsymbol{x}_n)$$
$$= \frac{1}{2}\sum_n e_n \left[\boldsymbol{\omega}\dot{\boldsymbol{x}}_n^2 - \boldsymbol{x}_n(\boldsymbol{x}_n \cdot \boldsymbol{\omega})\right] = \frac{1}{2}\underline{D} \cdot \boldsymbol{\omega}.$$

In the case of spherical symmetry,

$$\underline{I} = I\underline{1}, \quad \underline{D} = D\underline{1}, \quad \zeta = 0, \quad \underline{Q} = 0,$$

$$\boldsymbol{\mu} = \frac{1}{2}D\boldsymbol{\omega}, \quad \boldsymbol{S} = I\boldsymbol{\omega} + \frac{1}{2}D\boldsymbol{H} + \cdots,$$

$$\mathcal{H} = \frac{1}{2m}(\boldsymbol{p} - e\boldsymbol{A})^2 + e\phi + \frac{1}{2}I^{-1}\left(\boldsymbol{S} - \frac{1}{2}D\boldsymbol{H}\right)^2 + \cdots.$$

Ignoring terms indicated by ... up to now,

$$\dot{\boldsymbol{S}} = \boldsymbol{\mu} \times \boldsymbol{H} = \frac{1}{2}D\boldsymbol{\omega} \times \boldsymbol{H} = \frac{1}{2}DI^{-1}\left(\boldsymbol{S} - \frac{1}{2}D\boldsymbol{H}\right) \times \boldsymbol{H} = \frac{1}{2}DI^{-1}\boldsymbol{S} \times \boldsymbol{H},$$

$$\frac{\mathrm{d}}{\mathrm{d}t}S^2 = 2\boldsymbol{S} \cdot \dot{\boldsymbol{S}} = DI^{-1}\boldsymbol{S} \cdot (\boldsymbol{S} \times \boldsymbol{H}) = 0,$$

$$\mathcal{H} = \frac{1}{2m}(\boldsymbol{p} - e\boldsymbol{A})^2 + e\phi + \frac{1}{2}I^{-1}S^2 - \frac{1}{2}DI^{-1}\boldsymbol{S} \cdot \boldsymbol{H} + \frac{1}{8}I^{-1}D^2H^2.$$

The third term on the right hand side is a constant, inert term that affects neither the orbital nor the spin dynamics. The last term affects only the orbital dynamics.

We now make the transition to a spinning particle, i.e., $\varepsilon \longrightarrow 0$, where ε is the radius of the body. We have the finite quantities S, $\boldsymbol{\mu}$, $\boldsymbol{v}$, $m = m_c + m_p$, $e = e_c + e_p$, where

m_c = central mass, m_p = peripheral mass,
e_c = central charge, e_p = peripheral charge,
$\boldsymbol{v}$ = mean rotation velocity.

Now

$$v \sim \omega\varepsilon, \quad \omega \sim \frac{v}{\varepsilon} \longrightarrow \infty, \quad I \sim m_p\varepsilon^2, \quad D \sim e_p\varepsilon^2,$$

$$\mu \sim e_p v\varepsilon, \quad e_p \sim \tfrac{\mu}{v\varepsilon} \longrightarrow \pm\infty, \quad D \sim \tfrac{\mu}{v}\varepsilon \longrightarrow 0,$$
$$S \sim m_p v\varepsilon, \quad m_p \sim \tfrac{S}{v\varepsilon} \longrightarrow \infty, \quad I \sim \tfrac{S}{v}\varepsilon \longrightarrow 0,$$

$$m_c \longrightarrow -\infty, \quad e_c \longrightarrow \mp\infty, \quad I^{-1}S^2 \sim \frac{\nu S}{\varepsilon} \sim m_p \nu^2 \longrightarrow \infty,$$

$$DI^{-1}S \sim \mu, \quad I^{-1}D^2 \sim \frac{\mu^2}{\nu S}\varepsilon \longrightarrow 0.$$

Writing $DI^{-1} = ge/m$, we have

$$\boldsymbol{\mu} = \frac{ge}{2m}\boldsymbol{S},$$

and the limiting Hamiltonian becomes

$$\mathcal{H} = \frac{1}{2m}(\boldsymbol{p} - e\boldsymbol{A})^2 + e\phi + \frac{1}{2}I^{-1}S^2 - \frac{ge}{2m}\boldsymbol{S} \cdot \boldsymbol{H}.$$

A.2: Relativistic Spinning Body in Impressed Electromagnetic and Gravitational Fields

The metric is $g_{\mu\nu}$ and we use the notation

$$A \cdot B = g_{\mu\nu}A^\mu B^\nu, \quad A^2 = A \cdot A,$$

$$A_\mu = g_{\mu\nu}A^\nu, \quad A^\mu = g^{\mu\nu}A_\nu, \quad g_{\mu\sigma}g^{\sigma\nu} = \delta^\nu_{\ \mu},$$

$$\Gamma^\mu_{\nu\sigma} = g^{\mu\tau}\Gamma_{\nu\sigma\tau}, \quad \Gamma_{\nu\sigma\tau} = \frac{1}{2}(g_{\nu\tau,\sigma} + g_{\sigma\tau,\nu} - g_{\nu\sigma,\tau}),$$

$$R^\mu_{\nu\sigma\tau} = \Gamma^\mu_{\nu\sigma,\tau} - \Gamma^\mu_{\nu\tau,\sigma} + \Gamma^\rho_{\nu\sigma}\Gamma^\mu_{\rho\tau} - \Gamma^\rho_{\nu\tau}\Gamma^\mu_{\rho\sigma},$$

$$R_{\mu\nu\sigma\tau} = -R_{\nu\mu\sigma\tau} = R_{\sigma\tau\mu\nu}, \quad R_{\mu\nu\sigma\tau} + R_{\mu\sigma\tau\nu} + R_{\mu\tau\nu\sigma} = 0,$$

$$R_{\mu\nu} = R^\sigma_{\mu\nu\sigma}, \quad R = R^\mu_{\ \mu}.$$

Covariant differentiation is given by

$$A^\mu_{\cdot\nu} = A^\mu_{,\nu} + \Gamma^\mu_{\nu\sigma}A^\sigma, \quad A_{\mu\cdot\nu} = A_{\mu,\nu} - \Gamma^\sigma_{\mu\nu}A_\sigma,$$

and so on. The metric satisfies $g_{\mu\nu\cdot\sigma} = 0$. Then

$$\begin{aligned} A^\mu_{\cdot\nu\sigma} - A^\mu_{\cdot\sigma\nu} &= (A^\mu_{\cdot\nu})_{,\sigma} + \Gamma^\mu_{\sigma\tau}A^\tau_{\cdot\nu} - \Gamma^\tau_{\nu\sigma}A^\mu_{\cdot\tau} - (\nu \leftrightarrow \sigma) \\ &= A^\mu_{,\nu\sigma} + \Gamma^\mu_{\nu\tau,\sigma}A^\tau + \Gamma^\mu_{\nu\tau}A^\tau_{,\sigma} \\ &\quad + \Gamma^\mu_{\sigma\tau}A^\tau_{,\nu} + \Gamma^\mu_{\sigma\tau}\Gamma^\tau_{\nu\rho}A^\rho - (\nu \leftrightarrow \sigma) \\ &= R^\mu_{\tau\nu\sigma}A^\tau. \end{aligned}$$

We also have

$$R_{\mu\nu\sigma\tau\cdot\rho} + R_{\mu\nu\tau\rho\cdot\sigma} + R_{\mu\nu\rho\sigma\cdot\tau} = 0.$$

The world line of the center of energy is $x^\mu(\tau)$, with

$$\dot{x}^\mu = \frac{\mathrm{d}}{\mathrm{d}\tau}x^\mu.$$

Covariant differentiation with respect to τ is given by

$$\dot{A}^\mu = \frac{\mathrm{D}}{\mathrm{D}\tau}A^\mu = \frac{\mathrm{d}}{\mathrm{d}\tau}A^\mu + \Gamma^\mu_{\nu\sigma}A^\nu\dot{x}^\sigma,$$
$$\dot{A}_\mu = \frac{\mathrm{D}}{\mathrm{D}\tau}A_\mu = \frac{\mathrm{d}}{\mathrm{d}\tau}A_\mu - \Gamma^\nu_{\mu\sigma}A_\nu\dot{x}^\sigma,$$

and so on. If A^μ is a field then $\dot{A}^\mu = A^\mu_{;\nu}\dot{x}^\nu$. For example,

$$\dot{g}_{\mu\nu} = 0,$$

and

$$\ddot{x}^\mu = \frac{\mathrm{D}}{\mathrm{D}\tau}\dot{x}^\mu, \quad \dddot{x}^\mu = \frac{\mathrm{D}}{\mathrm{D}\tau}\ddot{x}^\mu,$$

and so on.

In a local rest frame $e^\mu_a(\tau)$, such that

$$e_a \cdot e_b = \delta_{ab}, \qquad e_a \cdot \dot{x} = 0,$$

we define the rest frame projection tensor

$$P^{\mu\nu} = e^\mu_a e^\nu_a = g^{\mu\nu} + (-\dot{x}^2)^{-1}\dot{x}^\mu\dot{x}^\nu,$$

and the frame rotation tensor

$$\Omega_{ab} = (-\dot{x}^2)^{-1/2}\dot{e}_a \cdot e_b = -\Omega_{ab},$$

whence

$$\dot{e}^\mu_a = (-\dot{x}^2)^{1/2}\Omega_{ab}e^\mu_b + (-\dot{x}^2)^{-1}(e_a \cdot \ddot{x})\dot{x}^\mu.$$

When $\Omega_{ab} = 0$, the vectors e^μ_a are generated by Fermi–Walker transport. However, Fermi–Walker transport is non-integrable, i.e., it depends globally on the world line and not merely locally. In order to have a local dependence, e.g., vectors e^μ_a propagated by Fermi–Walker transport along preselected paths and then boosted, we must allow $\Omega_{ab} \neq 0$.

A variation of the orbit is denoted by $\delta x^\mu(\tau)$. Covariant variations are given by

$$\bar{\delta}A^\mu = \delta A^\mu + \Gamma^\mu_{\nu\sigma}A^\nu\delta x^\sigma, \qquad \bar{\delta}A_\mu = \delta A_\mu - \Gamma^\nu_{\mu\sigma}A_\nu\delta x^\sigma,$$

and so on. If A^μ is a field, then $\bar{\delta}A^\mu = A^\mu_{;\nu}\delta x^\nu$. For example,

$$\bar{\delta}g_{\mu\nu} = 0, \quad \bar{\delta}\dot{x}^\mu = \delta\dot{x}^\mu + \Gamma^\mu_{\nu\sigma}\dot{x}^\nu\delta x^\sigma = \frac{\mathrm{D}}{\mathrm{D}\tau}\delta x^\mu.$$

Covariant variation and covariant differentiation with respect to τ are not generally commutative:

$$\begin{aligned}\bar{\delta}\dot{A}^\mu - \frac{\mathrm{D}}{\mathrm{D}\tau}\bar{\delta}A^\mu &= \delta\dot{A}^\mu + \Gamma^\mu_{\nu\sigma}\dot{A}^\nu\delta x^\sigma - \frac{\mathrm{d}}{\mathrm{d}\tau}\bar{\delta}A^\mu - \Gamma^\mu_{\nu\sigma}\bar{\delta}A^\nu\dot{x}^\sigma \\ &= \frac{\mathrm{d}}{\mathrm{d}\tau}\delta A^\mu + \delta\left(\Gamma^\mu_{\nu\sigma}A^\nu\dot{x}^\sigma\right) + \Gamma^\mu_{\nu\sigma}\left(\frac{\mathrm{d}}{\mathrm{d}\tau}A^\nu + \Gamma^\nu_{\rho\lambda}A^\rho\dot{x}^\lambda\right)\delta x^\sigma \\ &\quad - \frac{\mathrm{d}}{\mathrm{d}\tau}\delta A^\mu - \frac{\mathrm{d}}{\mathrm{d}\tau}(\Gamma^\mu_{\nu\sigma}A^\nu\delta x^\sigma) - \Gamma^\mu_{\nu\sigma}\left(\delta A^\nu + \Gamma^\nu_{\rho\lambda}A^\rho\delta x^\lambda\right)\dot{x}^\sigma \\ &= \left(\Gamma^\mu_{\nu\sigma,\tau} - \Gamma^\mu_{\nu\tau,\sigma} + \Gamma^\rho_{\nu\sigma}\Gamma^\mu_{\rho\tau} - \Gamma^\rho_{\nu\tau}\Gamma^\mu_{\rho\sigma}\right)A^\nu\dot{x}^\sigma\delta x^\tau \\ &= R^\mu_{\nu\sigma\tau}A^\nu\dot{x}^\sigma\delta x^\tau.\end{aligned}$$

Now

$$\delta\dot{x}^2 = 2\dot{x}\cdot\bar{\delta}\dot{x}, \quad \bar{\delta}e^\mu_a = \delta\lambda_{ab}e^\mu_b + (-\dot{x}^2)^{-1}(e_a\cdot\bar{\delta}\dot{x})\dot{x}^\mu,$$

for some antisymmetric $\delta\lambda_{ab}$. Also

$$\begin{aligned}\frac{\mathrm{D}}{\mathrm{D}\tau}\bar{\delta}e^\mu_a &= \delta\dot{\lambda}_{ab}e^\mu_b + \delta\lambda_{ab}\dot{e}^\mu_b + 2(-\dot{x}^2)^{-2}(\dot{x}\cdot\ddot{x})(e_a\cdot\bar{\delta}\dot{x})\dot{x}^\mu \\ &\quad + (-\dot{x}^2)^{-1}(\dot{e}_a\cdot\bar{\delta}\dot{x})\dot{x}^\mu + (-\dot{x}^2)^{-1}\left(e_a\cdot\frac{\mathrm{D}}{\mathrm{D}\tau}\bar{\delta}\dot{x}\right)\dot{x}^\mu \\ &\quad + (-\dot{x}^2)^{-1}(e_a\cdot\bar{\delta}\dot{x})\ddot{x}^\mu \\ &= \delta\dot{\lambda}_{ab}e^\mu_b + (-\dot{x}^2)^{1/2}\delta\lambda_{ab}\Omega_{bc}e^\mu_c + (-\dot{x}^2)^{-1}\delta\lambda_{ab}(e_b\cdot\ddot{x})\dot{x}^\mu \\ &\quad + 2(-\dot{x}^2)^{-2}(\dot{x}\cdot\ddot{x})(e_a\cdot\bar{\delta}\dot{x})\dot{x}^\mu + (-\dot{x}^2)^{-1/2}\Omega_{ab}(e_b\cdot\bar{\delta}\dot{x})\dot{x}^\mu \\ &\quad + (-\dot{x}^2)^{-2}(e_a\cdot\ddot{x})(\dot{x}\cdot\bar{\delta}\dot{x})\dot{x}^\mu + (-\dot{x}^2)^{-1}\left(e_a\cdot\frac{\mathrm{D}}{\mathrm{D}\tau}\bar{\delta}\dot{x}\right)\dot{x}^\mu \\ &\quad + (-\dot{x}^2)^{-1}(e_a\cdot\bar{\delta}\dot{x})\ddot{x}^\mu,\end{aligned}$$

$$\begin{aligned}\bar{\delta}\dot{e}^\mu_a &= -(-\dot{x}^2)^{-1/2}(\dot{x}\cdot\bar{\delta}\dot{x})\Omega_{ab}e^\mu_b + (-\dot{x}^2)^{1/2}\delta\Omega_{ab}e^\mu_b \\ &\quad + (-\dot{x}^2)^{1/2}\Omega_{ab}\delta\lambda_{bc}e^\mu_c + (-\dot{x}^2)^{-1/2}\Omega_{ab}(e_b\cdot\bar{\delta}\dot{x})\dot{x}^\mu \\ &\quad + 2(-\dot{x}^2)^{-2}(\dot{x}\cdot\bar{\delta}\dot{x})(e_a\cdot\ddot{x})\dot{x}^\mu + (-\dot{x}^2)^{-1}\delta\lambda_{ab}(e_b\cdot\ddot{x})\dot{x}^\mu \\ &\quad + (-\dot{x}^2)^{-2}(e_a\cdot\bar{\delta}\dot{x})(\dot{x}\cdot\ddot{x})\dot{x}^\mu + (-\dot{x}^2)^{-1}\left(e_a\cdot\frac{\mathrm{D}}{\mathrm{D}\tau}\bar{\delta}\dot{x}\right)\dot{x}^\mu \\ &\quad + (-\dot{x}^2)^{-1}R_{\rho\nu\sigma\tau}e^\rho_a\dot{x}^\nu\dot{x}^\sigma\delta x^\tau\dot{x}^\mu + (-\dot{x}^2)^{-1}(e_a\cdot\ddot{x})\bar{\delta}\dot{x}^\mu,\end{aligned}$$

$$\begin{aligned}0 &= \frac{\mathrm{D}}{\mathrm{D}\tau}\bar{\delta}e^\mu_a - \bar{\delta}\dot{e}^\mu_a + R^\mu_{\nu\sigma\tau}e^\nu_a\dot{x}^\sigma\delta x^\tau \\ &= \Big[\delta\dot{\lambda}_{ab} - (-\dot{x}^2)^{1/2}(\Omega_{ac}\delta\lambda_{cb} - \Omega_{bc}\delta\lambda_{ca}) \\ &\quad + (-\dot{x}^2)^{-1/2}\Omega_{ab}(\dot{x}\cdot\bar{\delta}\dot{x}) - (-\dot{x}^2)^{1/2}\delta\Omega_{ab}\Big]e^\mu_b \\ &\quad + (-\dot{x}^2)^{-1}P^\mu_\nu\ddot{x}^\nu(e_a\cdot\bar{\delta}\dot{x}) - (-\dot{x}^2)^{-1}P^\mu_\nu(e_a\cdot\ddot{x})\bar{\delta}\dot{x}^\nu + P^{\mu\rho}R_{\rho\nu\sigma\tau}e^\nu_a\dot{x}^\sigma\delta x^\tau,\end{aligned}$$

whence

$$\begin{aligned}0 &= \delta\dot{\lambda}_{ab} - (-\dot{x}^2)^{1/2}(\Omega_{ac}\delta\lambda_{cb} - \Omega_{bc}\delta\lambda_{ca}) + (-\dot{x}^2)^{-1/2}\Omega_{ab}(\dot{x}\cdot\overline{\delta}\dot{x}) \\ &\quad - (-\dot{x}^2)^{1/2}\delta\Omega_{ab} - (-\dot{x}^2)^{-1}\ddot{x}\cdot(e_a e_b - e_b e_a)\cdot\overline{\delta}\dot{x} \\ &\quad - e^\mu_a e^\nu_b R_{\mu\nu\sigma\tau}\dot{x}^\sigma \delta x^\tau.\end{aligned}$$

We now introduce the biscalar of geodetic interval $\sigma(z, x)$. We associate indices from the first part of the Greek alphabet with z and those from the middle of the alphabet with x. It is defined by the requirements

$$\sigma = \frac{1}{2}\sigma_{\cdot\mu}\sigma^\mu_{\cdot} = \frac{1}{2}\sigma_{\cdot\alpha}\sigma^\alpha_{\cdot} = \pm\frac{1}{2}s^2,$$

$$\lim_{z\to x}\sigma_{\cdot\mu} = 0, \quad \lim_{z\to x}\sigma_{\cdot\alpha} = 0,$$

and it has the properties

$$\sigma_{\cdot\mu} = \sigma_{\cdot\mu\tau}\sigma^\tau_{\cdot} = \sigma_{\cdot\mu\alpha}\sigma^\alpha_{\cdot},$$

$$\sigma_{\cdot\mu\nu} = \sigma_{\cdot\mu\tau\nu}\sigma^\tau_{\cdot} + \sigma_{\cdot\mu\tau}\sigma^\tau_{\cdot\nu} = \sigma_{\cdot\mu\nu\alpha}\sigma^\alpha_{\cdot} + \sigma_{\cdot\mu\alpha}\sigma^\alpha_{\cdot\nu},$$

$$\lim_{z\to x}\sigma_{\cdot\mu\nu} = g_{\mu\nu}, \quad \lim_{z\to x}\sigma_{\cdot\alpha\beta} = g_{\alpha\beta}, \quad \lim_{z\to x}\sigma_{\cdot\mu\alpha} = -g_{\mu\alpha},$$

$$\begin{aligned}\sigma_{\cdot\mu\nu\sigma} &= \sigma_{\cdot\mu\tau\nu\sigma}\sigma^\tau_{\cdot} + \sigma_{\cdot\mu\tau\nu}\sigma^\tau_{\cdot\sigma} + \sigma_{\cdot\mu\tau\sigma}\sigma^\tau_{\cdot\nu} + \sigma_{\cdot\mu\tau}\sigma^\tau_{\cdot\nu\sigma}, \\ 0 &= \lim_{z\to x}(-\sigma_{\cdot\mu\nu\sigma} + \sigma_{\cdot\mu\sigma\nu} + \sigma_{\cdot\mu\nu\sigma} + \sigma_{\cdot\nu\mu\sigma}) = 2\lim_{z\to x}\sigma_{\cdot\mu\nu\sigma}.\end{aligned}$$

We may expand

$$\sigma_{\cdot\mu\nu} = g_{\mu\nu} + A_{\mu\nu\sigma}\sigma^\sigma_{\cdot} + \frac{1}{2}A_{\mu\nu\sigma\tau}\sigma^\sigma_{\cdot}\sigma^\tau_{\cdot} + \cdots,$$

where $A_{\mu\nu\sigma}$, $A_{\mu\nu\sigma\tau}$, etc., are ordinary one-point tensors. We have

$$\sigma_{\cdot\mu\nu\sigma} = A_{\mu\nu\tau\cdot\sigma}\sigma^\tau_{\cdot} + A_{\mu\nu\tau}\sigma^\tau_{\cdot\sigma} + \cdots,$$

$$0 = \lim_{z\to x}\sigma_{\cdot\mu\nu\sigma} = A_{\mu\nu\sigma}.$$

Therefore

$$\sigma_{\cdot\mu\nu} = g_{\mu\nu} + O(s^2).$$

We also have

$$\begin{aligned}\sigma_{\cdot\mu\nu\alpha} &= \sigma_{\cdot\mu\tau\nu\alpha}\sigma^\tau_{\cdot} + \sigma_{\cdot\mu\tau\nu}\sigma^\tau_{\cdot\alpha} + \sigma_{\cdot\mu\tau\alpha}\sigma^\tau_{\cdot\nu} + \sigma_{\cdot\mu\tau}\sigma^\tau_{\cdot\nu\alpha}, \\ 0 &= \lim_{z\to x}(-\sigma_{\cdot\mu\nu\alpha} + \sigma_{\cdot\mu\nu\alpha} + \sigma_{\cdot\nu\mu\alpha}) = \lim_{z\to x}\sigma_{\cdot\mu\nu\alpha}.\end{aligned}$$

Therefore,

$$\sigma_{\cdot\mu\alpha}\sigma^\alpha_{\cdot\nu} = \sigma_{\cdot\mu\nu} - \sigma_{\cdot\mu\nu\alpha}\sigma^\alpha_{\cdot} = g_{\mu\nu} + O(s^2).$$

The world line of the nth component particle is denoted by $z_n^\mu(\tau)$. It is determined by

$$-\sigma_{\cdot\mu}(z_n(\tau), x(\tau)) = x_{na}(\tau)e_{a\mu}(\tau),$$

$$-\sigma_{\cdot\mu\alpha}\dot{z}_n^\alpha = \sigma_{\cdot\mu\nu}\dot{x}^\nu + \dot{x}_{na}e_{a\mu} + x_{na}\dot{e}_{a\mu},$$

$$\begin{aligned}-\dot{z}_n^2 &= -\dot{x}^2 - \dot{x}_{na}\dot{x}_{na} - 2x_{na}(\dot{e}_a \cdot \dot{x}) - 2x_{na}\dot{x}_{nb}(\dot{e}_a \cdot e_b) + O(s^2)\\ &= -\dot{x}^2 - \dot{x}_{na}\dot{x}_{na} + 2x_{na}(e_a \cdot \ddot{x}) - 2(-\dot{x}^2)^{1/2}x_{na}\dot{x}_{nb}\Omega_{ab} + O(s^2).\end{aligned}$$

The rotation velocity relative to the Fermi–Walker transported frame is

$$v_{na} = (-\dot{x}^2)^{-1/2}\dot{x}_{na} - \Omega_{ab}x_{nb},$$

and

$$\begin{aligned}-\dot{z}_n^2 &= \dot{x}^2 - (-\dot{x}^2)v_n^2 + 2x_{na}(e_a \cdot \ddot{x}) + O(s^2)\\ &= (-\dot{x}^2)(1 - v_n^2)\left[1 + 2(-\dot{x}^2)^{-1}(1 - v_n^2)^{-1}x_{na}(e_a \cdot \ddot{x})\right] + O(s^2),\\ (-\dot{z}_n^2)^{1/2} &= (-\dot{x}^2)^{1/2}(1 - v_n^2)^{1/2} + (-\dot{x}^2)^{-1/2}(1 - v_n^2)^{-1/2}x_{na}(e_a \cdot \ddot{x}) + O(s^2).\end{aligned}$$

The electromagnetic vector potential is A_μ. We try to expand $A_\alpha(z_n)\dot{z}_n^\alpha$ in powers of x_{na} and $\dot{x}_{na}$:

$$A_\alpha\dot{z}_n^\alpha = A \cdot \dot{x} + B_{na}\dot{x}_{na} + C_{na}x_{na} + D_{nab}x_{na}\dot{x}_{nb} + O(s^2).$$

Now

$$-\sigma_{\cdot\mu\alpha}\frac{\partial z_n^\alpha}{\partial x_{na}} = e_{a\mu}, \qquad -\sigma_{\cdot\mu\alpha}\frac{\partial \dot{z}_n^\alpha}{\partial \dot{x}_{na}} = e_{a\mu}.$$

Also, since $-\sigma_{\cdot\mu\alpha}$ differs from $g_{\mu\alpha}$ only by a quantity of order s^2,

$$-\sigma_{\cdot\mu\alpha}\frac{\partial \dot{z}_n^\alpha}{\partial x_{na}} = \dot{e}_{a\mu} + O(s).$$

Therefore,

$$B_{na} = \left(A_\alpha\frac{\partial \dot{z}_n^\alpha}{\partial \dot{x}_{na}}\right)_{x_n=0} = A \cdot e_a,$$

$$\begin{aligned}C_{na} + D_{nab}\dot{x}_{nb} &= \left(A_{\alpha\cdot\beta}\frac{\partial z_n^\beta}{\partial x_{na}}\dot{z}_n^\alpha + A_\alpha\frac{\partial \dot{z}_n^\alpha}{\partial x_{na}}\right)_{x_n=0}\\ &= A_{\mu\cdot\nu}e_a^\nu(\dot{x}^\mu + \dot{x}_{nb}e_b^\mu) + A \cdot \dot{e}_a,\end{aligned}$$

whence

$$\begin{aligned}A_\alpha\dot{z}_n^\alpha = {} & A \cdot \dot{x} + (A \cdot e_a)\dot{x}_{na} + (A \cdot \dot{e}_a)x_{na} + A_{\mu\cdot\nu}\dot{x}^\mu e_a^\nu x_{na}\\ & + A_{\mu\cdot\nu}e_a^\nu e_b^\mu x_{na}\dot{x}_{nb} + O(s^2).\end{aligned}$$

The action is

$$W = \int L d\tau,$$

with Lagrangian

$$\begin{aligned} L' &= -\sum_n m_n(-\dot{z}_n^2)^{1/2} + \sum_n e_n A_\alpha \dot{z}_n^\alpha \\ &= -(-\dot{x}^2)^{1/2} \sum_n m_n(1-v_n^2)^{1/2} + (-\dot{x}^2)^{-1/2} \sum_n m_n(1-v_n^2)^{-1/2} x_{na}(e_a \cdot \ddot{x}) \\ &\quad + eA \cdot \dot{x} + A \cdot \sum_n e_n(e_a \dot{x}_{na} + \dot{e}_a x_{na}) + A_{\mu \cdot \nu} \dot{x}^\mu e_a^\nu \sum_n e_n x_{na} \\ &\quad + A_{\mu \cdot \nu} e_a^\nu e_b^\mu \sum_n e_n x_{na} \dot{x}_{nb} + O(s^2) \\ &= L + \frac{\mathrm{d}}{\mathrm{d}\tau}\left[(A \cdot e_a) \sum_n e_n x_{na} + \frac{1}{2} A_{\mu \cdot \nu} e_a^\nu e_b^\mu \sum_n e_n x_{na} x_{nb} + O(s^2) \right], \end{aligned}$$

where

$$\begin{aligned} L &= -(-\dot{x}^2)^{1/2} \sum_n m_n(1-v_n^2)^{1/2} + (-\dot{x}^2)^{-1/2} \sum_n m_n(1-v_n^2)^{-1/2} x_{na}(e_a \cdot \ddot{x}) \\ &\quad + eA \cdot \dot{x} + (-\dot{x}^2)^{1/2}(\zeta_a E_a + \mu_a H_a) + O(s^2), \end{aligned}$$

the symbols e, ζ_a, μ_a, E_a, and H_a being defined as follows:

Total charge	$e = \sum_n e_n$
Electric dipole moment	$\zeta_a = \sum_n e_n x_{na}$
Magnetic dipole moment	$\mu_a = \frac{1}{2}\varepsilon_{abc} \sum_n e_n x_{nb} v_{nc}$
Electric field in local rest frame	$E_a = (-\dot{x}^2)^{-1/2} e_a^\nu F_{\nu\mu} \dot{x}^\mu$
Magnetic field in local rest frame	$H_a = \frac{1}{2}\varepsilon_{abc} F_{bc} \text{with} F_{ab} = e_a^\mu e_b^\nu F_{\mu\nu}$

Note that the term $\boldsymbol{\zeta} \cdot (\dot{\boldsymbol{x}} \times \boldsymbol{H})$ that appears in the Lagrangian in the nonrelativistic case is missing here. This is because the E_a are already the components of the electric field in the instantaneous rest frame of the center of energy of the body.

We shall now make four assumptions:

1. Nonrelativistic rotation: $v_n^2 \ll 1$. Although this assumption does not relieve us from the ultimate necessity of including the rotational energy in the total rest mass of the body, it does allow us to ignore the internal energy associated with strains produced in the body by centrifugal forces and special relativistic contraction effects. (These energies are of order $s^2 v_n^2$ and v_n^4, respectively.)

2. Formal or quasi-rigidity: $x_{na,i}R^i{}_b = \varepsilon_{abc}x_{nc}$. Then

$$\dot{q}^i = (-\dot{x}^2)^{1/2}R^i_a\omega_a, \qquad \dot{x}_{na}(-\dot{x}^2)^{1/2}\varepsilon_{abc}\omega_b x_{nc},$$
$$v_{na} = \varepsilon_{abc}(\omega_b + \Omega_b)x_{nc}, \qquad \Omega_a = \tfrac{1}{2}\varepsilon_{abc}\Omega_{bc}.$$

3. Inversion symmetry. By this we mean that for every component particle, located at x_n say, there is another particle located at $-x_n$ (relative to the center of energy) that has the same mass and charge. An immediate consequence of this assumption is $\zeta_a = 0$. Also, since v_n^2 is an even function of the x_n (by the quasi-rigidity assumption), we have

$$\sum_n m_n(1 - v_n^2)^{-1/2}x_{na} = 0,$$

so that the center of energy $x^\mu(\tau)$ is indeed located at the origin of local coordinates. Note that inversion symmetry is much less restrictive than spherical symmetry. Indeed, although we shall later make the spherical symmetry assumption in the present nonrelativistic case, the appropriate generalization of the spherical asymmetry assumption to the case of relativistic rotation would require the energies $m_n(1 - v_n^2)^{-1/2}$ rather than the rest masses m_n themselves to be spherically symmetrically distributed, an assumption that would be rather ad hoc and hard to justify.
4. Smallness. By this we mean that the body is sufficiently small that the strains (and associated internal energy) produced in the body by the curvature of spacetime and by the gradients of the electromagnetic field may be neglected. This will justify our neglecting terms of order s^2 in the Lagrangian, and we shall henceforth no longer indicate these terms.

We note that when all the above assumptions are valid, the departure from true rigidity of the formally rigid body is negligible.

The Lagrangian now takes the form

$$L = -m_0(-\dot{x}^2)^{1/2} + \frac{1}{2}(-\dot{x}^2)^{1/2}\sum_n m_n v_n^2 + eA\cdot\dot{x} + (-\dot{x}^2)^{1/2}\mu_a H_a,$$

where

$$m_0 = \sum_n m_n.$$

In calculating the dynamical equations for the internal (spin) motion of the body, we may set the parameter τ equal to proper time so that

$$-\dot{x}^2 = 1.$$

The internal motion is then described by the Lagrangian

$$L_s = \frac{1}{2}\sum_n m_n v_n^2 + \frac{1}{2}\varepsilon_{abc}H_a\sum_n e_n x_{nb}v_{nc},$$

where

$$v_{na} = x_{na,i}\dot{q}^i + \varepsilon_{abc}\Omega_b x_{nc} = \varepsilon_{abc}(\omega_b + \Omega_b)x_{nc}.$$

The conjugate momenta are

$$p_i = \frac{\partial L_s}{\partial \dot{q}^i} = \sum_n m_n v_{na} x_{na,i} + \frac{1}{2}\varepsilon_{abc} H_a \sum_n e_n x_{nb} x_{nc,i}.$$

The components of the spin angular momentum in the local rest frame are

$$\begin{aligned} S_a &= p_i R^i_a \\ &= \sum_n m_n \varepsilon_{bcd}(\omega_c + \Omega_c)x_{nd}\varepsilon_{bae}x_{ne} + \frac{1}{2}\varepsilon_{bcd} H_b \sum_n e_n x_{nc}\varepsilon_{dae}x_{ne} \\ &= I_{ab}(\omega_b + \Omega_b) + \frac{1}{2}D_{ab}H_b, \end{aligned}$$

where the moment of inertia tensor I_{ab} and charge moment tensor D_{ab} are as defined on p. 219.

Introducing a three-vector dyadic notation, we may write

$$\boldsymbol{\omega} + \boldsymbol{\Omega} = \underline{I}^{-1}\cdot\left(\boldsymbol{S} - \frac{1}{2}\underline{D}\cdot\boldsymbol{H}\right),$$

and the Hamiltonian for the internal motion may be expressed in the form

$$\begin{aligned} \mathcal{H}_s &= p_i\dot{q}^i - L_s \\ &= \frac{1}{2}\sum_n m_n v_n^2 - \varepsilon_{abc}\sum_n m_n v_{na}\Omega_b x_{nc} - \frac{1}{2}\varepsilon_{abc}\varepsilon_{cde}H_a\Omega_d \sum_n e_n x_{nb} x_{ne} \\ &= \frac{1}{2}(\boldsymbol{\omega}+\boldsymbol{\Omega})\cdot\underline{I}\cdot(\boldsymbol{\omega}+\boldsymbol{\Omega}) - \boldsymbol{\Omega}\cdot\underline{I}\cdot(\boldsymbol{\omega}+\boldsymbol{\Omega}) - \frac{1}{2}\boldsymbol{\Omega}\cdot\underline{D}\cdot\boldsymbol{H} \\ &= \frac{1}{2}\left(\boldsymbol{S} - \frac{1}{2}\boldsymbol{H}\cdot\underline{D}\right)\cdot\underline{I}^{-1}\cdot\left(\boldsymbol{S} - \frac{1}{2}\underline{D}\cdot\boldsymbol{H}\right) - \boldsymbol{\Omega}\cdot\left(\boldsymbol{S} - \frac{1}{2}\underline{D}\cdot\boldsymbol{H}\right) - \frac{1}{2}\boldsymbol{\Omega}\cdot\underline{D}\cdot\boldsymbol{H} \\ &= \frac{1}{2}\left(\boldsymbol{S} - \frac{1}{2}\boldsymbol{H}\cdot\underline{D}\right)\cdot\underline{I}^{-1}\cdot\left(\boldsymbol{S} - \frac{1}{2}\underline{D}\cdot\boldsymbol{H}\right) - \boldsymbol{\Omega}\cdot\boldsymbol{S}, \end{aligned}$$

$$\begin{aligned} \dot{S}_a &= (S_a, \mathcal{H}_s) \\ &= \frac{1}{2}(\omega_b + \Omega_b)D_{bc}\varepsilon_{acd}H_d + \varepsilon_{abc}\Omega_c S_b \\ &= \varepsilon_{abc}(\mu_b H_c - \Omega_b S_c) = F_{ab}\mu_b + \Omega_{ab}S_b, \end{aligned}$$

or

$$\dot{\boldsymbol{S}} = \boldsymbol{\mu}\times\boldsymbol{H} - \boldsymbol{\Omega}\times\boldsymbol{S}.$$

The spin angular momentum four-vector is

$$S^\mu = S_a e^\mu_a, \quad S\cdot\dot{x} = 0, \quad S_a = S\cdot e_a.$$

The magnetic dipole moment four-vector is

$$\mu^\mu = \mu_a e_a^\mu, \quad \mu \cdot \dot{x} = 0, \quad \mu_a = \mu \cdot e_a.$$

Then

$$\begin{aligned}\dot{S}^\mu &= \dot{S}_a e_a^\mu + S_a \dot{e}_a^\mu \\ &= e_a^\mu (F_{ab}\mu_b + \Omega_{ab} S_b) + S_a \left[\Omega_{ab} e_b^\mu + (e_a \cdot \ddot{x})\dot{x}^\mu\right] \\ &= e_a^\mu e_a^\nu e_b^\sigma F_{\nu\sigma}\mu_b + (S \cdot \ddot{x})\dot{x}^\mu \\ &= P^{\mu\nu} F_{\nu\sigma}\mu^\sigma - (\dot{S} \cdot \dot{x})\dot{x}^\mu,\end{aligned}$$

or

$$P^\mu_\nu (\dot{S}^\nu - F^\nu_\sigma \mu^\sigma) = 0.$$

The spin angular momentum tensor is

$$S^{\mu\nu} = S_{ab} e_a^\mu e_b^\nu, \quad S_{ab} = \varepsilon_{abc} S_c, \quad S_{\mu\nu}\dot{x}^\nu = 0.$$

The magnetic dipole moment tensor is

$$\mu^{\mu\nu} = \mu_{ab} e_a^\mu e_b^\nu, \quad \mu_{ab} = \varepsilon_{abc}\mu_c, \quad \mu_{\mu\nu}\dot{x}^\nu = 0.$$

Since

$$\dot{x}^\lambda \varepsilon_{\lambda\mu\nu\sigma} e_c^\sigma = g^{-1/2} e_{a\mu} e_{b\nu} \varepsilon_{abc}, \quad g = -\det(g_{\mu\nu}),$$

we may also write

$$\begin{aligned}S_{\mu\nu} &= g^{1/2}\dot{x}^\lambda \varepsilon_{\lambda\mu\nu\sigma} S^\sigma, \quad S_\mu = \tfrac{1}{2} g^{1/2}\dot{x}^\lambda \varepsilon_{\lambda\mu\nu\sigma} S^{\nu\sigma}, \\ \mu_{\mu\nu} &= g^{1/2}\dot{x}^\lambda \varepsilon_{\lambda\mu\nu\sigma}\mu^\sigma, \quad \mu_\mu = \tfrac{1}{2} g^{1/2}\dot{x}^\lambda \varepsilon_{\lambda\mu\nu\sigma}\mu^{\nu\sigma},\end{aligned}$$

Now, using

$$\begin{aligned}S_a \Omega_b &= \frac{1}{4}\varepsilon_{acd}\varepsilon_{bef} S_{cd}\Omega_{ef} \\ &= \frac{1}{4}(\delta_{ab}\delta_{ce}\delta_{df} + \delta_{ae}\delta_{cf}\delta_{db} + \delta_{af}\delta_{cb}\delta_{de} \\ &\quad - \delta_{ab}\delta_{cf}\delta_{de} - \delta_{af}\delta_{ce}\delta_{db} - \delta_{ae}\delta_{cb}\delta_{df}) S_{cd}\Omega_{ef} \\ &= \frac{1}{2}\delta_{ab} S_{cd}\Omega_{cd} + \Omega_{ac} S_{cb},\end{aligned}$$

whence

$$S_a \Omega_b - S_b \Omega_a = \Omega_{ac} S_{cb} - \Omega_{bc} S_{ca},$$

we find

$$\begin{aligned}\dot{S}_{ab} &= \varepsilon_{abc}\dot{S}_c = \varepsilon_{abc}\varepsilon_{cde}(\mu_d H_e + S_d \Omega_e)\\ &= \mu_a H_b - \mu_b H_a + S_a\Omega_b - S_b\Omega_a\\ &= F_{ac}\mu_{cb} - F_{bc}\mu_{ca} + \Omega_{ac}S_{cb} - \Omega_{bc}S_{ca},\end{aligned}$$

and hence

$$\begin{aligned}\dot{S}^{\mu\nu} &= \dot{S}_{ab}e^\mu_a e^\nu_b + S_{ab}\dot{e}^\mu_a e^\nu_b + S_{ab}e^\mu_a \dot{e}^\nu_b\\ &= (F_{ac}\mu_{cb} - F_{bc}\mu_{ca} + \Omega_{ac}S_{cb} - \Omega_{bc}S_{ca})e^\mu_a e^\nu_b\\ &\quad + S_{ab}\{[\Omega_{ac}e^\mu_c + (e_a\cdot\ddot{x})\dot{x}^\mu]e^\nu_b + e^\mu_a[\Omega_{bc}e^\nu_c + (e_b\cdot\ddot{x})\dot{x}^\nu]\}\\ &= P^{\mu\sigma}F_{\sigma\tau}\mu^{\tau\nu} - P^{\nu\sigma}F_{\sigma\tau}\mu^{\tau\mu} - \dot{x}^\mu\dot{x}^\sigma\dot{S}^\nu_\sigma + \dot{x}^\nu\dot{x}^\sigma\dot{S}^\mu_\sigma,\end{aligned}$$

or

$$P^\mu_\sigma P^\nu_\tau(\dot{S}^{\sigma\tau} - F^\sigma_\rho\mu^{\rho\tau} + F^\tau_\rho\mu^{\rho\sigma}) = 0.$$

Alternative expressions are

$$\begin{aligned}S_{ab} &= \varepsilon_{abc}S_c = \varepsilon_{abc}\left(\varepsilon_{cde}\sum_n m_n x_{nd}v_{ne} + \frac{1}{2}F_{de}\sum_n e_n x_{nd}\varepsilon_{ecf}x_{nf}\right) \quad \text{(see p. 231)}\\ &= \sum_n m_n(x_{na}v_{nb} - x_{nb}v_{na}) + \frac{1}{2}F_{ac}\sum_n e_n x_{nc}x_{nb} - \frac{1}{2}F_{bc}\sum_n e_n x_{nc}x_{na}\end{aligned}$$

and

$$\mu_{ab} = \varepsilon_{abc}\mu_c = \frac{1}{2}\sum_n e_n(x_{na}v_{nb} - x_{nb}v_{na}).$$

The rotational kinetic energy is

$$K = \frac{1}{2}\sum_n m_n v_n^2 = \mathcal{H}_s + \boldsymbol{\Omega}\cdot\boldsymbol{S}.$$

Hence,

$$\begin{aligned}\dot{K} &= \dot{\mathcal{H}}_s + \dot{\boldsymbol{\Omega}}\cdot\boldsymbol{S} + \boldsymbol{\Omega}\cdot\dot{\boldsymbol{S}}\\ &= -\frac{\partial L_s}{\partial\tau} + \dot{\boldsymbol{\Omega}}\cdot\boldsymbol{S} + \boldsymbol{\Omega}\cdot(\boldsymbol{\mu}\times\boldsymbol{H} - \boldsymbol{\Omega}\times\boldsymbol{S})\\ &= -\sum_n m_n\boldsymbol{v}_n\cdot(\dot{\boldsymbol{\Omega}}\times\boldsymbol{x}_n) - \frac{1}{2}\dot{F}_{ab}\mu_{ab} - \frac{1}{2}\boldsymbol{H}\cdot\sum_n e_n\boldsymbol{x}_n\times(\dot{\boldsymbol{\Omega}}\times\boldsymbol{x}_n)\\ &\quad + \dot{\boldsymbol{\Omega}}\cdot\boldsymbol{S} + \Omega_{ab}\mu_a H_b \qquad \text{(see p. 231)}\\ &= -\frac{1}{2}\mu_{ab}\frac{\mathrm{d}}{\mathrm{d}\tau}(e^\mu_a e^\nu_b F_{\mu\nu}) + \Omega_{ab}F_{ac}\mu_{cb}\end{aligned}$$

$$= -\mu_{ab}\Omega_{ac}F_{cb} - \mu_{ab}(e_a \cdot \ddot{x})\dot{x}^\mu F_{\mu\nu}e_b^\nu - \frac{1}{2}\mu^{\mu\nu}\dot{F}_{\mu\nu} + \Omega_{ab}F_{ac}\mu_{cb}$$

$$= \ddot{x}^\mu \mu_{\mu\nu}F^\nu_\sigma \dot{x}^\sigma - \frac{1}{2}\mu^{\mu\nu}\dot{F}_{\mu\nu}. \tag{A.1}$$

In calculating the dynamical equations for the orbital motion of the spinning body, we may set τ equal to the proper time only *after* performing the orbit variation $\delta x^\mu(\tau)$. We make use of the following relations:

$$\delta(-\dot{x}^2)^{1/2} = -\dot{x}\cdot\overline{\delta}\dot{x}, \quad \delta(-\dot{x}^2)^{-1/2} = \dot{x}\cdot\overline{\delta}\dot{x},$$

$$\overline{\delta}e_a^\mu = \delta\lambda_{ab}e_b^\mu + \dot{x}^\mu(e_a\cdot\overline{\delta}x) \qquad \text{(see p. 226)},$$

$$\begin{aligned}\delta\Omega_{ab} &= \delta\dot{\lambda}_{ab} - (\Omega_{ac}\delta\lambda_{cb} - \Omega_{bc}\delta\lambda_{ca}) + \Omega_{ab}(\dot{x}\cdot\overline{\delta}\dot{x})\\ &\quad - \ddot{x}\cdot(e_a e_b - e_b e_a)\cdot\overline{\delta}x - e_a^\mu e_b^\nu R_{\mu\nu\sigma\tau}\dot{x}^\sigma\delta x^\tau \qquad \text{(see p. 227)},\end{aligned}$$

$$\begin{aligned}\delta v_{na} &= (v_{na} + \Omega_{ab}x_{nb})(\dot{x}\cdot\overline{\delta}\dot{x}) - \delta\Omega_{ab}x_{nb} \qquad \text{(see p. 228)}\\ &= v_{na}(\dot{x}\cdot\overline{\delta}\dot{x}) - \Big[\delta\dot{\lambda}_{ab} - (\Omega_{ac}\delta\lambda_{cb} - \Omega_{bc}\delta\lambda_{ca})\\ &\quad -\ddot{x}\cdot(e_a e_b - e_b e_a)\cdot\overline{\delta}x - e_a^\mu e_b^\nu R_{\mu\nu\sigma\tau}\dot{x}^\sigma\delta x^\tau\Big]x_{nb},\end{aligned}$$

$$\begin{aligned}\delta(\mu_a H_a) &= \frac{1}{2}\delta\left(F_{\mu\nu}e_a^\mu e_b^\nu\sum_n e_n x_{na}v_{nb}\right)\\ &= \frac{1}{2}F_{\mu\nu\cdot\sigma}\mu^{\mu\nu}\delta x^\sigma + \delta\lambda_{ac}F_{cb}\mu_{ab} - \dot{x}^\mu F_{\mu\nu}\mu^\nu_\sigma\overline{\delta}\dot{x}^\sigma + \mu_a H_a(\dot{x}\cdot\overline{\delta}\dot{x})\\ &\quad - \frac{1}{2}F_{ab}\Big[\delta\dot{\lambda}_{bc} - (\Omega_{bd}\delta\lambda_{dc} - \Omega_{cd}\delta\lambda_{db}) - \ddot{x}\cdot(e_b e_c - e_c e_b)\cdot\overline{\delta}x\\ &\quad - e_b^\mu e_c^\nu R_{\mu\nu\sigma\tau}\dot{x}^\sigma\delta x^\tau\Big]\sum_n e_n x_{na}x_{nc},\end{aligned}$$

$$\begin{aligned}\delta L &= m_0(\dot{x}\cdot\overline{\delta}\dot{x}) - \frac{1}{2}\sum_n m_n v_n^2(\dot{x}\cdot\overline{\delta}\dot{x}) + \sum_n m_n v_n^2(\dot{x}\cdot\overline{\delta}\dot{x})\\ &\quad + \Big[\delta\dot{\lambda}_{ab} - (\Omega_{ac}\delta\lambda_{cb} - \Omega_{bc}\delta\lambda_{ca}) - \ddot{x}\cdot(e_a e_b - e_b e_a)\cdot\overline{\delta}x\\ &\quad - e_a^\mu e_b^\nu R_{\mu\nu\sigma\tau}\dot{x}^\sigma\delta x^\tau\Big]\sum_n m_n x_{na}v_{nb}\\ &\quad + eA_{\nu\cdot\mu}\dot{x}^\nu\delta x^\mu + eA\cdot\overline{\delta}\dot{x} - \mu_a H_a(\dot{x}\cdot\overline{\delta}\dot{x}) + \frac{1}{2}F_{\mu\nu\cdot\sigma}\mu^{\mu\nu}\delta x^\sigma\\ &\quad + \delta\lambda_{ac}F_{cb}\mu_{ab} - \dot{x}^\mu F_{\mu\nu}\mu^\nu_\sigma\overline{\delta}\dot{x}^\sigma + \mu_a H_a(\dot{x}\cdot\overline{\delta}\dot{x})\\ &\quad - \frac{1}{2}F_{ab}\Big[\delta\dot{\lambda}_{bc} - (\Omega_{bd}\delta\lambda_{dc} - \Omega_{cd}\delta\lambda_{db}) - \ddot{x}\cdot(e_b e_c - e_c e_b)\cdot\overline{\delta}x\\ &\quad - e_b^\mu e_c^\nu R_{\mu\nu\sigma\tau}\dot{x}^\sigma\delta x^\tau\Big]\sum_n e_n x_{na}x_{nc}\end{aligned}$$

$$
\begin{aligned}
&= (m_0 + K)(\dot{x} \cdot \overline{\delta}\dot{x}) + \frac{1}{2} S_{ab} \boldsymbol{\delta}\dot{\lambda}_{ab} + \Omega_{bc} \boldsymbol{\delta}\lambda_{ca} S_{ab} - \ddot{x}^{\nu} S_{\nu\mu} \overline{\delta}\dot{x}^{\mu} \\
&\quad + \frac{1}{2} R_{\mu\nu\sigma\tau} \dot{x}^{\nu} S^{\sigma\tau} \delta x^{\mu} + eA_{\nu \cdot \mu} \dot{x}^{\nu} \delta x^{\mu} + eA \cdot \overline{\delta}\dot{x} + \frac{1}{2} \mu^{\nu\sigma} F_{\nu\sigma \cdot \mu} \delta x^{\mu} \\
&\quad + \boldsymbol{\delta}\lambda_{ac} F_{cb} \mu_{ab} - \mu_{\mu\nu} F^{\nu}_{\sigma} \dot{x}^{\sigma} \overline{\delta}\dot{x}^{\mu} \\
&= \frac{\mathrm{d}}{\mathrm{d}\tau} \left[(m_0 + K)(\dot{x} \cdot \delta x) + \frac{1}{2} S_{ab} \boldsymbol{\delta}\lambda_{ab} + S_{\mu\nu} \ddot{x}^{\nu} \delta x^{\mu} + eA \cdot \delta x - \mu_{\mu\nu} F^{\nu}_{\sigma} \dot{x}^{\sigma} \delta x^{\mu} \right] \\
&\quad + \left\{ -\frac{\mathrm{D}}{\mathrm{D}\tau} \left[(m_0 + K) g_{\mu\nu} \dot{x}^{\nu} + S_{\mu\nu} \ddot{x}^{\nu} - \mu_{\mu\nu} F^{\nu}_{\sigma} \dot{x}^{\sigma} \right] + \frac{1}{2} R_{\mu\nu\sigma\tau} \dot{x}^{\nu} S^{\sigma\tau} \right. \\
&\quad \left. + eF_{\mu\nu} \dot{x}^{\nu} + \frac{1}{2} \mu^{\nu\sigma} F_{\nu\sigma \cdot \mu} \right\} \delta x^{\mu}.
\end{aligned}
$$

The action principle $\delta W = 0$ yields the dynamical equations

$$
\frac{\mathrm{D}}{\mathrm{D}\tau} \left[(m_0 + K)\dot{x}^{\mu} + S^{\mu}_{\nu} \ddot{x}^{\nu} - \mu^{\mu\nu} F_{\nu\sigma} \dot{x}^{\sigma} \right] = \frac{1}{2} R^{\mu}_{\nu\sigma\tau} \dot{x}^{\nu} S^{\sigma\tau} + eF^{\mu}_{\nu} \dot{x}^{\nu} + \frac{1}{2} \mu^{\nu\sigma} F^{\mu}_{\sigma\nu \cdot}. \tag{A.2}
$$

As a test for consistency, multiplication of the equation by $g_{\mu\nu}\dot{x}^{\nu}$ should yield an identity which holds independently of the equation itself:

$$
\begin{aligned}
&\left\{ -\frac{\mathrm{D}}{\mathrm{D}\tau} \left[(m_0 + K) g_{\mu\nu} \dot{x}^{\nu} + S_{\mu\nu} \ddot{x}^{\nu} - \mu_{\mu\nu} F^{\nu}_{\sigma} \dot{x}^{\sigma} \right] + \frac{1}{2} R_{\mu\nu\sigma\tau} \dot{x}^{\nu} S^{\sigma\tau} + eF_{\mu\nu} \dot{x}^{\nu} + \frac{1}{2} \mu^{\nu\sigma} F_{\nu\sigma \cdot \mu} \right\} \dot{x}^{\mu} \\
&= \dot{K} - \dot{x}^{\mu} \dot{S}_{\mu\nu} \ddot{x}^{\nu} + \dot{x}^{\mu} \dot{\mu}_{\mu\nu} F^{\nu}_{\sigma} \dot{x}^{\sigma} + \frac{1}{2} \mu^{\nu\sigma} \dot{F}_{\nu\sigma} = 0,
\end{aligned}
$$

by (A.1) on p. 234.

In the limit of a spherically symmetric point particle, we have

$$
\mu_{\mu\nu} = \frac{ge}{2m} S_{\mu\nu},
$$

$$
K = \frac{1}{2} I^{-1} S^2 - \frac{ge}{2m} \boldsymbol{S} \cdot \boldsymbol{H} = \frac{1}{2} I^{-1} S^2 - \frac{ge}{4m} S^{\mu\nu} F_{\mu\nu},
$$

$$
\frac{\mathrm{d}}{\mathrm{d}\tau} \left(\frac{1}{2} I^{-1} S^2 \right) = 0,
$$

$$
\begin{aligned}
\dot{K} &= -\frac{ge}{2m} (P^{\mu\sigma} F_{\sigma\tau} \mu^{\tau\nu} + \dot{x}^{\mu} \ddot{x}^{\sigma} S^{\nu}_{\sigma}) F_{\mu\nu} - \frac{ge}{4m} S^{\mu\nu} \dot{F}_{\mu\nu} \\
&= \ddot{x}^{\sigma} \mu_{\sigma\nu} F^{\nu}_{\mu} \dot{x}^{\mu} - \frac{1}{2} \mu^{\mu\nu} \dot{F}_{\mu\nu},
\end{aligned}
$$

and the dynamical equations take the form

$$
\begin{aligned}
&m\ddot{x}^{\mu} + \frac{\mathrm{D}}{\mathrm{D}\tau} \left(S^{\mu}_{\nu} \ddot{x}^{\nu} - \frac{ge}{4m} S^{\nu\sigma} F_{\nu\sigma} \dot{x}^{\mu} - \frac{ge}{2m} S^{\mu\nu} F_{\nu\sigma} \dot{x}^{\sigma} \right) \\
&= \frac{1}{2} R^{\mu}_{\nu\sigma\tau} \dot{x}^{\nu} S^{\sigma\tau} + eF^{\mu}_{\nu} \dot{x}^{\nu} + \frac{ge}{4m} S^{\nu\sigma} F^{\mu}_{\nu\sigma \cdot},
\end{aligned}
$$

where

$$m = m_0 + \frac{1}{2} I^{-1} S^2,$$

in which the second term on the right hand side is the spin energy. We note that the compensating central mass m_c, although again negatively infinite, differs from its nonrelativistic value only by a fraction of order v^2 (see p. 223).

We now consider conservation laws. Suppose the geometry of spacetime admits an isometry generated by a Killing vector ξ^μ,

$$\xi_{\mu \cdot \nu} + \xi_{\nu \cdot \mu} = 0.$$

Suppose furthermore that the Lie derivative $\mathcal{L}_\xi F_{\mu\nu}$ of the electromagnetic field tensor vanishes, i.e.,

$$\begin{aligned}
0 &= -F_{\mu\nu,\sigma}\xi^\sigma - F_{\sigma\nu}\xi^\sigma_{,\mu} - F_{\mu\sigma}\xi^\sigma_{,\nu} \\
&= -F_{\mu\nu\cdot\sigma}\xi^\sigma - F_{\sigma\nu}\xi^\sigma_{\cdot\mu} - F_{\mu\sigma}\xi^\sigma_{\cdot\nu} \\
&= -(A_{\nu\cdot\mu} - A_{\mu\cdot\nu})_{\cdot\sigma}\xi^\sigma - (A_{\nu\cdot\sigma} - A_{\sigma\cdot\nu})\xi^\sigma_{\cdot\mu} - (A_{\sigma\cdot\mu} - A_{\mu\cdot\sigma})\xi^\sigma_{\cdot\nu} \\
&= -(A_{\nu\cdot\sigma}\xi^\sigma + A_\sigma\xi^\sigma_{\cdot\nu})_{\cdot\mu} + (A_{\mu\cdot\sigma}\xi^\sigma + A_\sigma\xi^\sigma_{\cdot\mu})_{\cdot\nu} \\
&\quad + (-R^\tau_{\ \mu\sigma\nu} + R^\tau_{\ \nu\sigma\mu} + R^\tau_{\ \nu\mu\tau\sigma})A_\tau\xi^\sigma \\
&= -(A_{\nu\cdot\sigma}\xi^\sigma + A_\sigma\xi^\sigma_{\cdot\nu})_{\cdot\mu} + (A_{\mu\cdot\sigma}\xi^\sigma + A_\sigma\xi^\sigma_{\cdot\mu})_{\cdot\nu},
\end{aligned}$$

which implies

$$A_{\mu\cdot\nu}\xi^\nu + A_\nu\xi^\nu_{\cdot\mu} = B_{\cdot\mu},$$

for some B. (This assumes that spacetime is connected.) Then any orbit $x^\mu(\tau)$, whether or not it satisfies the dynamical equations, will encounter precisely the same physical environment after it has been displaced by an amount

$$\delta x^\mu(\tau) = \xi^\mu(x(\tau))$$

as it encountered before. The Lagrangian itself, in fact, will be left invariant under such a displacement provided the vector potential A_μ and the local frame vectors e^μ_a are properly chosen. Specifically, the vector potential must be chosen so as to satisfy $\mathcal{L}_\xi A_\mu = 0$, and the local frame vectors must be defined in terms of a field e^μ_a that has vanishing Lie derivative. That is

$$A_{\mu\cdot\nu}\xi^\nu + A_\nu\xi^\nu_{\cdot\mu} = 0$$

and

$$e^\mu_a(\tau) = e^\mu_a(x(\tau)),$$

where

$$e^\mu_{a\cdot\nu}\xi^\nu - e^\nu_a\xi^\mu_{\cdot\nu} = 0 \quad \forall a.$$

The required vector potential may be obtained from the earlier one by carrying out the gauge transformation

$$\overline{A}_\mu = A_\mu + \Lambda_{\cdot\mu},$$

where Λ is a solution of the equation

$$\Lambda_{\cdot\mu}\xi^\mu = -B,$$

and the field e_a^μ may be obtained by Lie displacement from a set of local frames (determined only up to an arbitrary rotation) assigned to an arbitrary two-parameter congruence of orbits. The local frames appropriate to the congruence obtained by Lie displacing any other two-parameter congruence of orbits may be obtained from these by local boosts.

Under the displacement $\delta x^\mu(\tau) = \xi^\mu(x(\tau))$, we now have

$$\delta\lambda_{ab} e_b^\mu + \dot{x}^\mu (e_a \cdot \dot{\xi}) = e^\mu_{a\cdot\nu}\xi^\nu = e_a^\nu \xi^\mu_{\cdot\nu},$$

whence

$$\delta\lambda_{ab} = e_a^\nu e_b^\mu \xi_{\mu\cdot\nu}.$$

Also

$$\delta L = 0.$$

For an orbit which satisfies the dynamical equations, the latter relation reduces (see p. 235) to the statement that the quantity

$$(m_0 + K)(\dot{x}\cdot\xi) + \frac{1}{2}S^{\mu\nu}\xi_{\nu\cdot\mu} + S_{\mu\nu}\ddot{x}^\nu\xi^\mu + eA\cdot\delta x - \mu_{\mu\nu}F^\nu_\sigma \dot{x}^\sigma \xi^\mu$$

is a constant of the motion. This statement may also be verified directly as follows:

$$\begin{aligned}
&\frac{\mathrm{d}}{\mathrm{d}\tau}\left[(m_0 + K)\dot{x}\cdot\xi + \frac{1}{2}S^{\mu\nu}\xi_{\nu\cdot\mu} + S_{\mu\nu}\ddot{x}^\nu\xi^\mu + eA\cdot\xi - \mu_{\mu\nu}F^\nu_\sigma\dot{x}^\sigma\xi^\mu\right]\\
&= \left(\frac{1}{2}R^\mu_{\nu\sigma\tau}\dot{x}^\nu S^{\sigma\tau} + eF^\mu_\nu\dot{x}^\nu + \frac{1}{2}\mu^{\nu\sigma}F^\mu_{\nu\sigma\cdot}\right)\xi_\mu + (m_0 + K)\dot{x}\cdot\dot{\xi}\\
&\quad + (P^{\mu\sigma}F_{\sigma\tau}\mu^{\tau\nu} + \dot{x}^\mu\ddot{x}^\sigma S^\nu_\sigma)\xi_{\nu\cdot\mu} + \frac{1}{2}S^{\mu\nu}\xi_{\nu\cdot\mu\sigma}\dot{x}^\sigma + S^\mu_\nu\ddot{x}^\nu\xi_{\mu\cdot\sigma}\dot{x}^\sigma\\
&\quad + eA_{\mu\cdot\nu}\dot{x}^\nu\xi^\mu + eA_\mu\xi^\mu_{\cdot\nu}\dot{x}^\nu - \xi_{\mu\cdot\tau}\mu^{\mu\nu}F_{\nu\sigma}\dot{x}^\sigma\dot{x}^\tau\\
&= \frac{1}{2}S^{\mu\nu}R_{\mu\nu\tau\sigma}\dot{x}^\sigma\xi^\tau - \mu^{\nu\sigma}F_{\tau\sigma}\xi^\tau_{\cdot\nu} + F^\mu_\tau\mu^{\tau\nu}\xi_{\nu\cdot\mu} + \dot{x}^\mu\dot{x}^\sigma F_{\sigma\tau}\mu^{\tau\nu}\xi_{\nu\cdot\mu}\\
&\quad - \frac{1}{2}S^{\mu\nu}\xi_{\sigma\cdot\nu\mu}\dot{x}^\sigma + \frac{1}{2}S^{\mu\nu}R_{\mu\sigma\nu\tau}\dot{x}^\sigma\xi^\tau - \xi_{\mu\cdot\tau}\mu^{\mu\nu}F_{\nu\sigma}\dot{x}^\sigma\dot{x}^\tau\\
&= 0.
\end{aligned}$$

In the case of a flat empty spacetime, we introduce Minkowski coordinates, and the general solution of Killing's equation becomes

$$\xi_\mu = \varepsilon_\mu + \varepsilon_{\mu\nu}x^\nu, \quad \varepsilon_{\mu\nu} = -\varepsilon_{\nu\mu},$$

The dynamical equations reduce to

$$\dot{S}^{\mu\nu} = \dot{x}^{\mu}\ddot{x}^{\sigma}S^{\nu}_{\sigma} - \dot{x}^{\nu}\ddot{x}^{\sigma}S^{\mu}_{\sigma},$$

$$\frac{\mathrm{d}}{\mathrm{d}\tau}(m\dot{x}^{\mu} + S^{\mu}_{\nu}\ddot{x}^{\nu}) = 0, \qquad m = m_0 + K, \qquad \dot{K} = 0,$$

and the conserved quantity takes the form

$$m\dot{x}^{\mu}(\varepsilon_{\mu} + \varepsilon_{\mu\nu}x^{\nu}) + \frac{1}{2}S^{\mu\nu}\varepsilon_{\nu\mu} + S^{\mu}_{\sigma}\ddot{x}^{\sigma}(\varepsilon_{\mu} + \varepsilon_{\mu\nu}x^{\nu}) = \varepsilon_{\mu}P^{\mu} - \frac{1}{2}\varepsilon_{\mu\nu}J^{\mu\nu},$$

where P^{μ} and $J^{\mu\nu}$ are the energy–momentum four-vector and the total angular momentum tensor, respectively:

$$P^{\mu} = m\dot{x}^{\mu} + S^{\mu}_{\nu}\ddot{x}^{\nu}, \qquad J^{\mu\nu} = x^{\mu}P^{\nu} - x^{\nu}P^{\mu} + S^{\mu\nu}.$$

The remarkable fact will be noted that the momentum of a relativistic spinning particle is not generally parallel to its velocity.

We note also that the dynamical equations are of the third differential order. This has the consequence that the particle can undergo non-uniform motion even in the absence of external fields. Suppose we pass to a Lorentz frame in which the particle comes to rest at some moment. Let us orient the spatial axes so that at that moment the only nonvanishing components of the spin angular momentum tensor are $S_{12} = -S_{21}$. Then, since

$$\dot{S}^{\mu}_{\nu}\ddot{x}^{\nu} = \dot{x}^{\mu}\ddot{x}^{\sigma}S_{\sigma\nu}\ddot{x}^{\nu} - (\ddot{x}\cdot\dot{x})\ddot{x}^{\sigma}S^{\mu}_{\sigma} = 0,$$

$$m\ddot{x}^{\mu} + S^{\mu}_{\nu}\dddot{x}^{\nu} = 0,$$

it follows that the acceleration can have nonvanishing components in the (1,2) plane. We may therefore study the motion in this plane, but instead of adopting a frame in which the particle comes momentarily to rest, we choose a frame in which the three-momentum vanishes. It is then not difficult to see that the motion must have the periodic form

$$x^1 = a\cos\omega x^0, \quad v_1 = \frac{\mathrm{d}x^1}{\mathrm{d}x^0} = -a\omega\sin\omega x^0,$$

$$x^2 = a\sin\omega x^0, \quad v_2 = \frac{\mathrm{d}x^2}{\mathrm{d}x^0} = a\omega\cos\omega x^0,$$

$$x^3 = 0, \quad v_3 = 0,$$

$$v^2 = a^2\omega^2, \quad \mathrm{d}\tau = (1 - v^2)^{1/2}\mathrm{d}x^0 = (1 - a^2\omega^2)^{1/2}\mathrm{d}x^0,$$

$$x^0 = \frac{\tau}{(1 - a^2\omega^2)^{1/2}}, \quad \omega x^0 = \overline{\omega}\tau, \quad \overline{\omega} = \frac{\omega}{(1 - a^2\omega^2)^{1/2}},$$

$$\begin{aligned}
& & \dot{x}^0 &= (1 - a^2\omega^2)^{-1/2}, & \ddot{x}^0 &= 0, \\
x^1 &= a\cos\varpi\tau, & \dot{x}^1 &= -a\varpi\sin\varpi\tau, & \ddot{x}^1 &= -a\varpi^2\cos\varpi\tau, \\
x^2 &= a\sin\varpi\tau, & \dot{x}^2 &= a\varpi\cos\varpi\tau, & \ddot{x}^2 &= -a\varpi^2\sin\varpi\tau, \\
x^3 &= 0, & \dot{x}^3 &= 0, & \ddot{x}^3 &= 0.
\end{aligned}$$

We set

$$S_{12} = \xi S, \quad S_{23} = 0 = S_{31},$$

where S is the magnitude of the spin angular momentum. From the conditions

$$\begin{aligned}
0 &= S_{01}\dot{x}^1 + S_{02}\dot{x}^2 + S_{03}\dot{x}^3, \\
0 &= S_{10}\dot{x}^0 + S_{12}\dot{x}^2 + S_{13}\dot{x}^3 = \dot{x}^0(S_{10} + \xi Sa\omega\cos\omega x^0), \\
0 &= S_{20}\dot{x}^0 + S_{21}\dot{x}^1 + S_{23}\dot{x}^3 = \dot{x}^0(S_{20} + \xi Sa\omega\sin\omega x^0), \\
0 &= S_{30}\dot{x}^0 + S_{31}\dot{x}^1 + S_{32}\dot{x}^2 = \dot{x}^0 S_{30},
\end{aligned}$$

it follows that

$$S_{10} = -\xi Sa\omega\cos\omega x^0, \quad S_{20} = -\xi Sa\omega\sin\omega x^0, \quad S_{30} = 0.$$

But since

$$S^2 = \frac{1}{2}S_{\mu\nu}S^{\mu\nu} = (S_{12})^2 - (S_{10})^2 - (S_{20})^2 = \xi^2 S^2(1 - a^2\omega^2),$$

we find

$$\xi = (1 - a^2\omega^2)^{-1/2},$$

$$S_{01} = Sa\varpi\cos\varpi\tau, \qquad S_{02} = Sa\varpi\sin\varpi\tau, \quad S_{03} = 0.$$

Finally, the vanishing momentum condition yields

$$\begin{aligned}
0 &= m\dot{x}^1 + S^1_0\ddot{x}^0 + S^1_2\ddot{x}^2 + S^1_3\ddot{x}^3 = -ma\varpi\sin\varpi\tau - \xi Sa\varpi^2\sin\varpi\tau, \\
0 &= m\dot{x}^2 + S^2_0\ddot{x}^0 + S^2_1\ddot{x}^1 + S^2_3\ddot{x}^3 = ma\varpi\cos\varpi\tau + \xi Sa\varpi^2\cos\varpi\tau, \\
0 &= m\dot{x}^3 + S^3_0\ddot{x}^0 + S^3_1\ddot{x}^1 + S^3_2\ddot{x}^2 = 0,
\end{aligned}$$

whence

$$\frac{\omega}{1 - a^2\omega^2} = \xi\varpi = -\frac{m}{S},$$

showing that the orbital motion must be retrograde to the spin and yielding the following relation between amplitude and frequency:

$$\frac{S\omega}{m} + 1 = a^2\omega^2, \quad a = \frac{1}{|\omega|}\sqrt{1 - \frac{S|\omega|}{m}}, \quad \xi = \sqrt{\frac{m}{S|\omega|}}.$$

The magnitude of the frequency can vary from 0 to m/S, the amplitude varying accordingly from ∞ to 0.

We thus see that there is a one-parameter family of allowed circular motions in flat empty space. In Lorentz frames in which the three-momentum is nonvanishing, these motions become spiral or cycloidal.

The energy, total angular momentum and absolute acceleration for the above motion are readily found:

$$\begin{aligned} P^0 &= m\dot{x}^0 + S^0_1\dddot{x}^1 + S^0_2\dddot{x}^2 + S^0_3\dddot{x}^3 \\ &= m\xi + Sa^2\overline{\omega}^3 = m\xi + \xi^3 Sa^2\omega^3 \\ &= m\left(\frac{m}{S|\omega|}\right)^{1/2} - \left(\frac{m}{S|\omega|}\right)^{3/2} S|\omega|\left(1 - \frac{S|\omega|}{m}\right) = \sqrt{mS|\omega|}, \end{aligned}$$

$$J_{12} = x_1P_2 - x_2P_1 + S_{12} = S_{12} = \xi S = \sqrt{\frac{mS}{|\omega|}},$$

$$\begin{aligned} J_{23} &= x_2P_3 - x_3P_2 + S_{23} = 0, \quad J_{31} = x_3P_1 - x_1P_3 + S_{31} = 0, \\ J_{01} &= x_0P_1 - x_1P_0 + S_{01} = a\sqrt{mS|\omega|}\cos\overline{\omega}\tau + S\xi a\omega\cos\overline{\omega}\tau = 0, \\ J_{02} &= x_0P_2 - x_2P_0 + S_{02} = a\sqrt{mS|\omega|}\sin\overline{\omega}\tau + S\xi a\omega\sin\overline{\omega}\tau = 0, \\ J_{03} &= x_0P_3 - x_3P_0 + S_{03} = 0, \end{aligned}$$

$$\begin{aligned} \ddot{x}^2 &= -(\ddot{x}^0)^2 + (\ddot{x}^1)^2 + (\ddot{x}^2)^2 + (\ddot{x}^3)^2 = a^2\overline{\omega}^4 \\ &= \xi^4 a^2\omega^4 = \frac{m^2}{S^2\omega^2}\frac{1}{\omega^2}\left(1 - \frac{S|\omega|}{m}\right)\omega^4 = \frac{m^2}{S^2}\left(1 - \frac{S|\omega|}{m}\right), \end{aligned}$$

$$(\ddot{x}^2)^{1/2} = \frac{m}{S}\sqrt{1 - \frac{S|\omega|}{m}},$$

The following table is instructive:

v	$\lvert\omega\rvert$	a	P^0	J_{12}	$(\ddot{x}^2)^{1/2}$
0	m/S	0	m	S	0
1	0	∞	0	∞	m/S

We observe the surprising and non-intuitive fact that the faster the particle circulates the smaller is the total energy, and the greater is the total angular momentum even though the circulation is retrograde to the spin.

A particle that is initially at rest will generally find itself executing this sort of Zitterbewegung after a field has acted upon it. Within the framework of the point particle limit, this result must be regarded as an anomaly. We may ask, however, how big the effect is in the case of a macroscopic spinning body. In this case we do not have a compensating central mass and the spin angular momentum is $m\varepsilon v$ (see p. 223 for the notation). Let ζ be the fractional loss of rest mass (defined by the squared four-momentum) that the body would be computed as suffering if idealized as a point particle. For the kinds of forces envisaged in macroscopic situations we shall have $\zeta \ll 1$.

In the frame in which the final three-momentum vanishes we have

$$(1-\zeta)m = P^0 = \sqrt{mS|\omega|} = \sqrt{m^2\varepsilon v|\omega|},$$

whence

$$\varepsilon v|\omega| = 1 - 2\zeta,$$

$$a = \frac{1}{|\omega|}\sqrt{1-\frac{S|\omega|}{m}} = \frac{\varepsilon v}{1-2\zeta}\sqrt{1-\frac{m\varepsilon v|\omega|}{m}} = \varepsilon v\sqrt{2\zeta} \ll \varepsilon,$$

$$J_{12} = \sqrt{\frac{mS}{|\omega|}} = \sqrt{\frac{m^2\varepsilon v}{1-2\zeta}\varepsilon v} = (1+\zeta)S, \quad a|\omega| = \sqrt{2\zeta} \ll 1,$$

$$(\ddot{x}^2)^{1/2} = \frac{m}{S}\sqrt{1-\frac{S|\omega|}{m}} = \frac{m}{m\varepsilon v}\sqrt{2\zeta} = \frac{\sqrt{2\zeta}}{\varepsilon v},$$

$$\frac{|\omega|}{\omega_{\rm rot}} = \frac{1-2\zeta}{\varepsilon v}\frac{\varepsilon}{v} = \frac{1}{v^2} \gg 1,$$

where $\omega_{\rm rot}$ is the angular velocity of rotation of the body. The radius of the Zitterbewegung circle is seen to be very small compared to the radius of the body, and the Zitterbewegung frequency is seen to be much greater than the rotation frequency of the body. The effect is therefore macroscopically unobservable. That it must, in fact, be spurious may be seen by remembering that a real rigid rotating body is an elastic medium and that the dynamical equations of an elastic medium are of only the second differential order even in relativity theory. The anomalous Zitterbewegung will be replaced by elastic internal vibrations arising from both relativistic effects and the fact that the field (gravitational and electric) is not generally uniform over the body and hence cannot be represented completely accurately by only one or two terms of a power series expansion.

A.3: Charge Current Density. Variation of Four-Vector Potential

Setting τ equal to the proper time after the variation, we have

$$\delta L = e\dot{x}^\mu \delta A_\mu + \frac{1}{2}\mu^{\mu\nu}\delta F_{\mu\nu} = e\dot{x}^\mu \delta A_\mu - \mu^{\mu\nu}\delta A_{\mu\cdot\nu},$$

$$j^\alpha \equiv \frac{\delta W}{\delta A_\alpha(z)} \equiv \int (e\underline{\delta}^\alpha_\mu \dot{x}^\mu - \underline{\delta}^\alpha_{\mu\cdot\nu}\mu^{\mu\nu})\mathrm{d}\tau,$$

where

$$\underset{-}{\delta}{}^{\alpha}_{\mu} \equiv \delta^{\alpha}_{\mu}\delta(z, x(\tau)), \quad \underset{-}{\delta}{}^{\alpha}_{\mu\cdot\alpha} \equiv -\delta_{\cdot\mu}(z, x(\tau)),$$

with $\delta(z, x(\tau))$ treated as a density of unit weight at z and a scalar at x, so that

$$\begin{aligned} j^{\alpha}_{\cdot\alpha} &\equiv \int \left[-e\delta_{\cdot\mu}(z, x(\tau))\dot{x}^{\mu} + \delta_{\cdot\mu\nu}(z, x(\tau))\mu^{\mu\nu}\right]\mathrm{d}\tau \\ &\equiv -e\int \frac{\mathrm{d}}{\mathrm{d}\tau}\delta(z, x(\tau))\mathrm{d}\tau \equiv 0. \end{aligned}$$

A.4: Energy–Momentum–Stress Density. Variation of Metric Tensor

Once again, setting τ equal to the proper time *after* all variations, we have

$$\delta e^{\mu}_{a} = \delta\rho_{ab}e^{\mu}_{b} - \frac{1}{2}(g^{\mu\nu} - \dot{x}^{\mu}\dot{x}^{\nu})e^{\sigma}_{a}\delta g_{\nu\sigma},$$

for some antisymmetric $\delta\rho_{ab}$. We check that

$$\begin{aligned} \delta e_a \cdot e_b + e_a \cdot \delta e_b &= -e^{\mu}_{a}e^{\nu}_{b}\delta g_{\mu\nu} \\ \dot{x} \cdot \delta e_a &= -\dot{x}^{\mu}e^{\nu}_{a}\delta g_{\mu\nu} \end{aligned}$$

which are consistent with $e_a \cdot e_b = \delta_{ab}$ and $\dot{x} \cdot e_a = 0$. Now

$$\begin{aligned} \delta\Gamma^{\mu}_{\nu\sigma} &= \frac{1}{2}\delta\left[g^{\mu\tau}(g_{\nu\tau,\sigma} + g_{\sigma\tau,\nu} - g_{\nu\sigma,\tau})\right] \\ &= -\frac{1}{2}g^{\mu\rho}g^{\tau\lambda}\delta g_{\rho\lambda}(g_{\nu\tau,\sigma} + g_{\sigma\tau,\nu} - g_{\nu\sigma,\tau}) + \frac{1}{2}g^{\mu\tau}(\delta g_{\nu\tau,\sigma} + \delta g_{\sigma\tau,\nu} - \delta g_{\nu\sigma,\tau}) \\ &= -g^{\mu\rho}\Gamma^{\lambda}_{\nu\sigma}\delta g_{\rho\lambda} + \frac{1}{2}g^{\mu\tau}(\delta g_{\nu\tau\cdot\sigma} + \delta g_{\sigma\tau\cdot\nu} - \delta g_{\nu\sigma\cdot\tau}) \\ &\quad + \frac{1}{2}g^{\mu\tau}(\Gamma^{\rho}_{\nu\sigma}\delta g_{\rho\tau} + \Gamma^{\rho}_{\tau\sigma}\delta g_{\nu\rho} + \Gamma^{\rho}_{\sigma\nu}\delta g_{\rho\tau} + \Gamma^{\rho}_{\tau\nu}\delta g_{\sigma\rho} \\ &\quad -\Gamma^{\rho}_{\nu\tau}\delta g_{\rho\sigma} - \Gamma^{\rho}_{\sigma\tau}\delta g_{\nu\rho}) \\ &= \frac{1}{2}g^{\mu\tau}(\delta g_{\nu\tau\cdot\sigma} + \delta g_{\sigma\tau\cdot\nu} - \delta g_{\nu\sigma\cdot\tau}). \end{aligned}$$

Variation of the metric and covariant differentiation with respect to τ are not generally commutative:

$$\begin{aligned} \delta\dot{A}^{\mu} - \frac{\mathrm{D}}{\mathrm{D}\tau}\delta A^{\mu} &= \delta\left(\frac{\mathrm{d}}{\mathrm{d}\tau}A^{\mu} + \Gamma^{\mu}_{\nu\sigma}A^{\nu}\dot{x}^{\sigma}\right) - \frac{\mathrm{d}}{\mathrm{d}\tau}\delta A^{\mu} - \Gamma^{\mu}_{\nu\sigma}\dot{x}^{\sigma}\delta A^{\nu} \\ &= \delta\Gamma^{\mu}_{\nu\sigma}A^{\nu}\dot{x}^{\sigma}, \end{aligned}$$

$$\begin{aligned}
\frac{\mathrm{D}}{\mathrm{D}\tau}\delta e_a^\mu &= \delta\dot\rho_{ab}e_b^\mu + \delta\rho_{ab}\dot e_b^\mu + \frac{1}{2}\ddot x^\mu \dot x^\nu e_a^\sigma \delta g_{\nu\sigma} + \frac{1}{2}\dot x^\mu \ddot x^\nu e_a^\sigma \delta g_{\nu\sigma} \\
&\quad - \frac{1}{2}(g^{\mu\nu} - \dot x^\mu \dot x^\nu)\dot e_a^\sigma \delta g_{\nu\sigma} - \frac{1}{2}(g^{\mu\nu} - \dot x^\mu \dot x^\nu) e_a^\sigma \frac{\mathrm{D}}{\mathrm{D}\tau}\delta g_{\nu\sigma} \\
&= \delta\dot\rho_{ab}e_b^\mu + \delta\rho_{ab}\Omega_{bc}e_c^\mu + \delta\rho_{ab}(e_b\cdot\ddot x)\dot x^\mu \\
&\quad + \frac{1}{2}\ddot x^\mu \dot x^\nu e_a^\sigma \delta g_{\nu\sigma} + \frac{1}{2}\dot x^\mu \ddot x^\nu e_a^\sigma \delta g_{\nu\sigma} \\
&\quad - \frac{1}{2}(g^{\mu\nu} - \dot x^\mu \dot x^\nu)\left[\Omega_{ab}e_b^\sigma \delta g_{\nu\sigma} + (e_a\cdot\ddot x)\dot x^\sigma \delta g_{\nu\sigma} + e_a^\sigma \delta g_{\nu\sigma\cdot\tau}\dot x^\tau\right],
\end{aligned}$$

$$\begin{aligned}
\delta\dot e_a^\mu &= -\frac{1}{2}\Omega_{ab}e_b^\mu \dot x^\nu \dot x^\sigma \delta g_{\nu\sigma} + \delta\Omega_{ab}e_b^\mu + \Omega_{ab}\delta e_b^\mu \\
&\quad + \frac{1}{2}(e_a\cdot\ddot x)\dot x^\mu \dot x^\nu \dot x^\sigma \delta g_{\nu\sigma} + (\delta e_a\cdot\ddot x)\dot x^\mu + \dot x^\mu e_a^\nu \ddot x^\sigma \delta g_{\nu\sigma} + (e_a\cdot\delta\ddot x)\dot x^\mu \\
&= -\frac{1}{2}\Omega_{ab}e_b^\mu \dot x^\nu \dot x^\sigma \delta g_{\nu\sigma} + \delta\Omega_{ab}e_b^\mu + \Omega_{ab}\delta\rho_{bc}e_c^\mu \\
&\quad - \frac{1}{2}(g^{\mu\nu} - \dot x^\mu \dot x^\nu)\Omega_{ab}e_b^\sigma \delta g_{\nu\sigma} + (e_a\cdot\ddot x)\dot x^\mu \dot x^\nu \dot x^\sigma \delta g_{\nu\sigma} \\
&\quad + \delta\rho_{ab}(e_b\cdot\ddot x)\dot x^\mu + \frac{1}{2}\dot x^\mu \ddot x^\nu e_a^\sigma \delta g_{\nu\sigma} + \dot x^\mu \dot x^\nu \dot x^\sigma e_{a\tau}\delta\Gamma^\tau_{\nu\sigma},
\end{aligned}$$

$$\begin{aligned}
0 &= \frac{\mathrm{D}}{\mathrm{D}\tau}\delta e_a^\mu - \delta\dot e_a^\mu + e_a^\nu \dot x^\sigma \delta\Gamma^\mu_{\nu\sigma} \\
&= \left[\delta\dot\rho_{ab} - (\Omega_{ac}\delta\rho_{cb} - \Omega_{bc}\delta\rho_{ca}) + \frac{1}{2}\Omega_{ab}\dot x^\nu \dot x^\sigma \delta g_{\nu\sigma} - \delta\Omega_{ab}\right]e_b^\mu \\
&\quad + \frac{1}{2}\ddot x^\mu \dot x^\nu e_a^\sigma \delta g_{\nu\sigma} - \frac{1}{2}P^{\mu\nu}(e_a\cdot\ddot x)\dot x^\sigma \delta g_{\nu\sigma} + \frac{1}{2}P^{\mu\nu}e_a^\sigma \dot x^\tau(\delta g_{\nu\tau\cdot\sigma} - \delta g_{\sigma\tau\cdot\nu}),
\end{aligned}$$

since

$$\begin{aligned}
&-\frac{1}{2}(g^{\mu\nu} - \dot x^\mu \dot x^\nu)e_a^\sigma \dot x^\tau \delta g_{\nu\sigma\cdot\tau} - \dot x^\mu \dot x^\nu e_{a\sigma}\dot x^\tau \delta\Gamma^\sigma_{\nu\tau} + e_a^\sigma \dot x^\tau \delta\Gamma^\mu_{\sigma\tau} \\
&\quad = \frac{1}{2}g^{\mu\nu}e_a^\sigma \dot x^\tau(-\delta g_{\nu\sigma\cdot\tau} + \delta g_{\sigma\nu\cdot\tau} + \delta g_{\tau\nu\cdot\sigma} - \delta g_{\sigma\tau\cdot\nu}) \\
&\qquad + \frac{1}{2}\dot x^\mu \dot x^\nu e_a^\sigma \dot x^\tau(\delta g_{\nu\sigma\cdot\tau} - \delta g_{\nu\sigma\cdot\tau} - \delta g_{\tau\sigma\cdot\nu} + \delta g_{\nu\tau\cdot\sigma}).
\end{aligned}$$

From this it follows that

$$\begin{aligned}
0 &= \delta\dot\rho_{ab} - (\Omega_{ac}\delta\rho_{cb} - \Omega_{bc}\delta\rho_{ca}) + \frac{1}{2}\Omega_{ab}\dot x^\nu \dot x^\sigma \delta g_{\nu\sigma} - \delta\Omega_{ab} \\
&\quad + \frac{1}{2}\dot x^\mu\left[e_a^\nu(e_b\cdot\ddot x) - e_b^\nu(e_a\cdot\ddot x)\right]\delta g_{\mu\nu} - \frac{1}{2}\dot x^\mu\left(e_a^\nu e_b^\sigma - e_b^\nu e_a^\sigma\right)\delta g_{\mu\nu\cdot\sigma},
\end{aligned}$$

$$\begin{aligned}
\delta v_{na} &= \frac{1}{2}\dot x_{na}\dot x^\mu \dot x^\nu \delta g_{\mu\nu} - \delta\Omega_{ab}x_{nb} \\
&= \frac{1}{2}(v_{na} + \Omega_{ab}x_{nb})\dot x^\mu \dot x^\nu \delta g_{\mu\nu} - \delta\dot\rho_{ab}x_{nb} + (\Omega_{ac}\delta\rho_{cb} - \Omega_{bc}\delta\rho_{ca})x_{nb}
\end{aligned}$$

$$
\begin{aligned}
&-\frac{1}{2}\Omega_{ab}x_{nb}\dot{x}^{\mu}\dot{x}^{\nu}\boldsymbol{\delta}g_{\mu\nu}-\frac{1}{2}\dot{x}^{\mu}\left[e_a^{\nu}(e_b\cdot\ddot{x})-e_b^{\nu}(e_a\cdot\ddot{x})\right]x_{nb}\boldsymbol{\delta}g_{\mu\nu}\\
&+\frac{1}{2}\dot{x}^{\mu}\left(e_a^{\nu}e_b^{\sigma}-e_b^{\nu}e_a^{\sigma}\right)x_{nb}\boldsymbol{\delta}g_{\mu\nu\cdot\sigma},
\end{aligned}
$$

$$
\begin{aligned}
\boldsymbol{\delta}\mu_a=\frac{1}{2}\varepsilon_{abc}\sum_n e_n x_{nb}\Bigg[&\frac{1}{2}v_{nc}\dot{x}^{\mu}\dot{x}^{\nu}\boldsymbol{\delta}g_{\mu\nu}-\boldsymbol{\delta}\dot{\rho}_{cd}x_{nd}+(\Omega_{ce}\boldsymbol{\delta}\rho_{ed}-\Omega_{de}\boldsymbol{\delta}\rho_{ec})x_{nd}\\
&-\frac{1}{2}\dot{x}^{\mu}\left[e_c^{\nu}(e_d\cdot\ddot{x})-e_d^{\nu}(e_c\cdot\ddot{x})\right]x_{nd}\boldsymbol{\delta}g_{\mu\nu}\\
&+\frac{1}{2}\dot{x}^{\mu}\left(e_c^{\nu}e_d^{\sigma}-e_d^{\nu}e_c^{\sigma}\right)x_{nb}\boldsymbol{\delta}g_{\mu\nu\cdot\sigma}\Bigg],
\end{aligned}
$$

$$
\begin{aligned}
\boldsymbol{\delta}L=&\frac{1}{2}(m_0-K)\dot{x}^{\mu}\dot{x}^{\nu}\boldsymbol{\delta}g_{\mu\nu}\\
&+\sum_n m_n v_{na}\Bigg[\frac{1}{2}v_{na}\dot{x}^{\mu}\dot{x}^{\nu}\boldsymbol{\delta}g_{\mu\nu}-\boldsymbol{\delta}\dot{\rho}_{ab}x_{nb}+(\Omega_{ac}\boldsymbol{\delta}\rho_{cb}-\Omega_{bc}\boldsymbol{\delta}\rho_{ca})x_{nb}\\
&-\frac{1}{2}\dot{x}^{\mu}\left[e_a^{\nu}(e_b\cdot\ddot{x})-e_b^{\nu}(e_a\cdot\ddot{x})\right]x_{nb}\boldsymbol{\delta}g_{\mu\nu}\\
&+\frac{1}{2}\dot{x}^{\mu}\left(e_a^{\nu}e_b^{\sigma}-e_b^{\nu}e_a^{\sigma}\right)x_{nb}\boldsymbol{\delta}g_{\mu\nu\cdot\sigma}\Bigg]\\
&-\frac{1}{4}\mu_{ab}F_{ab}\dot{x}^{\mu}\dot{x}^{\nu}\boldsymbol{\delta}g_{\mu\nu}+\mu_{ab}\left[\boldsymbol{\delta}\rho_{ac}e_c^{\mu}-\frac{1}{2}(g^{\mu\sigma}-\dot{x}^{\mu}\dot{x}^{\sigma})e_a^{\tau}\boldsymbol{\delta}g_{\sigma\tau}\right]e_b^{\nu}F_{\mu\nu}\\
&+\frac{1}{2}F_{ab}\sum_n e_n x_{na}\Bigg[\frac{1}{2}v_{nb}\dot{x}^{\mu}\dot{x}^{\nu}\boldsymbol{\delta}g_{\mu\nu}-\boldsymbol{\delta}\dot{\rho}_{bc}x_{nc}+(\Omega_{bd}\boldsymbol{\delta}\rho_{dc}-\Omega_{cd}\boldsymbol{\delta}\rho_{db})x_{nc}\\
&-\frac{1}{2}\dot{x}^{\mu}\left[e_b^{\nu}(e_c\cdot\ddot{x})-e_c^{\nu}(e_b\cdot\ddot{x})\right]x_{nc}\boldsymbol{\delta}g_{\mu\nu}\\
&+\frac{1}{2}\dot{x}^{\mu}\left(e_b^{\nu}e_c^{\sigma}-e_c^{\nu}e_b^{\sigma}\right)x_{nc}\boldsymbol{\delta}g_{\mu\nu\cdot\sigma}\Bigg]\\
=&\frac{1}{2}(m_0+K)\dot{x}^{\mu}\dot{x}^{\nu}\boldsymbol{\delta}g_{\mu\nu}+\frac{1}{2}S_{ab}\boldsymbol{\delta}\dot{\rho}_{ab}+\Omega_{ac}S_{cb}\boldsymbol{\delta}\rho_{ab}+F_{ac}\mu_{cb}\boldsymbol{\delta}\rho_{ab}\\
&-\frac{1}{2}\dot{x}^{\mu}\ddot{x}^{\sigma}S_{\sigma}^{\nu}\boldsymbol{\delta}g_{\mu\nu}-\frac{1}{2}\dot{x}^{\mu}S^{\nu\sigma}\boldsymbol{\delta}g_{\mu\nu\cdot\sigma}-\frac{1}{2}(g^{\mu\sigma}-\dot{x}^{\mu}\dot{x}^{\sigma})\mu^{\tau\nu}F_{\mu\nu}\boldsymbol{\delta}g_{\sigma\tau}\\
=&\frac{1}{2}(m_0+K)\dot{x}^{\mu}\dot{x}^{\nu}\boldsymbol{\delta}g_{\mu\nu}+\frac{1}{2}\frac{\mathrm{d}}{\mathrm{d}\tau}(S_{ab}\boldsymbol{\delta}\rho_{ab})+\frac{1}{2}\dot{x}^{\mu}S_{\sigma}^{\nu}\ddot{x}^{\sigma}\boldsymbol{\delta}g_{\mu\nu}\\
&-\frac{1}{2}\dot{x}^{\mu}S^{\nu\sigma}\boldsymbol{\delta}g_{\mu\nu\cdot\sigma}+\frac{1}{2}\mu^{\nu\sigma}F_{\sigma\tau}(g^{\tau\mu}-\dot{x}^{\tau}\dot{x}^{\mu})\boldsymbol{\delta}g_{\mu\nu},
\end{aligned}
$$

where we have used the relations on p. 229. We may choose the proper time interval in the action integral $W=\int L\mathrm{d}\tau$ so that $\boldsymbol{\delta}\rho_{ab}$ vanishes at the endpoints when $\boldsymbol{\delta}g_{\mu\nu}$ has compact support. Then introducing

$$
\underline{\delta}_{\mu\nu}^{\alpha\beta}=\frac{1}{2}\left(\boldsymbol{\delta}_{\mu}^{\alpha}\boldsymbol{\delta}_{\nu}^{\beta}+\boldsymbol{\delta}_{\nu}^{\alpha}\boldsymbol{\delta}_{\mu}^{\beta}\right)\delta(z,x(\tau)),
$$

we have

$$
\begin{aligned}
T^{\alpha\beta} &\equiv 2\frac{\delta W}{\delta g_{\alpha\beta}(z)} \\
&= \int \Big[(m_0 + K)\underline{\delta}^{\alpha\beta}_{\mu\nu}\dot{x}^{\mu}\dot{x}^{\nu} + \underline{\delta}^{\alpha\beta}_{\mu\nu}\dot{x}^{\mu}S^{\nu}_{\sigma}\ddot{x}^{\sigma} - \underline{\delta}^{\alpha\beta}_{\mu\nu\cdot\sigma}\dot{x}^{\mu}S^{\nu\sigma} \\
&\quad + \underline{\delta}^{\alpha\beta}_{\mu\nu}\mu^{\nu\sigma}F_{\sigma\tau}(g^{\tau\mu} - \dot{x}^{\tau}\dot{x}^{\mu}) \Big] \mathrm{d}\tau.
\end{aligned}
$$

Finally, let us consider conservation of energy–momentum. Let $g_{\alpha\beta}$, A_{α} and the dynamical variables suffer an infinitesimal coordinate transformation $\delta x^{\mu}(\tau)$, δq^{i},

$$
\begin{aligned}
\delta g_{\alpha\beta} &= -g_{\alpha\beta,\gamma}\delta\xi^{\gamma} - g_{\gamma\beta}\delta\xi^{\gamma}_{,\alpha} - g_{\alpha\gamma}\delta\xi^{\gamma}_{,\beta} = -\delta\xi_{\alpha\cdot\beta} - \delta\xi_{\beta\cdot\alpha}, \\
\delta A_{\alpha} &= -A_{\alpha,\beta}\delta\xi^{\beta} - A_{\beta}\delta\xi^{\beta}_{,\alpha} = -A_{\alpha\cdot\beta}\delta\xi^{\beta} - A_{\beta}\delta\xi^{\beta}_{\cdot\alpha}.
\end{aligned}
$$

Because of the coordinate invariance of the action, we have

$$
\begin{aligned}
0 = \delta W &= \int \left(\frac{\delta W}{\delta g_{\alpha\beta}}\delta g_{\alpha\beta} + \frac{\delta W}{\delta A_{\beta}}\delta A_{\beta} \right) \mathrm{d}^4 z + \int \left(\frac{\delta W}{\delta x^{\mu}}\delta x^{\mu} + \frac{\delta W}{\delta q^{i}}\delta q^{i} \right) \mathrm{d}\tau \\
&= \int \Big[-T^{\alpha\beta}\delta\xi_{\alpha\cdot\beta} - j^{\beta}(A_{\beta\cdot\alpha}\delta\xi^{\alpha} + A_{\alpha}\delta\xi^{\alpha}_{\cdot\beta}) \Big] \mathrm{d}^4 z \\
&= \int \left(T^{\alpha\beta}_{\cdot\beta} - F^{\alpha}_{\beta}j^{\beta} \right) \delta\xi_{a}\mathrm{d}^4 z,
\end{aligned}
$$

where the penultimate line follows by virtue of the dynamical equations and the last by virtue of the charge conservation law $j^{\beta}_{\cdot\beta} = 0$. Since $\delta\xi_{a}$ is arbitrary, we must have

$$
T^{\alpha\beta}_{\cdot\beta} = F^{\alpha}_{\beta}j^{\beta}.
$$

Let us now check this. Making use of the identity

$$
\underline{\delta}^{\alpha\beta}_{\mu\nu\cdot\beta} = -\frac{1}{2}(\underline{\delta}^{\alpha}_{\mu\cdot\nu} + \underline{\delta}^{\alpha}_{\nu\cdot\mu}),
$$

we find

$$
\begin{aligned}
T^{\alpha\beta}_{\cdot\beta} - F^{\alpha}_{\beta}j^{\beta} &= \int \Big[-(m_0 + K)\underline{\delta}^{\alpha}_{\mu\cdot\nu}\dot{x}^{\mu}\dot{x}^{\nu} - \frac{1}{2}(\underline{\delta}^{\alpha}_{\mu\cdot\nu} + \underline{\delta}^{\alpha}_{\nu\cdot\mu})\dot{x}^{\mu}S^{\nu}_{\sigma}\ddot{x}^{\sigma} \\
&\quad + \frac{1}{2}(\underline{\delta}^{\alpha}_{\mu\cdot\nu\sigma} + \underline{\delta}^{\alpha}_{\nu\cdot\mu\sigma})\dot{x}^{\mu}S^{\nu\sigma} - \frac{1}{2}(\underline{\delta}^{\alpha}_{\mu\cdot\nu} + \underline{\delta}^{\alpha}_{\nu\cdot\mu})\mu^{\nu\sigma}F_{\sigma\tau}(g^{\tau\mu} - \dot{x}^{\tau}\dot{x}^{\mu}) \\
&\quad - eF^{\alpha}_{\beta}\underline{\delta}^{\beta}_{\mu}\dot{x}^{\mu} + eF^{\alpha}_{\beta}\underline{\delta}^{\beta}_{\mu\cdot\nu}\mu^{\mu\nu} \Big] \mathrm{d}\tau \\
&= \int \Big\{ \underline{\delta}^{\alpha}_{\mu}\frac{\mathrm{D}}{\mathrm{D}\tau}[(m_0 + K)\dot{x}^{\mu}] - \frac{1}{2}\underline{\delta}^{\alpha}_{\mu\cdot\nu}\dot{x}^{\mu}S^{\nu}_{\sigma}\ddot{x}^{\sigma} + \frac{1}{2}\underline{\delta}^{\alpha}_{\mu}\frac{\mathrm{D}}{\mathrm{D}\tau}(S^{\mu}_{\nu}\ddot{x}^{\nu}) \\
&\quad + \frac{1}{4}R^{\tau}_{\nu\sigma\mu}\underline{\delta}^{\alpha}_{\tau}\dot{x}^{\mu}S^{\nu\sigma} + \frac{1}{2}R^{\tau}_{\mu\sigma\nu}\underline{\delta}^{\alpha}_{\tau}\dot{x}^{\mu}S^{\nu\sigma}
\end{aligned}
$$

$$
\begin{aligned}
&-\frac{1}{2}\underline{\delta}^{\alpha}_{\mu\cdot\nu}(P^{\mu\sigma}F_{\sigma\tau}\mu^{\tau\nu} - P^{\nu\sigma}F_{\sigma\tau}\mu^{\tau\mu} - \dot{x}^{\mu}S^{\nu}_{\sigma}\ddot{x}^{\sigma} + \dot{x}^{\nu}S^{\mu}_{\sigma}\ddot{x}^{\sigma}) \\
&-\frac{1}{2}(\underline{\delta}^{\alpha}_{\mu\cdot\nu} + \underline{\delta}^{\alpha}_{\nu\cdot\mu})\mu^{\nu\sigma}F^{\mu}_{\sigma} + \frac{1}{2}\underline{\delta}^{\alpha}_{\mu\cdot\nu}\mu^{\nu\sigma}F_{\sigma\tau}\dot{x}^{\tau}\dot{x}^{\mu} \\
&-\frac{1}{2}\underline{\delta}^{\alpha}_{\mu}\frac{\mathrm{D}}{\mathrm{D}\tau}(\mu^{\mu\nu}F_{\nu\sigma}\dot{x}^{\sigma}) - e\underline{\delta}^{\alpha}_{\mu}F^{\mu}_{\nu}\dot{x}^{\nu} + (\underline{\delta}^{\alpha}_{\sigma}F^{\sigma}_{\mu})_{\cdot\nu}\mu^{\mu\nu}\Big\}\mathrm{d}\tau \\
&= \int \underline{\delta}^{\alpha}_{\mu}\Big\{\frac{\mathrm{D}}{\mathrm{D}\tau}[(m_0 + K)\dot{x}^{\mu} + S^{\mu}_{\nu}\ddot{x}^{\nu} - \mu^{\mu\nu}F_{\nu\sigma}\dot{x}^{\sigma}] \\
&-\frac{1}{2}R^{\mu}_{\nu\sigma\tau}\dot{x}^{\nu}S^{\sigma\tau} - eF^{\mu}_{\nu}\dot{x}^{\nu} - \frac{1}{2}\mu^{\nu\sigma}F^{\mu}_{\nu\sigma\cdot}\Big\}\mathrm{d}\tau \\
&= 0,
\end{aligned}
$$

by virtue of the dynamical equations (A.2) on p. 235.

Appendix B
Weak Field Gravitational Wave

$$\begin{aligned}
R_{\mu\nu\sigma\tau} &= -\frac{1}{2}(h_{\mu\sigma,\nu\tau} + h_{\nu\tau,\mu\sigma} - h_{\mu\tau,\nu\sigma} - h_{\nu\sigma,\mu\tau}),\\
\Box R_{\mu\nu\sigma\tau} &= R^{\rho}_{\ \mu\nu\sigma\tau,\rho} = -R^{\rho}_{\ \mu\nu\tau\rho,\sigma} - R^{\rho}_{\ \mu\nu\rho\sigma,\tau}\\
&= R^{\rho}_{\ \nu\tau\rho,\sigma\mu} + R^{\rho}_{\ \mu\tau\rho,\sigma\nu} + R^{\rho}_{\ \nu\rho\sigma,\tau\mu} + R^{\rho}_{\ \mu\rho\sigma,\tau\nu}\\
&= R_{\mu\sigma,\nu\tau} + R_{\nu\tau,\mu\sigma} - R_{\mu\tau,\nu\sigma} - R_{\nu\sigma,\mu\tau}\\
&= 0
\end{aligned}$$

in empty space because $R_{\mu\nu} = 0$ in empty space.

A plane gravitational wave is given by

$$R_{\mu\nu\sigma\tau} = a_{\mu\nu\sigma\tau}e^{ip\cdot x} + \text{c.c.}, \quad p^2 = 0,$$

where the amplitude $a_{\mu\nu\sigma\tau}$ satisfies

$$a_{\mu\nu\sigma\tau} = -a_{\mu\nu\tau\sigma} = a_{\sigma\tau\mu\nu}, \quad a_{\mu\nu\sigma\tau} + a_{\mu\sigma\tau\nu} + a_{\mu\tau\nu\sigma} = 0,$$

$$a_{\mu\nu\sigma\tau}p_\rho + a_{\mu\nu\tau\rho}p_\sigma + a_{\mu\nu\rho\sigma}p_\tau = 0.$$

We introduce three vectors e_1, e_2 and n satisfying

$$e_a \cdot e_b = \delta_{ab}, \quad e_a \cdot n = 0, \quad n^2 = -1, \quad e_a \cdot p = 0.$$

Define

$$\overline{p} \equiv p + 2(n \cdot p)n.$$

Then

$$n \cdot \overline{p} = -n \cdot p, \quad p \cdot \overline{p} = 2(n \cdot p)^2,$$

$$\overline{p}^2 = 4(n \cdot p)^2 - 4(n \cdot p)^2 = 0, \quad e_a \cdot \overline{p} = 0.$$

Let $T_{\mu\nu}$ be any antisymmetric tensor satisfying

$$T_{\mu\nu}p_\sigma + T_{\nu\sigma}p_\mu + T_{\sigma\mu}p_\nu = 0.$$

$T_{\mu\nu}$ must have the form

$$T_{\mu\nu} = a_\mu p_\nu - a_\nu p_\mu,$$

for some a_μ. It suffices to choose a_μ in the form

$$a_\mu = a_a e_{a\mu} + a\overline{p}_\mu,$$

whence

$$T_{\mu\nu} = a_a(e_{a\mu}p_\nu - e_{a\nu}p_\mu) + a(\overline{p}_\mu p_\nu - \overline{p}_\nu p_\mu).$$

It is clear similarly that $a_{\mu\nu\sigma\tau}$ must have the form

$$\begin{aligned} a_{\mu\nu\sigma\tau} = {} & A_{ab}(e_{a\mu}p_\nu - e_{a\nu}p_\mu)(e_{b\sigma}p_\tau - e_{b\tau}p_\sigma) + B_a(e_{a\mu}p_\nu - e_{a\nu}p_\mu)(p_\sigma\overline{p}_\tau - p_\tau\overline{p}_\sigma) \\ & + B_a(e_{a\sigma}p_\tau - e_{a\tau}p_\sigma)(p_\mu\overline{p}_\nu - p_\nu\overline{p}_\mu) + C(p_\mu\overline{p}_\nu - p_\nu\overline{p}_\mu)(p_\sigma\overline{p}_\tau - p_\tau\overline{p}_\sigma), \end{aligned}$$

with $A_{ab} = -A_{ba}$, in which we have used the algebraic symmetries

$$a_{\mu\nu\sigma\tau} = -a_{\mu\nu\tau\sigma} = a_{\sigma\tau\mu\nu}.$$

But we also have

$$\begin{aligned} 0 &= a_{\mu\nu\sigma\tau} + a_{\mu\sigma\tau\nu} + a_{\mu\tau\nu\sigma} \\ &= A_{ab}\big[(e_{a\mu}p_\nu - e_{a\nu}p_\mu)(e_{b\sigma}p_\tau - e_{b\tau}p_\sigma) + (e_{a\mu}p_\sigma - e_{a\sigma}p_\mu)(e_{b\tau}p_\nu - e_{b\nu}p_\tau) \\ &\quad +(e_{a\mu}p_\tau - e_{a\tau}p_\mu)(e_{b\nu}p_\sigma - e_{b\sigma}p_\nu)\big] \\ &\quad + B_a\big[(e_{a\mu}p_\nu - e_{a\nu}p_\mu)(p_\sigma\overline{p}_\tau - p_\tau\overline{p}_\sigma) + (e_{a\mu}p_\sigma - e_{a\sigma}p_\mu)(p_\tau\overline{p}_\nu - p_\nu\overline{p}_\tau) \\ &\quad +(e_{a\mu}p_\tau - e_{a\tau}p_\mu)(p_\nu\overline{p}_\sigma - p_\sigma\overline{p}_\nu)\big] \\ &\quad + B_a\big[(p_\mu\overline{p}_\nu - p_\nu\overline{p}_\mu)(e_{a\sigma}p_\tau - e_{a\tau}p_\sigma) + (p_\mu\overline{p}_\sigma - p_\sigma\overline{p}_\mu)(e_{a\tau}p_\nu - e_{a\nu}p_\tau) \\ &\quad +(p_\mu\overline{p}_\tau - p_\tau\overline{p}_\mu)(e_{a\nu}p_\sigma - e_{a\sigma}p_\nu)\big] \\ &\quad + C\big[(p_\mu\overline{p}_\nu - p_\nu\overline{p}_\mu)(p_\sigma\overline{p}_\tau - p_\tau\overline{p}_\sigma) + (p_\mu\overline{p}_\sigma - p_\sigma\overline{p}_\mu)(p_\tau\overline{p}_\nu - p_\nu\overline{p}_\tau) \\ &\quad +(p_\mu\overline{p}_\tau - p_\tau\overline{p}_\mu)(p_\nu\overline{p}_\sigma - p_\sigma\overline{p}_\nu)\big] \\ &= 0. \end{aligned}$$

Finally, using $R_{\mu\nu} = 0$ (empty space), we have

$$\begin{aligned} 0 &= a^\sigma_{\mu\sigma\nu} \\ &= A_{ab}(e_{a\mu}p_\sigma - e_{a\sigma}p_\mu)(e_{b\nu}p^\sigma - e^\sigma_b p_\nu) + B_a(e_{a\mu}p_\sigma - e_{a\sigma}p_\mu)(p_\nu\overline{p}^\sigma - p^\sigma\overline{p}_\nu) \\ &\quad + B_a(e_{a\nu}p^\sigma - e^\sigma_a p_\nu)(p_\mu\overline{p}_\sigma - p_\sigma\overline{p}_\mu) + C(p_\mu\overline{p}_\sigma - p_\sigma\overline{p}_\mu)(p_\nu\overline{p}^\sigma - p^\sigma\overline{p}_\nu) \\ &= A_{aa}p_\mu p_\nu + (p\cdot\overline{p})B_a(e_{a\mu}p_\nu + e_{a\nu}p_\mu) - (p\cdot\overline{p})C(\overline{p}_\mu p_\nu + \overline{p}_\nu p_\mu), \end{aligned}$$

which implies

$$B_a = 0, \quad C = 0, \quad A_{aa} = 0,$$

whence, writing

$$A_+ \equiv A_{11} = -A_{22}, \quad A_\times \equiv A_{12} = A_{21},$$

we have

$$a_{\mu\nu\sigma\tau} = H_{\mu\sigma}p_\nu p_\tau + H_{\nu\tau}p_\mu p_\sigma - H_{\mu\tau}p_\nu p_\sigma - H_{\nu\sigma}p_\mu p_\tau,$$

where

$$H_{\mu\nu} = A_+(e_{1\mu}e_{1\nu} - e_{2\mu}e_{2\nu}) + A_\times(e_{1\mu}e_{2\nu} + e_{1\nu}e_{2\mu}).$$

The equation of geodesic deviation in the rest frame of a particle pair, with $n = (1, 0, 0, 0)$ in this frame, is

$$\begin{aligned}\frac{\mathrm{d}^2\eta_i}{\mathrm{d}t^2} &= -R_{i0j0}\eta_j = -\left(H_{ij}p_0p_0\mathrm{e}^{\mathrm{i}p\cdot x} + \mathrm{c.c.}\right)\eta_j \\ &= -2\left[A_+(e_{1i}e_{1j} - e_{2i}e_{2j}) + A_\times(e_{1i}e_{2j} + e_{2i}e_{1j})\right]\omega^2\cos\omega t\eta_j,\end{aligned}$$

setting the particle pair at the origin and writing $x^0 = t$. For $|A_+|$, $|A_\times| \ll 1$, we have (with $t = 0$ when $\dot{\eta}_i = 0$)

$$\eta_i(t) = \left\{\delta_{ij} + 2\left[A_+(e_{1i}e_{1j} - e_{2i}e_{2j}) + A_\times(e_{1i}e_{2j} + e_{2i}e_{1j})\right]\cos\omega t\right\}\eta_j(0).$$

Appendix C
Stationary Spherically (or Rotationally) Symmetric Metric

The metric is given by

$$ds^2 = -F(r)dt^2 + 2E(r)\boldsymbol{x}\cdot d\boldsymbol{x}dt + D(r)(\boldsymbol{x}\cdot d\boldsymbol{x})^2 + C(r)d\boldsymbol{x}^2,$$

where $r = (\boldsymbol{x}\cdot\boldsymbol{x})^{1/2}$, $t = x^0$. Note that

$$dr = r^{-1}\boldsymbol{x}\cdot d\boldsymbol{x},$$

whence

$$\boldsymbol{x}\cdot d\boldsymbol{x} = rdr.$$

Introducing spherical coordinates r, θ, ϕ, where

$$x^1 = r\sin\theta\cos\phi, \quad x^2 = r\sin\theta\sin\phi, \quad x^3 = r\cos\phi,$$

we get

$$ds^2 = -Fdt^2 + 2rEdrdt + r^2Ddr^2 + C(dr^2 + r^2d\theta^2 + r^2\sin^2\theta d\phi^2).$$

Let

$$t' = t - \int \frac{rE}{F}dr.$$

Then

$$dt' = dt - \frac{rE}{F}dr, \quad Fdt'^2 = Fdt^2 - 2rEdrdt + \frac{r^2E^2}{F}dr^2,$$

and

$$ds^2 = -Fdt'^2 + Gdr^2 + C(dr^2 + r^2d\theta^2 + r^2\sin^2\theta d\phi^2),$$

where

$$G = r^2\left(D + \frac{E^2}{F}\right).$$

Let $r' = C^{1/2}r$. Then

$$\mathrm{d}r' = C^{1/2}\mathrm{d}r + \frac{1}{2}rC^{-1/2}C'\mathrm{d}r,$$

or

$$\mathrm{d}r = C^{-1/2}\left(1 + \frac{1}{2}r\frac{C'}{C}\right)^{-1}\mathrm{d}r',$$

and

$$\begin{aligned}\mathrm{d}s^2 &= -F\mathrm{d}t'^2 + GC^{-1}\left(1 + \frac{1}{2}r\frac{C'}{C}\right)^{-2}\mathrm{d}r'^2 + \left(1 + \frac{1}{2}r\frac{C'}{C}\right)^{-2}\mathrm{d}r'^2 \\ &\quad + r'^2(\mathrm{d}\theta^2 + \sin^2\theta\mathrm{d}\phi^2) \\ &= -F\mathrm{d}t'^2 + A\mathrm{d}r'^2 + r'^2(\mathrm{d}\theta^2 + \sin^2\theta\mathrm{d}\phi^2),\end{aligned}$$

where

$$A = \left(1 + \frac{G}{C}\right)\left(1 + \frac{1}{2}r\frac{C'}{C}\right)^{-2}.$$

We now drop the primes and write

$$\mathrm{d}s^2 = -\mathrm{e}^{2\Phi}\mathrm{d}t^2 + \mathrm{e}^{2\Lambda}\mathrm{d}r^2 + r^2(\mathrm{d}\theta^2 + \sin^2\theta\mathrm{d}\phi^2),$$

where

$$\Phi = \frac{1}{2}\ln F, \quad \Lambda = \frac{1}{2}\ln A.$$

C.1: Introducing an Orthonormal Frame Field $\{e_\alpha\}$ and Its Dual $\{e^\alpha\}$

We have

$$g^{-1} = \eta^{\alpha\beta}e_\alpha \otimes e_\beta, \quad g = \eta_{\alpha\beta}e^\alpha \otimes e^\beta = g_{\mu\nu}\mathrm{d}x^\mu \otimes \mathrm{d}x^\nu.$$

In the present case,

$$\begin{aligned}g &= -\mathrm{e}^{2\Phi}\mathrm{d}t \otimes \mathrm{d}t + \mathrm{e}^{2\Lambda}\mathrm{d}r \otimes \mathrm{d}r + r^2(\mathrm{d}\theta \otimes \mathrm{d}\theta + \sin^2\theta\mathrm{d}\phi \otimes \mathrm{d}\phi) \\ &= -e^t \otimes e^t + e^r \otimes e^r + e^\theta \otimes e^\theta + e^\phi \otimes e^\phi,\end{aligned}$$

where

$$e^t = \mathrm{e}^\Phi\mathrm{d}t, \quad e^r = \mathrm{e}^\Lambda\mathrm{d}r, \quad e^\theta = r\mathrm{d}\theta, \quad e^\phi = r\sin\theta\mathrm{d}\phi,$$

$$\mathrm{d}e^t = \mathrm{e}^\Phi\Phi'\mathrm{d}r \wedge \mathrm{d}t = \mathrm{e}^{-\Lambda}\Phi'e^r \wedge e^t,$$

$$de^r = e^{\Lambda}\Lambda' dr \wedge dr = 0, \quad de^\theta = dr \wedge d\theta = r^{-1}e^{-\Lambda}e^r \wedge e^\theta,$$

$$de^\phi = \sin\theta dr \wedge d\phi + r\cos\theta d\theta \wedge d\phi = r^{-1}e^{-\Lambda}e^r \wedge e^\phi + r^{-1}\cot\theta e^\theta \wedge e^\phi.$$

From the equation

$$de^\alpha = -\omega^\alpha_\beta \wedge e^\beta$$

and the antisymmetry of the connection one-form $\omega^{\alpha\beta}$, we may infer that the nonvanishing components of $\omega^\alpha{}_\beta$ are

$$\begin{aligned}
\omega^t_r &= e^{-\Lambda}\Phi' e^t = e^{\Phi-\Lambda}\Phi' dt,\\
\omega^r_t &= e^{-\Lambda}\Phi' e^t = e^{\Phi-\Lambda}\Phi' dt,\\
\omega^r_\theta &= -r^{-1}e^{-\Lambda}e^\theta = -e^{-\Lambda}d\theta,\\
\omega^r_\phi &= -r^{-1}e^{-\Lambda}e^\phi = -e^{-\Lambda}\sin\theta d\phi,\\
\omega^\theta_r &= r^{-1}e^{-\Lambda}e^\theta = e^{-\Lambda}d\theta,\\
\omega^\theta_\phi &= -r^{-1}\cot\theta e^\phi = -\cos\theta d\phi,\\
\omega^\phi_r &= r^{-1}e^{-\Lambda}e^\phi = e^{-\Lambda}\sin\theta d\phi,\\
\omega^\phi_\theta &= r^{-1}\cot\theta e^\phi = \cos\theta d\phi.
\end{aligned}$$

C.2: Computing the Curvature Tensor

The curvature two-form is

$$\Omega^\alpha_\beta = d\omega^\alpha_\beta + \omega^\alpha_\varepsilon \wedge \omega^\varepsilon_\beta.$$

Hence,

$$\begin{aligned}
\Omega^t_r &= d\omega^t_r = e^{\Phi-\Lambda}[\Phi'' + \Phi'(\Phi' - \Lambda')]dr \wedge dt\\
&= e^{-2\Lambda}[\Phi'' + \Phi'(\Phi' - \Lambda')]e^r \wedge e^t,
\end{aligned}$$

$$\begin{aligned}
\Omega^t_\theta &= \omega^t_r \wedge \omega^r_\theta = -r^{-1}e^{-2\Lambda}\Phi' e^t \wedge e^\theta,\\
\Omega^t_\phi &= \omega^t_r \wedge \omega^r_\phi = -r^{-1}e^{-2\Lambda}\Phi' e^t \wedge e^\phi,
\end{aligned}$$

$$\begin{aligned}
\Omega^r_\theta &= d\omega^r_\theta + \omega^r_\phi \wedge \omega^\phi_\theta = e^{-\Lambda}\Lambda' dr \wedge d\theta = r^{-1}e^{-2\Lambda}\Lambda' e^r \wedge e^\theta,\\
\Omega^r_\phi &= d\omega^r_\phi + \omega^r_\theta \wedge \omega^\theta_\phi\\
&= e^{-\Lambda}\Lambda' \sin\theta dr \wedge d\phi - e^{-\Lambda}\cos\theta d\theta \wedge d\phi + e^{-\Lambda}\cos\theta d\theta \wedge d\phi\\
&= r^{-1}e^{-2\Lambda}\Lambda' e^r \wedge e^\phi,
\end{aligned}$$

$$\begin{aligned}
\Omega^\theta_\phi &= d\omega^\theta_\phi + \omega^\theta_r \wedge \omega^r_\phi = \sin\theta d\theta \wedge d\phi - e^{-2\Lambda}\sin\theta d\theta \wedge d\phi\\
&= r^{-2}\left(1 - e^{-2\Lambda}\right)e^\theta \wedge e^\phi.
\end{aligned}$$

From the equation

$$\Omega^{\alpha}_{\beta} = \frac{1}{2} R^{\alpha}_{\beta\gamma\delta} e^{\gamma} \wedge e^{\delta},$$

we may infer that the nonvanishing components of the curvature tensor are

$$R_{trtr} = \mathrm{e}^{-2\Lambda}[\Phi'' + \Phi'(\Phi' - \Lambda')], \quad R_{t\theta t\theta} = r^{-1}\mathrm{e}^{-2\Lambda}\Phi',$$

$$R_{t\phi t\phi} = r^{-1}\mathrm{e}^{-2\Lambda}\Phi', \quad R_{r\theta r\theta} = r^{-1}\mathrm{e}^{-2\Lambda}\Lambda',$$
$$R_{r\phi r\phi} = r^{-1}\mathrm{e}^{-2\Lambda}\Lambda', \quad R_{\theta\phi\theta\phi} = r^{-2}\left(1 - \mathrm{e}^{-2\Lambda}\right),$$

together with the components obtained from these by using the antisymmetry of $R_{\alpha\beta\gamma\delta}$ in its first pair of indices and in its last pair.

The nonvanishing components of the Ricci tensor

$$R_{\alpha\beta} = \eta^{\gamma\delta} R_{\alpha\gamma\beta\delta}$$

are

$$R_{tt} = R_{trtr} + R_{t\theta t\theta} + R_{t\phi t\phi} = \mathrm{e}^{-2\Lambda}\left[\Phi'' + \Phi'(\Phi' - \Lambda') + 2r^{-1}\Phi'\right],$$
$$R_{rr} = -R_{rtrt} + R_{r\theta r\theta} + R_{r\phi r\phi} = \mathrm{e}^{-2\Lambda}\left[-\Phi'' - \Phi'(\Phi' - \Lambda') + 2r^{-1}\Lambda'\right],$$
$$R_{\theta\theta} = -R_{\theta t\theta t} + R_{\theta r\theta r} + R_{\theta\phi\theta\phi} = r^{-1}\mathrm{e}^{-2\Lambda}\left[-\Phi' + \Lambda' + r^{-1}\left(\mathrm{e}^{2\Lambda} - 1\right)\right],$$
$$R_{\phi\phi} = -R_{\phi t\phi t} + R_{\phi r\phi r} + R_{\phi\theta\phi\theta} = r^{-1}\mathrm{e}^{-2\Lambda}\left[-\Phi' + \Lambda' + r^{-1}\left(\mathrm{e}^{2\Lambda} - 1\right)\right].$$

C.3: Vacuum Solution

In this case, $R_{\alpha\beta} = 0$, whence

$$\Phi' + \Lambda' = 0.$$

The boundary condition is

$$\lim_{r\to\infty} \Phi = 0 = \lim_{r\to\infty} \Lambda.$$

Hence,

$$\Phi = -\Lambda,$$

and we then have

$$2\Lambda' + r^{-1}\left(\mathrm{e}^{2\Lambda} - 1\right) = 0,$$

so that

$$\frac{2\mathrm{e}^{-2\Lambda}\Lambda'}{1 - \mathrm{e}^{-2\Lambda}} = -\frac{1}{r}, \quad \frac{\mathrm{d}}{\mathrm{d}r}\ln\left(1 - \mathrm{e}^{-2\Lambda}\right) = -\frac{1}{r}.$$

Therefore

$$\ln\left(1-\mathrm{e}^{-2\Lambda}\right) = -\ln r + \text{const.},$$

and

$$1-\mathrm{e}^{-2\Lambda} = \frac{C}{r}, \quad \mathrm{e}^{-2\Lambda} = \mathrm{e}^{2\Phi} = 1-\frac{C}{r}, \quad \Phi = \frac{1}{2}\ln\left(1-\frac{C}{r}\right).$$

We check as follows:

$$\Phi' = \frac{C}{2}\frac{1/r^2}{1-C/r} = \frac{C}{2}\frac{1}{r^2-Cr},$$
$$\Phi'' = -\frac{C}{2}\frac{2r-C}{(r^2-Cr)^2},$$

$$\begin{aligned}\Phi'' + \Phi'(\Phi'-\Lambda') + 2r^{-1}\Phi' &= \Phi'' + 2\Phi'^2 + 2r^{-1}\Phi' \\ &= -\frac{C}{2}\frac{2r-C}{(r^2-Cr)^2} + \frac{C^2}{2}\frac{1}{(r^2-Cr)^2} + C\frac{1}{r(r^2-Cr)} \\ &= \frac{C}{r(r^2-Cr)^2}\left(-r^2+\frac{1}{2}Cr+\frac{1}{2}Cr+r^2-Cr\right) = 0.\end{aligned}$$

Now

$$\mathrm{d}s^2 = -\left(1-\frac{C}{r}\right)\mathrm{d}t^2 + \left(1-\frac{C}{r}\right)^{-1}\mathrm{d}r^2 + r^2(\mathrm{d}\theta^2 + \sin^2\theta\mathrm{d}\phi^2).$$

The quantity $h_{00} = C/r$ is -2 times Newton's potential energy, i.e.,

$$h_{00} = \frac{C}{r} = \frac{2MG}{r}.$$

Hence, $C = 2MG$ and

$$\mathrm{d}s^2 = -\left(1-\frac{2MG}{r}\right)\mathrm{d}t^2 + \left(1-\frac{2MG}{r}\right)^{-1}\mathrm{d}r^2 + r^2(\mathrm{d}\theta^2 + \sin^2\theta\mathrm{d}\phi^2).$$

For a radially infalling photon,

$$0 = g_{\mu\nu}\frac{\mathrm{d}x^\mu}{\mathrm{d}t}\frac{\mathrm{d}x^\nu}{\mathrm{d}t} = -\left(1-\frac{2MG}{r}\right) + \left(1-\frac{2MG}{r}\right)^{-1}\left(\frac{\mathrm{d}r}{\mathrm{d}t}\right)^2,$$
$$\frac{\mathrm{d}r}{\mathrm{d}t} = -\left(1-\frac{2MG}{r}\right).$$

The Eddington time coordinate is

$$t^* = t + 2MG\ln\left|\frac{r}{2MG}-1\right|, \quad \mathrm{d}t^* = \mathrm{d}t + \frac{\mathrm{d}r}{\frac{r}{2MG}-1},$$

whence

$$\frac{dt^*}{dr} = -\frac{1}{1-\frac{2MG}{r}} + \frac{1}{\frac{r}{2MG}-1} = \frac{-1+\frac{2MG}{r}}{1-\frac{2MG}{r}} = -1,$$

and

$$ds^2 = -\left(1-\frac{2MG}{r}\right)\left(dt^* - \frac{dr}{\frac{r}{2MG}-1}\right)^2 + \left(1-\frac{2MG}{r}\right)^{-1} dr^2 + r^2(d\theta^2 + \sin^2\theta d\phi^2) \tag{C.1}$$

$$\begin{aligned} ds^2 &= -\left(1-\frac{2MG}{r}\right)dt^{*2} + \frac{4MG}{r}dt^*dr \\ &\quad + \left(1-\frac{2MG}{r}\right)^{-1}\left[1-\left(\frac{2MG}{r}\right)^2\right]dr^2 + r^2(d\theta^2+\sin^2\theta d\phi^2) \\ &= -\left(1-\frac{2MG}{r}\right)dt^{*2} + \frac{4MG}{r}dt^*dr + \left(1+\frac{2MG}{r}\right)dr^2 \\ &\quad + r^2(d\theta^2+\sin^2\theta d\phi^2) \\ &= -dt^{*2} + dr^2 + r^2(d\theta^2+\sin^2\theta d\phi^2) + \frac{2MG}{r}(dt^*+dr)^2. \end{aligned} \tag{C.2}$$

Returning to Cartesian coordinates, we have

$$ds^2 = -dt^{*2} + d\boldsymbol{x}^2 + \frac{2MG}{r}\left(dt^* + \frac{\boldsymbol{x}\cdot d\boldsymbol{x}}{r}\right)^2, \quad r = (\boldsymbol{x}\cdot x)^{1/2},$$

or

$$ds^2 = (\eta_{\mu\nu} + l_\mu l_\nu)dx^\mu dx^\nu, \quad dx^0 = dt^*,$$

where

$$(l_\mu) = \left(\frac{2MG}{r}\right)^{1/2}\left(1, \frac{\boldsymbol{x}}{r}\right), \quad \eta^{\mu\nu} l_\mu l_\nu = 0.$$

C.4: Metrics of the Form $g_{\mu\nu} = \eta_{\mu\nu} + l_\mu l_\nu$, with $\eta^{\mu\nu} l_\mu l_\nu = 0$

Define $l^\mu \equiv \eta^{\mu\nu} l_\nu$. Then $g^{\mu\nu} = \eta^{\mu\nu} - l^\mu l^\nu$. This is proved by

$$g^{\mu\sigma} g_{\sigma\nu} = (\eta^{\mu\sigma} - l^\mu l^\sigma)(\eta_{\sigma\nu} + l_\sigma l_\nu) = \delta^\mu_\nu + l^\mu l_\nu - l^\mu l_\nu = \delta^\mu_\nu,$$

since $l^\sigma l_\sigma = 0$. Note also that

$$g^{\mu\nu} l_\nu = l^\mu, \quad l^\mu l_{\mu,\nu} = l^\mu l_{\mu;\nu} = 0,$$

$$\delta \ln g = g^{\mu\nu}\delta g_{\mu\nu} = 2(\eta^{\mu\nu} - l^{\mu}l^{\nu})l_{\mu}\delta l_{\nu} = 2l^{\nu}\delta l_{\nu} = 0,$$

whence

$$\det(g_{\mu\nu}) = -g = -1, \quad g = 1, \quad \Gamma^{\nu}_{\nu\mu} = 0.$$

C.4.1: Ricci Tensor

The Ricci tensor is

$$\begin{aligned} R_{\mu\nu} = R^{\sigma}_{\mu\sigma\nu} &= \Gamma^{\sigma}_{\mu\nu,\sigma} - \Gamma^{\sigma}_{\mu\sigma,\nu} + \Gamma^{\sigma}_{\sigma\tau}\Gamma^{\tau}_{\nu\mu} - \Gamma^{\sigma}_{\nu\tau}\Gamma^{\tau}_{\sigma\mu} \\ &= \Gamma^{\sigma}_{\mu\nu,\sigma} - \Gamma^{\sigma}_{\tau\mu}\Gamma^{\tau}_{\sigma\nu}. \end{aligned}$$

All equations must hold equally well if we rescale l_{μ} and write

$$g_{\mu\nu} = \eta_{\mu\nu} + \alpha l_{\mu}l_{\nu}, \quad g^{\mu\nu} = \eta^{\mu\nu} - \alpha l^{\mu}l^{\nu},$$

Equations must hold for arbitrary constant α. We have

$$\Gamma_{\mu\nu\sigma} = \frac{1}{2}\alpha\left[(l_{\mu}l_{\nu})_{,\sigma} + (l_{\mu}l_{\sigma})_{,\nu} - (l_{\nu}l_{\sigma})_{,\mu}\right],$$

$$\begin{aligned} \Gamma^{\mu}_{\nu\sigma} &= \frac{1}{2}\alpha(\eta^{\mu\tau} - \alpha l^{\mu}l^{\tau})\left[(l_{\tau}l_{\nu})_{,\sigma} + (l_{\tau}l_{\sigma})_{,\nu} - (l_{\nu}l_{\sigma})_{,\tau}\right] \\ &= \frac{1}{2}\alpha\left[(l^{\mu}l_{\nu})_{,\sigma} + (l^{\mu}l_{\sigma})_{,\nu} - \eta^{\mu\tau}(l_{\nu}l_{\sigma})_{,\tau}\right] + \frac{1}{2}\alpha^2 l^{\mu}l^{\tau}(l_{\nu}l_{\sigma})_{,\tau}, \end{aligned}$$

$$\begin{aligned} R_{\mu\nu} &= \frac{1}{2}\alpha\left[(l^{\sigma}l_{\mu})_{,\nu\sigma} + (l^{\sigma}l_{\nu})_{,\mu\sigma} - \eta^{\sigma\tau}(l_{\mu}l_{\nu})_{,\sigma\tau}\right] \\ &\quad + \frac{1}{2}\alpha^2(l^{\sigma}l^{\tau})_{,\sigma}(l_{\mu}l_{\nu})_{,\tau} + \frac{1}{2}\alpha^2 l^{\sigma}l^{\tau}(l_{\mu}l_{\nu})_{,\sigma\tau} \\ &\quad - \left\{\frac{1}{2}\alpha\left[(l^{\sigma}l_{\tau})_{,\mu} + (l^{\sigma}l_{\mu})_{,\tau} - \eta^{\sigma\rho}(l_{\tau}l_{\mu})_{,\rho}\right] + \frac{1}{2}\alpha^2 l^{\sigma}l^{\rho}(l_{\tau}l_{\mu})_{,\rho}\right\} \\ &\quad \times \left\{\frac{1}{2}\alpha\left[(l^{\tau}l_{\sigma})_{,\nu} + (l^{\tau}l_{\nu})_{,\sigma} - \eta^{\tau\lambda}(l_{\sigma}l_{\nu})_{,\lambda}\right] + \frac{1}{2}\alpha^2 l^{\tau}l^{\lambda}(l_{\sigma}l_{\nu})_{,\lambda}\right\} \\ &= \frac{1}{2}\alpha\left[(l^{\sigma}l_{\mu})_{,\nu\sigma} + (l^{\sigma}l_{\nu})_{,\mu\sigma} - \eta^{\sigma\tau}(l_{\mu}l_{\nu})_{,\sigma\tau}\right] \\ &\quad + \frac{1}{2}\alpha^2\left[(l^{\sigma}l^{\tau})_{,\sigma}(l_{\mu}l_{\nu})_{,\tau} + l^{\sigma}l^{\tau}(l_{\mu}l_{\nu})_{,\sigma\tau}\right. \\ &\quad - \frac{1}{2}l^{\sigma}l_{\tau,\mu}l^{\tau}_{,\sigma}l_{\nu} + \frac{1}{2}l^{\sigma}_{,\mu}l^{\lambda}l_{\sigma,\lambda}l_{\nu} - \frac{1}{2}l^{\sigma}_{,\tau}l_{\mu}l^{\tau}l_{\sigma,\nu} \\ &\quad - \frac{1}{2}(l^{\sigma}l_{\mu})_{,\tau}(l^{\tau}l_{\nu})_{,\sigma} + \frac{1}{2}\eta^{\tau\lambda}l^{\sigma}_{,\tau}l_{\mu}l_{\sigma,\lambda}l_{\nu} \end{aligned}$$

$$+\frac{1}{2}l_{\tau,\rho}l_\mu l^\tau_{,\nu}l^\rho+\frac{1}{2}\eta^{\sigma\rho}l_{\tau,\rho}l_\mu l^\tau_{,\sigma}l_\nu-\frac{1}{2}(l^\lambda l_\mu)_{,\rho}(l^\rho l_\nu)_{,\lambda}\Big]$$
$$-\frac{1}{4}\alpha^3\left(l^\tau l^\lambda l^\sigma_{,\tau}l_\mu l_{\sigma,\lambda}l_\nu+l^\sigma l^\rho l_{\tau,\rho}l_\mu l^\tau_{,\sigma}l_\nu\right). \tag{C.3}$$

In empty spacetime, we have $R_{\mu\nu}=0$, and since α is arbitrary the three terms above must vanish separately. If $l_\mu \neq 0$, the third term implies

$$a^2=0, \quad \text{where } a_\mu=l^\nu l_{\mu,\nu}=l^\nu(l_{\mu;\nu}+\Gamma^\sigma_{\mu\nu}l^\mu)=l^\nu l_{\mu;\nu},$$

since $l^\mu l^\nu\Gamma^\sigma_{\mu\nu}=0$. We also have $a\cdot l=0$, whence

$$a_\mu=Al_\mu \text{ for some scalar function} A.$$

Now

$$a^\mu=g^{\mu\nu}a_\nu=\eta^{\mu\nu}a_\nu=l^\nu l^\mu_{;\nu}=l^\nu l^\mu_{,\nu}.$$

Define $B\equiv l^\mu_{,\mu}$. Then the first term implies

$$\Box(l_\mu l_\nu)=\left[(A+B)l_\mu\right]_{,\nu}+[(A+B)l_\nu]_{,\mu},$$

where

$$\Box\equiv\eta^{\mu\nu}\frac{\partial^2}{\partial x^\mu\partial x^\nu}.$$

Contracting this equation with $\eta^{\mu\nu}$ and dividing by 2, we get

$$[(A+B)l^\mu]_{,\mu}=0.$$

An alternative version of the uncontracted equation is

$$l_\mu\Box l_\nu+l_\nu\Box l_\mu+2\eta^{\sigma\tau}l_{\mu,\sigma}l_{\nu,\tau}=l_\mu(A+B)_{,\nu}+l_\nu(A+B)_{,\mu}+(A+B)(l_{\mu,\nu}+l_{\nu,\mu}).$$

Multiplication with l_μ and removal of a common factor l_ν yields

$$\begin{aligned}l^\mu\Box l_\mu&=l^\mu(A+B)_{,\mu}+A(A+B)\\&=[l^\mu(A+B)]_{,\mu}-B(A+B)+A(A+B)=A^2-B^2.\end{aligned}$$

But

$$0=\Box(l^\mu l_\mu)=2(l^\mu\Box l_\mu+\eta^{\sigma\tau}l^\mu_{,\sigma}l_{\mu,\tau}),$$

whence

$$\eta^{\sigma\tau}l^\mu_{,\sigma}l_{\mu,\tau}=B^2-A^2.$$

We now prove that the term in α^2 in the expression (C.3) automatically vanishes when the terms in α and α^3 do:

$$(l^\sigma l^\tau)_{,\sigma}(l_\mu l_\nu)_{,\tau} = (A+B)l^\tau(l_\mu l_\nu)_{,\tau} = 2A(A+B)l_\mu l_\nu$$

$$\begin{aligned} l^\sigma l^\tau l_{\mu,\sigma\tau} &= l^\sigma(l^\tau l_{\mu,\tau})_{,\sigma} - l^\sigma l^\tau_{,\sigma} l_{\mu,\tau} = l^\sigma(Al_\mu)_{,\sigma} - Al^\tau l_{\mu,\tau} \\ &= l^\sigma A_{,\sigma} l_\mu + A^2 l_\mu - A^2 l_\mu = l^\sigma A_{,\sigma} l_\mu, \end{aligned}$$

$$\begin{aligned} l^\sigma l^\tau(l_\mu l_\nu)_{,\sigma\tau} &= l^\sigma l^\tau(l_\mu l_{\nu,\sigma\tau} + l_\nu l_{\mu,\sigma\tau} + l_{\mu,\sigma} l_{\nu,\tau} + l_{\mu,\tau} l_{\nu,\sigma}) \\ &= 2(l^\sigma A_{,\sigma} + A^2) l_\mu l_\nu \end{aligned}$$

$$\begin{aligned} (l^\sigma l_\mu)_{,\tau}(l^\tau l_\nu)_{,\sigma} &= (l_\mu l^\sigma_{,\tau} + l^\sigma l_{\mu,\tau})(l_\nu l^\tau_{,\sigma} + l^\tau l_{\nu,\sigma}) \\ &= l_\mu l_\nu\left[(l^\sigma_{,\tau} l^\tau)_{,\sigma} - l^\sigma_{,\tau\sigma} l^\tau\right] + Al_\mu l^\sigma l_{\nu,\sigma} + Al_\nu l^\tau l_{\mu,\tau} + A^2 l_\mu l_\nu \\ &= l_\mu l_\nu\left[(Al^\sigma)_{,\sigma} - B_{,\tau} l^\tau + 3A^2\right] \end{aligned}$$

$$\eta^{\tau\lambda} l^\sigma_{,\tau} l_\mu l_{\sigma,\lambda} l_\nu = l_\mu l_\nu (B^2 - A^2)$$

Removing the factor $\alpha^2 l_\mu l_\nu/2$ from the term in α^2, we now get

$$\begin{aligned} &2A(A+B) + 2(A_{,\sigma} l^\sigma + A^2) - (Al^\sigma)_{,\sigma} + B_{,\sigma} l^\sigma - 3A^2 + B^2 - A^2 \\ &= (Al^\sigma)_{,\sigma} + (Bl^\sigma)_{,\sigma} = 0, \end{aligned}$$

as claimed. From now on, we set $\alpha = 1$.

C.4.2: Stationary Case

We assume that the l_μ are independent of the coordinate x^0. Write $l_\mu = l(1, \boldsymbol{\lambda})$. Since l_μ is null, $\boldsymbol{\lambda}$ is a unit three-vector, and the equation

$$\Box(l_\mu l_\nu) = [(A+B)l_\mu]_{,\nu} + [(A+B)l_\nu]_{,\mu}$$

decomposes into

$$\nabla^2(l^2) = 0, \quad \nabla^2(l^2\lambda_i) = [(A+B)l]_{,i}, \tag{C.4}$$

$$\nabla^2(l^2\lambda_i\lambda_j) = [(A+B)l\lambda_i]_{,j} + [(A+B)l\lambda_j]_{,i}.$$

The first equation allows the second to be rewritten in the form

$$l^2\lambda_{i,kk} + 2(l^2)_{,k}\lambda_{i,k} = [(A+B)l]_{,i}.$$

This, together with the first equation, allows the third equation to be rewritten in the form

$$
\begin{aligned}
0 &= l^2\lambda_{i,kk}\lambda_j + l^2\lambda_i\lambda_{j,kk} + 2l^2\lambda_{i,k}\lambda_{i,k} + 2(l^2)_{,k}\lambda_i\lambda_{j,k} + 2(l^2)_{,k}\lambda_{i,k}\lambda_j \\
&\quad - [(A+B)l]_{,j}\lambda_i - [(A+B)l]_{,i}\lambda_j - (A+B)l(\lambda_{i,j} + \lambda_{j,i}) \\
&= 2l^2\lambda_{i,k}\lambda_{i,k} - (A+B)l(\lambda_{i,j} + \lambda_{j,i}),
\end{aligned}
$$

or

$$M + M^{\mathrm{T}} - \frac{1}{8}MM^{\mathrm{T}} = 0,$$

where

$$M \equiv (\lambda_{i,j}), \quad p = \frac{A+B}{2l} \quad (\text{assuming} A + B \neq 0).$$

Since $\boldsymbol{\lambda}^2 = 1$, it follows that $\lambda_j\lambda_{j,i} = 0$, or $M^{\mathrm{T}}\boldsymbol{\lambda} = 0$. From $Al_\mu = l^\nu l_{\mu,\nu}$, we also have

$$Al = l^\nu l_{,\nu} = l\lambda_i l_{,i}, \quad \text{whence } \lambda_i l_{,i} = A,$$

and

$$Al\lambda_i = l^\nu(l\lambda_i)_{,\nu} = l\lambda_j(l\lambda_i)_{,j} = Al\lambda_i + l^2\lambda_j\lambda_{i,j},$$

or

$$M\boldsymbol{\lambda} = 0,$$

assuming that $l^2 \neq 0$. Evidently there exists an orthogonal matrix O such that

$$O\boldsymbol{\lambda} = \begin{pmatrix} 1 \\ 0 \\ 0 \end{pmatrix}, \quad OMO^{\mathrm{T}} = \mathrm{diag}(0, N),$$

where N is a 2×2 matrix satisfying

$$N + N^{\mathrm{T}} - \frac{1}{p}NN^{\mathrm{T}} = 0.$$

Note that $O_{1i}\lambda_i = 1$ and $O_{1i}O_{1i} = 1$, whence it follows that

$$O_{1i} = \lambda_i.$$

Note also that

$$\left(1 - \frac{1}{p}N\right)\left(1 - \frac{1}{p}N^{\mathrm{T}}\right) = 1 - \frac{1}{p}\left(N + N^{\mathrm{T}} - \frac{1}{p}NN^{\mathrm{T}}\right) = 1,$$

which implies that $1 - N/p$ is a 2×2 orthogonal matrix and hence that

$$1 - \frac{1}{p}N = \begin{pmatrix} \cos\theta & -\sin\theta \\ \sin\theta & \cos\theta \end{pmatrix} \quad \text{or} \quad \begin{pmatrix} \cos\theta & -\sin\theta \\ -\sin\theta & -\cos\theta \end{pmatrix},$$

for some θ. Choosing the first possibility, we have

$$N = p\begin{pmatrix} 1-\cos\theta & \sin\theta \\ -\sin\theta & 1-\cos\theta \end{pmatrix}$$

and

$$\begin{aligned}
\lambda_{i,j} &= O_{\cdot i}\begin{pmatrix} 0 & 0 & 0 \\ 0 & p(1-\cos\theta) & p\sin\theta \\ 0 & -p\sin\theta & p(1-\cos\theta) \end{pmatrix} O_{\cdot j} \\
&= p(1-\cos\theta)(O_{2i}O_{2j} + O_{3i}O_{3j}) + p\sin\theta(O_{2i}O_{3j} - O_{3i}O_{2j}) \\
&= p(1-\cos\theta)(\delta_{ij} - O_{1i}O_{1j}) + p\sin\theta\varepsilon_{ijk}O_{1k} \\
&= \alpha(\delta_{ij} - \lambda_i\lambda_j) + \beta\varepsilon_{ijk}\lambda_k,
\end{aligned}$$

where

$$\alpha \equiv p(1-\cos\theta), \quad \beta \equiv p\sin\theta.$$

Now

$$\nabla\cdot\boldsymbol{\lambda} = 2\alpha, \quad \nabla\times\boldsymbol{\lambda} = -2\beta\boldsymbol{\lambda},$$

and

$$\nabla\times(\nabla\times\boldsymbol{\lambda}) = \nabla(\nabla\cdot\boldsymbol{\lambda}) - \nabla^2\boldsymbol{\lambda},$$

whence

$$\begin{aligned}
\nabla^2\boldsymbol{\lambda} &= \nabla(\nabla\cdot\boldsymbol{\lambda}) - \nabla\times(\nabla\times\boldsymbol{\lambda}) = 2\nabla\alpha + 2\nabla\times(\beta\boldsymbol{\lambda}) \\
&= 2\nabla\alpha - 2\boldsymbol{\lambda}\times\nabla\beta + 2\beta\nabla\times\boldsymbol{\lambda} = 2\nabla\alpha - 2\boldsymbol{\lambda}\times\nabla\beta - 4\beta^2\boldsymbol{\lambda}.
\end{aligned}$$

But also

$$\lambda_{i,jj} = \alpha_{,j}(\delta_{ij} - \lambda_i\lambda_j) - \alpha\lambda_{i,j}\lambda_j - \alpha\lambda_i\lambda_{j,j} + \varepsilon_{ijk}(\beta\lambda_k)_{,j},$$

or

$$\nabla^2\boldsymbol{\lambda} = \nabla\alpha - \boldsymbol{\lambda}(\boldsymbol{\lambda}\cdot\nabla\alpha) - 2(\alpha^2+\beta^2)\boldsymbol{\lambda} - \boldsymbol{\lambda}\times\nabla\beta. \tag{C.5}$$

Subtracting one equation from the other, we get

$$0 = \nabla\alpha + \boldsymbol{\lambda}(\boldsymbol{\lambda}\cdot\nabla\alpha) + 2(\alpha^2-\beta^2)\boldsymbol{\lambda} - \boldsymbol{\lambda}\times\nabla\beta,$$

whence

$$\boldsymbol{\lambda}\cdot\nabla\alpha = \beta^2 - \alpha^2, \quad \nabla\alpha = (\beta^2-\alpha^2)\boldsymbol{\lambda} + \boldsymbol{\lambda}\times\nabla\beta,$$

and (see below)

$$\boldsymbol{\lambda}\times\nabla\alpha = \boldsymbol{\lambda}\times(\boldsymbol{\lambda}\times\nabla\beta) = \boldsymbol{\lambda}(\boldsymbol{\lambda}\cdot\nabla\beta) - \nabla\beta = -2\alpha\beta\boldsymbol{\lambda} - \nabla\beta.$$

Now

$$0 = -\frac{1}{2}\nabla \cdot (\nabla \times \boldsymbol{\lambda}) = \nabla \cdot (\beta \boldsymbol{\lambda}) = \boldsymbol{\lambda} \cdot \nabla \beta + 2\alpha\beta,$$

whence

$$\boldsymbol{\lambda} \cdot \nabla \beta = -2\alpha\beta, \quad \nabla \beta = -2\alpha\beta\boldsymbol{\lambda} - \boldsymbol{\lambda} \times \nabla \alpha.$$

Let $\gamma \equiv \alpha + \mathrm{i}\beta$. Then

$$\begin{aligned} \boldsymbol{\lambda} \cdot \nabla \gamma &= \boldsymbol{\lambda} \cdot (\nabla \alpha + \mathrm{i}\nabla \beta) = \beta^2 - \alpha^2 - 2\mathrm{i}\alpha\beta = -\gamma^2, \\ \nabla \gamma &= \nabla \alpha + \mathrm{i}\nabla \beta = (\beta^2 - \alpha^2 - \mathrm{i}\alpha\beta)\boldsymbol{\lambda} + \boldsymbol{\lambda} \times (\nabla \beta - \mathrm{i}\nabla \alpha) \\ &= -\gamma^2\boldsymbol{\lambda} - \mathrm{i}\boldsymbol{\lambda} \times \nabla \gamma, \end{aligned}$$

and finally,

$$\begin{aligned} \nabla^2 \gamma &= -(\gamma^2 \lambda_i)_{,i} - \mathrm{i}\varepsilon_{ijk}(\lambda_j \gamma_{,a})_{,i} \\ &= -2\gamma(\boldsymbol{\lambda} \cdot \nabla \gamma) - \gamma^2 \nabla \cdot \boldsymbol{\lambda} - \mathrm{i}(\nabla \gamma) \cdot (\nabla \times \boldsymbol{\lambda}) \\ &= 2\gamma^3 - 2\alpha\gamma^2 - 2\mathrm{i}\beta\gamma^2 = 2\gamma^2(\gamma - \alpha - \mathrm{i}\beta) = 0, \end{aligned}$$

$$\begin{aligned} (\nabla \gamma)^2 &= \gamma^4 - (\boldsymbol{\lambda} \times \nabla \gamma)^2 = \gamma^4 - \varepsilon_{ijk}\varepsilon_{imn}\lambda_j \gamma_{,k} \lambda_m \gamma_{,n} \\ &= \gamma^4 - (\nabla \gamma)^2 - (\boldsymbol{\lambda} \cdot \nabla \gamma)^2 = \frac{1}{2}(\gamma^4 + \gamma^4) = \gamma^4. \end{aligned}$$

Let $\omega = 1/\gamma$. Then

$$\nabla \omega = -\frac{\nabla \gamma}{\gamma^2} = \boldsymbol{\lambda} + \mathrm{i}\boldsymbol{\lambda} \times \frac{\nabla \gamma}{\gamma^2} = \boldsymbol{\lambda} - \mathrm{i}\boldsymbol{\lambda} \times \nabla \omega,$$

$$\boldsymbol{\lambda} \cdot \nabla \omega = -\frac{1}{\gamma^2}\boldsymbol{\lambda} \cdot \nabla \gamma = 1, \quad (\nabla \omega)^2 = 1,$$

$$\begin{aligned} \nabla \omega \times \nabla \omega^* &= (\boldsymbol{\lambda} - \mathrm{i}\boldsymbol{\lambda} \times \nabla \omega) \times (\boldsymbol{\lambda} + \mathrm{i}\boldsymbol{\lambda} \times \nabla \omega^*) \\ &= \mathrm{i}\boldsymbol{\lambda}(\boldsymbol{\lambda} \cdot \nabla \omega^*) - \mathrm{i}(\nabla \omega^*)\boldsymbol{\lambda} \cdot \boldsymbol{\lambda} - \mathrm{i}(\nabla \omega)\boldsymbol{\lambda} \cdot \boldsymbol{\lambda} + \mathrm{i}\boldsymbol{\lambda}(\boldsymbol{\lambda} \cdot \nabla \omega) \\ &\quad + \boldsymbol{\lambda}(\nabla \omega^*) \cdot (\boldsymbol{\lambda} \times \nabla \omega) \\ &= -\mathrm{i}(\nabla \omega + \nabla \omega^*) + \boldsymbol{\lambda}[(\nabla \omega^*) \cdot (\boldsymbol{\lambda} \times \nabla \omega) + 2\mathrm{i}], \end{aligned}$$

$$0 = \nabla \omega \cdot (\nabla \omega \times \nabla \omega^*) = -\mathrm{i} - \mathrm{i}(\nabla \omega) \cdot (\nabla \omega^*) + (\nabla \omega^*) \cdot (\boldsymbol{\lambda} \times \nabla \omega) + 2\mathrm{i},$$

$$\nabla \omega \times \nabla \omega^* = -\mathrm{i}(\nabla \omega + \nabla \omega^*) + \mathrm{i}\boldsymbol{\lambda}[1 + (\nabla \omega) \cdot (\nabla \omega^*)],$$

$$\boldsymbol{\lambda} = \frac{\nabla \omega + \nabla \omega^* - \mathrm{i}\nabla \omega \times \nabla \omega^*}{1 + \nabla \omega \cdot \nabla \omega^*}.$$

Note that $\nabla^2 \alpha = 0$ and $\nabla^2 \beta = 0$. Let $l^2 = f\alpha$. Then l^2 and hence f will be determined by the two equations in (C.4) on p. 259. The first of these equations yields

$$0 = \nabla^2(l^2) = \alpha\nabla^2 f + 2\nabla\alpha \cdot \nabla f.$$

By noting that

$$\begin{aligned}\alpha^2 + \beta^2 &= p^2(1 - 2\cos\theta + \cos^2\theta + \sin^2\theta)\\ &= 2p^2(1 - \cos\theta) = 2\alpha p = \frac{\alpha}{l}(A + B),\end{aligned}$$

whence

$$l(A + B) = \frac{l^2}{\alpha}(\alpha^2 + \beta^2) = f(\alpha^2 + \beta^2),$$

we may rewrite the second equation of (C.4) in the form

$$\begin{aligned}0 &= \nabla^2(l^2\lambda_i) - [(A + B)l]_{,i}\\ &= \nabla^2(f\alpha\lambda_i) - \left[f(\alpha^2 + \beta^2)\right]_{,i}\\ &= (\alpha\nabla^2 f + 2\nabla\alpha \cdot \nabla f)\lambda_i + 2(f\alpha)_{,j}\lambda_{i,j} + f\alpha\nabla^2\lambda_i - \left[f(\alpha^2 + \beta^2)\right]_{,i}\\ &= 2\alpha(f\alpha)_{,j}(\delta_{ij} - \lambda_i\lambda_j) + 2\beta(f\alpha)_{,j}\varepsilon_{ijk}\lambda_k + f\alpha\nabla^2\lambda_i - \left[f(\alpha^2 + \beta^2)\right]_{,i}.\end{aligned}$$

Using (C.5) for $\nabla^2\boldsymbol{\lambda}$ on p. 261, we get

$$\begin{aligned}0 &= 2\alpha^2\nabla f + 2f\alpha\nabla\alpha - 2\alpha^2\boldsymbol{\lambda}(\boldsymbol{\lambda}\cdot\nabla f) - 2f\alpha\boldsymbol{\lambda}(\boldsymbol{\lambda}\cdot\nabla\alpha) - 2\alpha\beta\boldsymbol{\lambda}\times\nabla f\\ &\quad - 2f\beta\boldsymbol{\lambda}\times\nabla\alpha + f\alpha\nabla\alpha - f\alpha\boldsymbol{\lambda}(\boldsymbol{\lambda}\cdot\nabla\alpha) - 2f\alpha(\alpha^2 + \beta^2)\boldsymbol{\lambda}\\ &\quad - f\alpha\boldsymbol{\lambda}\times\nabla\beta - (\alpha^2 + \beta^2)\nabla f - 2f\alpha\nabla\alpha - 2f\beta\nabla\beta\\ &= (\alpha^2 - \beta^2)\nabla f - 2\alpha^2\boldsymbol{\lambda}(\boldsymbol{\lambda}\cdot\nabla f) - 2\alpha\beta\boldsymbol{\lambda}\times\nabla f + f\alpha\nabla\alpha - 3f\alpha(\beta^2 - \alpha^2)\boldsymbol{\lambda}\\ &\quad + 2f\beta(2\alpha\beta\boldsymbol{\lambda} + \nabla\beta) - 2f\alpha(\alpha^2 + \beta^2)\boldsymbol{\lambda} - f\alpha\left[\nabla\alpha - (\beta^2 - \alpha^2)\boldsymbol{\lambda}\right] - 2f\beta\nabla\beta\\ &= (\alpha^2 - \beta^2)\nabla f - 2\alpha^2\boldsymbol{\lambda}(\boldsymbol{\lambda}\cdot\nabla f) - 2\alpha\beta\boldsymbol{\lambda}\times\nabla f.\end{aligned}$$

Whence, dotting and crossing with $\boldsymbol{\lambda}$,

$$0 = -(\alpha^2 + \beta^2)\boldsymbol{\lambda}\cdot\nabla f, \quad \text{or } \boldsymbol{\lambda}\cdot\nabla f = 0,$$

$$0 = 2\alpha\beta\nabla f + (\alpha^2 - \beta^2)\boldsymbol{\lambda}\times\nabla f.$$

Since

$$\det\begin{pmatrix}\alpha^2 - \beta^2 & -2\alpha\beta\\ 2\alpha\beta & \alpha^2 - \beta^2\end{pmatrix} = (\alpha^2 - \beta^2)^2 + 4\alpha^2\beta^2 = (\alpha^2 + \beta^2)^2 \neq 0,$$

it follows that

$$\nabla f = 0 \quad \text{and} \quad \boldsymbol{\lambda}\times\nabla f = 0.$$

That is, f is a constant and $l^2 = \text{const.} \times \alpha$.

C.5: Kerr Metric

In this case,

$$\gamma = \frac{1}{\sqrt{(\boldsymbol{x} + \mathrm{i}\boldsymbol{a})^2}}, \qquad \omega = \sqrt{(\boldsymbol{x} + \mathrm{i}\boldsymbol{a})^2} \equiv \rho + \mathrm{i}\sigma,$$

$$\rho^2 - \sigma^2 + 2\mathrm{i}\rho\sigma = \omega^2 = (\boldsymbol{x} + \mathrm{i}\boldsymbol{a})^2 = x^2 - a^2 + 2\mathrm{i}\boldsymbol{a} \cdot \boldsymbol{x}.$$

We write

$$r = \sqrt{\boldsymbol{x}^2}, \quad a = \sqrt{\boldsymbol{a}^2}.$$

Then

$$\rho^2 - \sigma^2 = r^2 - a^2, \quad \rho\sigma = \boldsymbol{a} \cdot \boldsymbol{x},$$

$$\rho^2 - \frac{(\boldsymbol{a} \cdot \boldsymbol{x})^2}{\rho^2} = r^2 - a^2, \quad \rho^4 - (r^2 - a^2)\rho^2 - (\boldsymbol{a} \cdot \boldsymbol{x})^2 = 0,$$

$$\begin{aligned}\rho^2 &= \frac{1}{2}\left[r^2 - a^2 + \sqrt{(r^2 - a^2)^2 + 4(\boldsymbol{a} \cdot \boldsymbol{x})^2}\right] \\ &= \frac{1}{2}\left[r^2 - a^2 + \sqrt{(r^2 + a^2)^2 - 4(\boldsymbol{a} \times \boldsymbol{x})^2}\right],\end{aligned}$$

$$\sigma = \frac{\boldsymbol{a} \cdot \boldsymbol{x}}{\rho}, \quad \gamma = \frac{1}{\omega} = \frac{1}{\rho + \mathrm{i}\sigma} = \frac{\rho - \mathrm{i}\sigma}{\rho^2 + \sigma^2},$$

$$\rho^2 + \sigma^2 = \rho^2 + \frac{(\boldsymbol{a} \cdot \boldsymbol{x})^2}{\rho^2} = \frac{\rho^4 + (\boldsymbol{a} \cdot \boldsymbol{x})^2}{\rho^2},$$

$$\alpha = \frac{\rho}{\rho^2 + \sigma^2} = \frac{\rho^3}{\rho^4 + (\boldsymbol{a} \cdot \boldsymbol{x})^2}, \quad \beta = -\frac{\sigma}{\rho^2 + \sigma^2} = -\frac{\rho \boldsymbol{a} \cdot \boldsymbol{x}}{\rho^4 + (\boldsymbol{a} \cdot \boldsymbol{x})^2},$$

$$\nabla\omega = \frac{1}{\omega}(\boldsymbol{x} + \mathrm{i}\boldsymbol{a}), \quad \nabla\omega^* = \frac{1}{\omega^*}(\boldsymbol{x} - \mathrm{i}\boldsymbol{a}),$$

$$\begin{aligned}1 + \nabla\omega \cdot \nabla\omega^* &= 1 + \frac{(\boldsymbol{x} + \mathrm{i}\boldsymbol{a}) \cdot (\boldsymbol{x} + \mathrm{i}\boldsymbol{a})}{\omega\omega^*} = 1 + \frac{r^2 + a^2}{\rho^2 + \sigma^2} \\ &= \frac{\rho^2 + \sigma^2 + r^2 + a^2}{\rho^2 + \sigma^2} = \frac{\rho^2 + (\rho^2 - r^2 + a^2) + r^2 + a^2}{\rho^2 + \sigma^2} \\ &= \frac{2(\rho^2 + a^2)}{\rho^2 + \sigma^2},\end{aligned}$$

$$\begin{aligned}\nabla\omega + \nabla\omega^* - \mathrm{i}\nabla\omega \times \nabla\omega^* &= \left(\frac{1}{\omega} + \frac{1}{\omega^*}\right)\boldsymbol{x} + \mathrm{i}\left(\frac{1}{\omega} - \frac{1}{\omega^*}\right)\boldsymbol{a} \\ &\quad - \frac{\mathrm{i}}{\omega\omega^*}(\boldsymbol{x} + \mathrm{i}\boldsymbol{a}) \times (\boldsymbol{x} - \mathrm{i}\boldsymbol{a}) \\ &= \frac{1}{\omega\omega^*}[(\omega + \omega^*)\boldsymbol{x} - \mathrm{i}(\omega - \omega^*)\boldsymbol{a} + 2\boldsymbol{a} \times \boldsymbol{x}] \\ &= \frac{2}{\rho^2 + \sigma^2}(\rho\boldsymbol{x} + \sigma\boldsymbol{a} + \boldsymbol{a} \times \boldsymbol{x}) \\ &= \frac{2\rho}{\rho^2 + \sigma^2}\left[\boldsymbol{x} + \frac{1}{\rho^2}\boldsymbol{a}(\boldsymbol{a}\cdot\boldsymbol{x}) + \frac{1}{\rho}\boldsymbol{a} \times \boldsymbol{x}\right],\end{aligned}$$

$$\boldsymbol{\lambda} = \frac{\nabla\omega + \nabla\omega^* - \mathrm{i}\nabla\omega \times \nabla\omega^*}{1 + \nabla\omega\cdot\nabla\omega^*} = \frac{\rho}{\rho^2 + a^2}\left[\boldsymbol{x} + \frac{1}{\rho^2}\boldsymbol{a}(\boldsymbol{a}\cdot\boldsymbol{x}) + \frac{1}{\rho}\boldsymbol{a} \times \boldsymbol{x}\right].$$

Choose axes so that $\boldsymbol{a} = (0, 0, a)$ and $\boldsymbol{x} = (x, y, z)$. Then

$$\boldsymbol{a}\cdot\boldsymbol{x} = az, \quad \boldsymbol{a} \times \boldsymbol{x} = (-ay, ax, 0),$$

$$\boldsymbol{\lambda} = \frac{\rho}{\rho^2 + a^2}\left(x - \frac{ay}{\rho}, y + \frac{ax}{\rho}, z + \frac{a^2 z}{\rho^2}\right) = \left(\frac{\rho x - ay}{\rho^2 + a^2}, \frac{\rho y + ax}{\rho^2 + a^2}, \frac{z}{\rho}\right),$$

$$\begin{aligned}(l_\mu \mathrm{d}x^\mu)^2 &= l^2(\mathrm{d}x^0 + \boldsymbol{\lambda}\cdot\mathrm{d}\boldsymbol{x})^2 \\ &= \frac{C\rho^3}{\rho^4 + a^2 z^2}\left(\mathrm{d}x^0 + \frac{\rho x - ay}{\rho^2 + a^2}\mathrm{d}x + \frac{\rho y + ax}{\rho^2 + a^2}\mathrm{d}y + \frac{z}{\rho}\mathrm{d}z\right)^2.\end{aligned}$$

Setting $x^0 = t$, we have finally, for the arc length,

$$\begin{aligned}\mathrm{d}s^2 &= \eta_{\mu\nu}\mathrm{d}x^\mu \mathrm{d}x^\nu + (l_\mu \mathrm{d}x^\mu)^2 \\ &= -\mathrm{d}t^2 + \mathrm{d}\boldsymbol{x}^2 + \frac{2GM\rho^3}{\rho^4 + a^2 z^2}\left[\mathrm{d}t + \frac{\rho}{\rho^2 + a^2}(x\mathrm{d}x + y\mathrm{d}y)\right. \\ &\quad \left. + \frac{a}{\rho^2 + a^2}(x\mathrm{d}y - y\mathrm{d}x) + \frac{1}{\rho}z\mathrm{d}z\right],\end{aligned}$$

where we have set $C = 2GM$ so that the metric reduces to that of the Eddington form (C.2) of the Schwarzschild metric on p. 256 when $a = 0$.

Alternative coordinates are t, ρ, θ, ϕ such that

$$\cos\theta = \frac{z}{\rho}, \quad (\rho + \mathrm{i}a)\mathrm{e}^{\mathrm{i}\phi}\sin\theta = x + \mathrm{i}y,$$

$$\mathrm{d}z = \mathrm{d}(\rho\cos\theta) = \cos\theta\mathrm{d}\rho - \rho\sin\theta\mathrm{d}\theta, \quad \frac{1}{\rho}z\mathrm{d}z = \cos^2\theta\mathrm{d}\rho - \rho\sin\theta\cos\theta\mathrm{d}\theta,$$

$$\mathrm{d}z^2 = \cos^2\theta\mathrm{d}\rho^2 + \rho^2\sin^2\theta\mathrm{d}\theta^2 - 2\rho\sin\theta\cos\theta\mathrm{d}\rho\mathrm{d}\theta,$$

$$
\begin{aligned}
\mathrm{d}x^2+\mathrm{d}y^2 &= |\mathrm{d}(x+\mathrm{i}y)|^2 = \left|\mathrm{d}\left[(\rho+\mathrm{i}a)\mathrm{e}^{\mathrm{i}\phi}\sin\theta\right]\right|^2 \\
&= \left|\mathrm{e}^{\mathrm{i}\phi}\sin\theta\mathrm{d}\rho+(\rho+\mathrm{i}a)\mathrm{e}^{\mathrm{i}\phi}\cos\theta\mathrm{d}\theta+\mathrm{i}(\rho+\mathrm{i}a)\mathrm{e}^{\mathrm{i}\phi}\sin\theta\mathrm{d}\phi\right|^2 \\
&= |\sin\theta\cos\phi\mathrm{d}\rho+(\rho\cos\phi-a\sin\phi)\cos\theta\mathrm{d}\theta-(\rho\sin\phi+a\cos\phi)\sin\theta\mathrm{d}\phi \\
&\quad +\mathrm{i}[\sin\theta\sin\phi\mathrm{d}\rho+(\rho\sin\phi+a\cos\phi)\cos\theta\mathrm{d}\theta+(\rho\cos\phi-a\sin\phi)\sin\theta\mathrm{d}\phi]|^2 \\
&= [\sin\theta\cos\phi\mathrm{d}\rho+(\rho\cos\phi-a\sin\phi)\cos\theta\mathrm{d}\theta-(\rho\sin\phi+a\cos\phi)\sin\theta\mathrm{d}\phi]^2 \\
&\quad +[\sin\theta\sin\phi\mathrm{d}\rho+(\rho\sin\phi+a\cos\phi)\cos\theta\mathrm{d}\theta+(\rho\cos\phi-a\sin\phi)\sin\theta\mathrm{d}\phi]^2 \\
&= \sin^2\theta\mathrm{d}\rho^2+2\rho\sin\theta\cos\theta\mathrm{d}\rho\mathrm{d}\theta-2a\sin^2\theta\mathrm{d}\rho\mathrm{d}\phi+\rho^2\cos^2\theta\mathrm{d}\theta^2 \\
&\quad -2a\rho\sin\theta\cos\theta\mathrm{d}\theta\mathrm{d}\phi+a^2\cos^2\theta\mathrm{d}\theta^2+2a\rho\sin\theta\cos\theta\mathrm{d}\theta\mathrm{d}\phi+\rho^2\sin^2\theta\mathrm{d}\phi^2 \\
&\quad +a^2\sin^2\theta\mathrm{d}\phi^2,
\end{aligned}
$$

$$
\mathrm{d}\boldsymbol{x}^2 = \mathrm{d}\rho^2+(\rho^2+a^2\cos^2\theta)\mathrm{d}\theta^2+(\rho^2+a^2)\sin^2\theta\mathrm{d}\phi^2-2a\sin^2\theta\mathrm{d}\rho\mathrm{d}\phi,
$$

$$
\begin{aligned}
x\mathrm{d}x+y\mathrm{d}y &= \frac{1}{2}\mathrm{d}|x+\mathrm{i}y|^2 = \frac{1}{2}\mathrm{d}\left[(\rho^2+a^2)\sin^2\theta\right] \\
&= \rho\sin^2\theta\mathrm{d}\rho+(\rho^2+a^2)\sin\theta\cos\theta\mathrm{d}\theta,
\end{aligned}
$$

$$
\begin{aligned}
x\mathrm{d}y-y\mathrm{d}x &= \mathrm{Im}[(x-\mathrm{i}y)\mathrm{d}(x+\mathrm{i}y)] \\
&= \mathrm{Im}\left\{(\rho-\mathrm{i}a)\mathrm{e}^{-\mathrm{i}\phi}\sin\theta\left[\mathrm{e}^{\mathrm{i}\phi}\sin\theta\mathrm{d}\rho+(\rho+\mathrm{i}a)\mathrm{e}^{\mathrm{i}\phi}\cos\theta\mathrm{d}\theta\right.\right. \\
&\quad \left.\left.+\mathrm{i}(\rho+\mathrm{i}a)\mathrm{e}^{\mathrm{i}\phi}\sin\theta\mathrm{d}\phi\right]\right\} \\
&= \mathrm{Im}\left[(\rho-\mathrm{i}a)\sin^2\theta\mathrm{d}\rho+(\rho^2+a^2)\sin\theta\cos\theta\mathrm{d}\theta+\mathrm{i}(\rho^2+a^2)\sin^2\theta\mathrm{d}\phi\right] \\
&= -a\sin^2\theta\mathrm{d}\rho+(\rho^2+a^2)\sin^2\theta\mathrm{d}\phi,
\end{aligned}
$$

$$
\begin{aligned}
\mathrm{d}s^2 &= -\mathrm{d}t^2+\mathrm{d}\rho^2+(\rho^2+a^2\cos^2\theta)\mathrm{d}\theta^2+(\rho^2+a^2)\sin^2\theta\mathrm{d}\phi^2-2a\sin^2\theta\mathrm{d}\rho\mathrm{d}\phi \\
&\quad +\frac{2GM\rho}{\rho^2+a^2\cos^2\theta}\left[\mathrm{d}t+\frac{\rho^2}{\rho^2+a^2}\sin^2\theta\mathrm{d}\rho+\rho\sin\theta\cos\theta\mathrm{d}\theta-\frac{a^2}{\rho^2+a^2}\sin^2\theta\mathrm{d}\rho\right. \\
&\quad \left.+a\sin^2\theta\mathrm{d}\phi+\cos^2\theta\mathrm{d}\rho-\rho\sin\theta\cos\theta\mathrm{d}\theta\right]^2 \\
&= -\mathrm{d}t^2+\mathrm{d}\rho^2+(\rho^2+a^2\cos^2\theta)\mathrm{d}\theta^2+(\rho^2+a^2)\sin^2\theta\mathrm{d}\phi^2-2a\sin^2\theta\mathrm{d}\rho\mathrm{d}\phi \\
&\quad +\frac{2GM\rho}{\rho^2+a^2\cos^2\theta}\left(\mathrm{d}t+\frac{\rho^2+a^2\cos 2\theta}{\rho^2+a^2}\mathrm{d}\rho+a\sin^2\theta\mathrm{d}\phi\right)^2.
\end{aligned}
$$

We introduce new variables t' and ϕ' given by

$$
t = t' + \int\frac{2GM\rho}{\rho^2+a^2-2GM\rho}\mathrm{d}\rho,
$$

$$
\phi = \phi' + \int\left(\frac{2a}{\rho^2+a^2}-\frac{a}{\rho^2+a^2-2GM\rho}\right)\mathrm{d}\rho.
$$

Then

$$
\begin{aligned}
ds^2 = &-\left(dt' + \frac{2GM\rho}{\rho^2 + a^2 - 2GM\rho} d\rho\right)^2 + d\rho^2 + (\rho^2 + a^2\cos^2\theta)d\theta^2 \\
&+ (\rho^2 + a^2)\sin^2\theta\left[d\phi' + \left(\frac{2a}{\rho^2 + a^2} - \frac{a}{\rho^2 + a^2 - 2GM\rho}\right)d\rho\right]^2 \\
&- 2a\sin^2\theta d\rho\left[d\phi' + \left(\frac{2a}{\rho^2 + a^2} - \frac{a}{\rho^2 + a^2 - 2GM\rho}\right)d\rho\right] \\
&+ \frac{2GM\rho}{\rho^2 + a^2\cos^2\theta}\left[dt' + \left(\frac{2GM\rho}{\rho^2 + a^2 - 2GM\rho} + \frac{\rho^2 - a^2}{\rho^2 + a^2}\sin^2\theta + \cos^2\theta\right.\right. \\
&\left.\left.+ \frac{2a^2}{\rho^2 + a^2}\sin^2\theta - \frac{a^2}{\rho^2 + a^2 - 2GM\rho}\sin^2\theta\right)d\rho + a\sin^2\theta d\phi'\right]^2.
\end{aligned}
$$

Everything in the round brackets of the last two lines amounts to

$$
\frac{\rho^2 + a^2\cos^2\theta}{\rho^2 + a^2 - 2GM\rho}.
$$

Hence,

$$
\begin{aligned}
ds^2 = &-\left(1 - \frac{2GM\rho}{\rho^2 + a^2\cos^2\theta}\right)dt'^2 + d\rho^2 \\
&+ \frac{2GM\rho}{\rho^2 + a^2 - 2GM\rho}\left\{\frac{2GM\rho}{\rho^2 + a^2 - 2GM\rho}\right. \\
&+ \frac{1}{2GM\rho}\left[\frac{4a^2(\rho^2 + a^2 - 2GM\rho)}{\rho^2 + a^2} - 4a^2\right. \\
&+ \frac{a^2(\rho^2 + a^2)}{\rho^2 + a^2 - 2GM\rho} \\
&\left.- \frac{4a^2(\rho^2 + a^2 - 2GM\rho)}{\rho^2 + a^2} + 2a^2\right]\sin^2\theta \\
&\left.+ \frac{\rho^2 + a^2\cos^2\theta}{\rho^2 + a^2 - 2GM\rho}\right\}d\rho^2 \\
&+ (\rho^2 + a^2\cos^2\theta)d\theta^2 + \left(\rho^2 + a^2 + \frac{2a^2GM\rho\sin^2\theta}{\rho^2 + a^2\cos^2\theta}\right)\sin^2\theta d\phi'^2 \\
&+ \left(-\frac{4GM\rho}{\rho^2 + a^2 - 2GM\rho} + \frac{4GM\rho}{\rho^2 + a^2 - 2GM\rho}\right)dt' d\rho \\
&+ 2\left[2a - \frac{a(\rho^2 + a^2)}{\rho^2 + a^2 - 2GM\rho} - a + a\frac{2GM\rho}{\rho^2 + a^2 - 2GM\rho}\right]\sin^2\theta d\rho d\phi' \\
&+ \frac{4aGM\rho\sin^2\theta}{\rho^2 + a^2\cos^2\theta}dt' d\phi'.
\end{aligned}
$$

The terms in curly brackets here amount to

$$
\begin{aligned}
&-\frac{2GM\rho}{\rho^2+a^2-2GM\rho}+\frac{1}{2GM\rho}\frac{-2a^2(\rho^2+a^2-2GM\rho)+a^2(\rho^2+a^2)}{\rho^2+a^2-2GM\rho}\sin^2\theta \\
&\quad+\frac{\rho^2+a^2\cos^2\theta}{\rho^2+a^2-2GM\rho} \\
&=\frac{1}{2GM\rho(\rho^2+a^2-2GM\rho)} \\
&\quad\times\left[-(2GM\rho)^2-a^2(\rho^2+a^2-4GM\rho)\sin^2\theta+2GM\rho(\rho^2+a^2\cos^2\theta)\right] \\
&=\frac{1}{2GM\rho(\rho^2+a^2-2GM\rho)} \\
&\quad\times\left[-(2GM\rho)^2-a^2(\rho^2+a^2-2GM\rho)\sin^2\theta+2GM\rho(\rho^2+a^2)\right] \\
&=\frac{1}{2GM\rho}(2GM\rho-a^2\sin^2\theta).
\end{aligned}
$$

Also

$$1+\frac{2GM\rho-a^2\sin^2\theta}{\rho^2+a^2-2GM\rho}=\frac{\rho^2+a^2\cos^2\theta}{\rho^2+a^2-2GM\rho}.$$

Hence,

$$
\begin{aligned}
\mathrm{d}s^2=&-\left(1-\frac{2GM\rho}{\rho^2+a^2\cos^2\theta}\right)\mathrm{d}t'^2+\frac{\rho^2+a^2\cos^2\theta}{\rho^2+a^2-2GM\rho}\mathrm{d}\rho^2+(\rho^2+a^2\cos^2\theta)\mathrm{d}\theta^2 \\
&+\left(\rho^2+a^2+\frac{2a^2GM\rho\sin^2\theta}{\rho^2+a^2\cos^2\theta}\right)\sin^2\theta\mathrm{d}\phi'^2+\frac{4aGM\rho\sin^2\theta}{\rho^2+a^2\cos^2\theta}\mathrm{d}t'\mathrm{d}\phi'.
\end{aligned}
$$

A common alternative notation is

$$t'\longrightarrow t,\quad \rho\longrightarrow r,\quad \theta\longrightarrow\theta,\quad \phi'\longrightarrow\phi,\quad a\longrightarrow -a,$$

$$\rho\equiv\sqrt{r^2+a^2\cos^2\theta},\quad \Delta\equiv r^2-2GMr+a^2,$$

whence

$$
\begin{aligned}
\mathrm{d}s^2=&-\left(1-\frac{2GMr}{\rho^2}\right)\mathrm{d}t^2+\frac{\rho^2}{\Delta}\mathrm{d}r^2+\rho^2\mathrm{d}\theta^2 \\
&+\left(r^2+a^2+\frac{2a^2GMr\sin^2\theta}{\rho^2}\right)\sin^2\theta\mathrm{d}\phi^2-\frac{4aGMr\sin^2\theta}{\rho^2}\mathrm{d}t\mathrm{d}\phi.
\end{aligned}
$$

Note that

$$\frac{1}{\rho^2}(\Delta-a^2\sin^2\theta)=\frac{1}{r^2+a^2\cos^2\theta}(r^2+a^2\cos^2\theta-2GMr)=1-\frac{2GMr}{\rho^2},$$

$$\frac{1}{\rho^2}\left[(r^2+a^2)^2-a^2\Delta\sin^2\theta\right]=\frac{1}{\rho^2}\left[(r^2+a^2)^2-a^2(r^2+a^2)\sin^2\theta+2a^2GMr\sin^2\theta\right]$$
$$=r^2+a^2+\frac{2a^2GMr\sin^2\theta}{\rho^2},$$

$$\frac{1}{\rho^2}\left[2a\Delta-2a(r^2+a^2)\right]=-\frac{4aGMr}{\rho^2}.$$

Hence we may write ds^2 in the alternative form

$$ds^2=-\frac{\Delta}{\rho^2}(dt-a\sin^2\theta d\phi)^2+\frac{\sin^2\theta}{\rho^2}\left[(r^2+a^2)d\phi-adt\right]^2+\frac{\rho^2}{\Delta}dr^2+\rho^2d\theta^2.$$

Choose $a\geq 0$. Independent Killing vector fields ξ_t and ξ_ϕ are

$$(\xi_t^\mu)=(1,0,0,0),\quad (\xi_\phi^\mu)=(0,0,0,1),$$

with

$$\xi_t\cdot\xi_t=-\left(1-\frac{2GMr}{\rho^2}\right),\quad \xi_\phi\cdot\xi_\phi=\left(r^2+a^2+\frac{2a^2GMr\sin^2\theta}{\rho^2}\right)\sin^2\theta,$$

$$\xi_t\cdot\xi_\phi=-\frac{2aGMr\sin^2\theta}{\rho^2},$$

and

$$\begin{aligned}(\xi_t\cdot\xi_\phi)^2-\xi_t^2\xi_\phi^2=&\frac{1}{\rho^4}\{4a^2G^2M^2r^2\sin^4\theta\\&+(\rho^2-2GMr)\left[\rho^2(r^2+a^2)+2a^2GMr\sin^2\theta\right]\sin^2\theta\}\\=&\frac{\sin^2\theta}{\rho^4}(4a^2G^2M^2r^2\sin^2\theta+\rho^4r^2+\rho^4a^2-2GM\rho^2r^3-2GM\rho^2a^2r\\&+2a^2GMr\rho^2\sin^2\theta-4a^2G^2M^2r^2\sin^2\theta)\\=&\frac{\sin^2\theta}{\rho^2}\left[\rho^2r^2+\rho^2a^2-2GMr(r^2+a^2\cos^2\theta)\right]\\=&(r^2-2GMr+a^2)\sin^2\theta=\Delta\sin^2\theta.\end{aligned}$$

Appendix D
Kerr Metric Subcalculations

$$\begin{aligned}
\rho^2 &= \frac{1}{2}\left[r^2 - a^2 + \sqrt{(r^2-a^2)^2 + 4(\boldsymbol{a}\cdot\boldsymbol{x})^2}\right] \\
&= \frac{1}{2}(r^2 - a^2)\left[1 + \sqrt{1 + \frac{4(\boldsymbol{a}\cdot\boldsymbol{x})^2}{(r^2-a^2)^2}}\right] \\
&= \frac{1}{2}(r^2 - a^2)\left[2 + \frac{2(\boldsymbol{a}\cdot\boldsymbol{x})^2}{(r^2-a^2)^2} + \cdots\right] \\
&= r^2 - a^2 + \frac{(\boldsymbol{a}\cdot\boldsymbol{x})^2}{r^2-a^2} + \cdots = r^2\left[1 - \frac{a^2}{r^2} + \frac{(\boldsymbol{a}\cdot\boldsymbol{x})^2}{r^4} + \cdots\right] \\
&= r^2\left[1 + O\left(\frac{a^2}{r^2}\right)\right],
\end{aligned}$$

$$\rho = r\left[1 + O\left(\frac{a^2}{r^2}\right)\right],$$

$$\rho^2 + a^2 = r^2\left[1 + \frac{(\boldsymbol{a}\cdot\boldsymbol{x})^2}{r^4} + \cdots\right] = r^2\left[1 + O\left(\frac{a^2}{r^2}\right)\right],$$

$$\frac{\rho^2 + a^2}{\rho} = \left[1 + O\left(\frac{a^2}{r^2}\right)\right],$$

$$\frac{\rho}{\rho^2 + a^2} = r\frac{1}{r}\left[1 + O\left(\frac{a^2}{r^2}\right)\right],$$

$$\begin{aligned}
\boldsymbol{\lambda} &= \frac{\rho}{\rho^2 + a^2}\left[\boldsymbol{x} + \frac{1}{\rho^2}\boldsymbol{a}(\boldsymbol{a}\cdot\boldsymbol{x}) - \frac{1}{\rho}\boldsymbol{a}\times\boldsymbol{x}\right] \quad (a \to -a) \\
&= \frac{1}{r}\left[\boldsymbol{x} - \frac{\boldsymbol{a}\times\boldsymbol{r}}{r} + O\left(\frac{a^2}{r}\right)\right],
\end{aligned}$$

$$l^2 = 2Ma = \frac{\rho^3}{\rho^4 + (\boldsymbol{a}\cdot\boldsymbol{x})^2},$$

$$\rho^4 + (\boldsymbol{a}\cdot\boldsymbol{x})^2 = r^4\left[1 + O\left(\frac{a^2}{r^2}\right)\right] + r^4 O\left(\frac{a^2}{r^2}\right) = r^4\left[1 + O\left(\frac{a^2}{r^2}\right)\right],$$

$$\frac{\rho^3}{\rho^4 + (\boldsymbol{a}\cdot\boldsymbol{x})^2} = \frac{1}{r}\left[1 + O\left(\frac{a^2}{r^2}\right)\right],$$

$$l^2 = \frac{2M}{r}\left[1 + O\left(\frac{a^2}{r^2}\right)\right],$$

$$l^0 = -\left(\frac{2M}{r}\right)^{1/2}\left[1 + O\left(\frac{a^2}{r^2}\right)\right],$$

$$l^i = \frac{(2M)^{1/2}}{r^{3/2}}(x_i - \varepsilon_{ijk}a_j\hat{x}_k)\left[1 + O\left(\frac{a^2}{r^2}\right)\right],$$

$$l^{00} = l^0 l^0 \sim \frac{2M}{r},$$

$$l^{0i} = l^0 l^i \sim -\frac{2M}{r^2}(x_i - \varepsilon_{imn}a_m\hat{x}_n),$$

$$l^{ij} = l^i l^j \sim \frac{2M}{r^3}(x_i x_j - x_i\varepsilon_{jmn}a_m\hat{x}_n - x_j\varepsilon_{imn}a_m\hat{x}_n),$$

$$\begin{aligned} H^{0i0j} &= -(l^{00}\eta^{ij} + l^{ij}\eta^{00} - l^{0j}\eta^{i0} - l^{i0}\eta^{0j}) \\ &= -\delta_{ij}l^{00} + l^{ij}, \end{aligned}$$

$$\begin{aligned} H^{ij0k} &= -(l^{i0}\eta^{jk} + l^{jk}\eta^{i0} - l^{ik}\eta^{j0} - l^{j0}\eta^{ik}) \\ &= -\delta_{jk}l^{i0} + \delta_{ik}l^{j0}, \end{aligned}$$

$$\begin{aligned} P^0 &= \lim_{S\to\infty}\frac{1}{16\pi}\int_S H^{0i0j}_{,i}\mathrm{d}^2S_j = \lim_{S\to\infty}\frac{1}{16\pi}\int_S \left(-\delta_{ij}l^{00} + l^{ij}\right)_{,i}\mathrm{d}^2S_j \\ &= \lim_{S\to\infty}\frac{1}{16\pi}\int_S \left(-l^{00}_{,j} + l^{ij}_{,i}\right)\mathrm{d}^2S_j \\ &= \frac{2M}{16\pi}\int_{4\pi}\left[-\left(\frac{1}{r}\right)_{,j} + \left(\frac{1}{r^3}x_i x_j - \frac{1}{r^4}\varepsilon_{jmn}a_m x_i x_n - \frac{1}{r^4}\varepsilon_{imn}a_m x_j x_n\right)_{,i}\right] r^2\hat{x}_j\mathrm{d}^2\Omega \\ &= \frac{2M}{16\pi}\int_{4\pi}\left(\hat{x}_j - 3\hat{x}_i\hat{x}_i\hat{x}_j + 3\hat{x}_j + \hat{x}_j\right)\hat{x}_j\mathrm{d}^2\Omega \\ &= \frac{4M}{16\pi}\int_{4\pi}\mathrm{d}^2\Omega = M, \end{aligned}$$

$$P^i = \lim_{S\to\infty}\frac{1}{16\pi}\int_S H^{ij0k}_{\ ,j}\mathrm{d}^2S_k = \lim_{S\to\infty}\frac{1}{16\pi}\int_S\left(-\delta_{jk}l^{i0}+\delta_{ik}l^{j0}\right)_{,j}\mathrm{d}^2S_k$$
$$= \lim_{S\to\infty}\frac{1}{16\pi}\int_S\left(-l^{i0}_{,k}+\delta_{ik}l^{j0}_{,j}\right)\mathrm{d}^2S_k$$
$$= \frac{2M}{16\pi}\int_{4\pi}\left[\left(\frac{x_i}{r^2}\right)_{,k}-\delta_{ik}\left(\frac{x_j}{r^2}\right)_{,j}\right]r^2\hat{x}_k\mathrm{d}^2\Omega$$
$$= \frac{2M}{16\pi}\int_{4\pi}\left(\delta_{ik}-2\hat{x}_i\hat{x}_k-3\delta_{ik}+2\delta_{ik}\hat{x}_i\hat{x}_j\hat{x}_j\right)\hat{x}_k\mathrm{d}^2\Omega$$
$$= -\frac{4M}{16\pi}\int_{4\pi}\hat{x}_i\mathrm{d}^2\Omega = 0,$$

$$S_{ij} = J^{ij} = \lim_{S\to\infty}\frac{1}{16\pi}\int_S\left(x^iH^{jk0l}_{\ ,k}-x^jH^{ik0l}_{\ ,k}+H^{il0j}-H^{jl0i}\right)\mathrm{d}^2S_l$$
$$= \lim_{S\to\infty}\frac{1}{16\pi}\int_S\Big[x^i(-\delta_{kl}l^{j0}+\delta_{jl}l^{k0})_{,k}-x^j(-\delta_{kl}l^{i0}+\delta_{il}l^{k0})_{,k}$$
$$-\delta_{lj}l^{i0}+\delta_{ij}l^{l0}+\delta_{li}l^{j0}-\delta_{ji}l^{l0}\Big]\mathrm{d}^2S_l$$
$$= \frac{2M}{16\pi}\int_{4\pi}\Big\{x_i\left[\delta_{kl}\left(\frac{x_j}{r^2}-\varepsilon_{jmn}a_m\frac{x_n}{r^3}\right)-\delta_{jl}\left(\frac{x_k}{r^2}-\varepsilon_{kmn}a_m\frac{x_n}{r^3}\right)\right]_{,k}$$
$$-x_j\left[\delta_{kl}\left(\frac{x_i}{r^2}-\varepsilon_{imn}a_m\frac{x_n}{r^3}\right)-\delta_{il}\left(\frac{x_k}{r^2}-\varepsilon_{kmn}a_m\frac{x_n}{r^3}\right)\right]_{,k}$$
$$+\delta_{lj}\left(\frac{x_i}{r^2}-\varepsilon_{imn}a_m\frac{x_n}{r^3}\right)-\delta_{ij}\left(\frac{x_l}{r^2}-\varepsilon_{lmn}a_m\frac{x_n}{r^3}\right)$$
$$-\delta_{li}\left(\frac{x_j}{r^2}-\varepsilon_{jmn}a_m\frac{x_n}{r^3}\right)+\delta_{ji}\left(\frac{x_l}{r^2}-\varepsilon_{lmn}a_m\frac{x_n}{r^3}\right)\Big\}r^2\hat{x}_l\mathrm{d}^2\Omega$$
$$+2\hat{x}_j\hat{x}_i\hat{x}_l+3\hat{x}_j\delta_{il}-2\hat{x}_j\delta_{il}+\hat{x}_i\delta_{lj}-\hat{x}_j\delta_{li})$$
$$-\varepsilon_{jmn}a_m\delta_{kl}\delta_{nk}\hat{x}_i+3\varepsilon_{jmn}a_m\delta_{kl}\hat{x}_n\hat{x}_k\hat{x}_i+\varepsilon_{kmn}a_m\delta_{jl}\delta_{nk}\hat{x}_i$$
$$-3\varepsilon_{kmn}a_m\delta_{jl}\hat{x}_n\hat{x}_k\hat{x}_i+\varepsilon_{imn}a_m\delta_{kl}\delta_{nk}\hat{x}_j-3\varepsilon_{imn}a_m\delta_{kl}\hat{x}_n\hat{x}_k\hat{x}_j$$
$$-\varepsilon_{kmn}a_m\delta_{il}\delta_{nk}\hat{x}_j+3\varepsilon_{kmn}a_m\delta_{il}\hat{x}_n\hat{x}_k\hat{x}_j$$
$$-\varepsilon_{imn}a_m\delta_{lj}\hat{x}_n+\varepsilon_{jmn}a_m\delta_{li}\hat{x}_n\big]\hat{x}_l\mathrm{d}^2\Omega$$
$$= \frac{2M}{16\pi}\int_{4\pi}\big[\varepsilon_{jmn}a_m(-\hat{x}_n\hat{x}_i+3\hat{x}_n\hat{x}_i+\hat{x}_n\hat{x}_i)$$
$$+\varepsilon_{imn}a_m(\hat{x}_n\hat{x}_j-3\hat{x}_n\hat{x}_j-\hat{x}_n\hat{x}_j)\big]\mathrm{d}^2\Omega$$
$$= \frac{2M}{16\pi}\int_{4\pi}(\varepsilon_{jmi}-\varepsilon_{imj})a_m\mathrm{d}^2\Omega = M\varepsilon_{ijm}a_m,$$

$$S = Ma,$$

$$h_{00} = \frac{2M}{r} + \cdots, \quad h_{0i} = 2M\frac{x_i}{r^2} + 2S_{ij}\frac{x_j}{r^3} + \cdots,$$

$$h_{ij} = 2M\frac{x_i x_j}{r^3} + 2\frac{S_{ik}x_k x_j}{r^4} + 2\frac{S_{jk}x_k x_i}{r^4} + \cdots \quad (\text{since } l = 0).$$

Let

$$\xi_0 = 2M \ln r, \quad \xi_i = -M\frac{x_i}{r} - S_{ij}\frac{x_j}{r^2},$$

so that

$$\xi_{0,i} = 2M\frac{x_i}{r^2}, \quad \xi_{i,j} = -M\frac{\delta_{ij}}{r} + M\frac{x_i x_j}{r^3} - \frac{S_{ij}}{r^2} + 2\frac{S_{ik}x_k x_j}{r^4}.$$

Then

$$\overline{h}_{00} = h_{00} = \frac{2M}{r} + \cdots, \quad \overline{h}_{0i} = h_{0i} - \xi_{0,i} = 2S_{ij}\frac{x_j}{r^3} + \cdots,$$

$$\overline{h}_{ij} = h_{ij} - \xi_{i,j} - \xi_{j,i} = 2M\frac{\delta_{ij}}{r} + \cdots,$$

$$\overline{h} = \eta^{\mu\nu}\overline{h}_{\mu\nu} = -\overline{h}_{00} + \overline{h}_{ii} = \frac{4M}{r} + \cdots,$$

$$\overline{l}_{00} = \overline{h}_{00} - \frac{1}{2}\eta_{00}\overline{h} = \overline{h}_{00} + \frac{1}{2}\overline{h} = \frac{4M}{r} + \cdots,$$

$$\overline{l}_{0i} = \overline{h}_{0i} = 2S_{ij}\frac{x_j}{r^3} + \cdots, \quad \overline{l}_{ij} = \overline{h}_{ij} - \frac{1}{2}\delta_{ij}\overline{h} = O\left(\frac{1}{r^3}\right).$$

Appendix E
Friedmann Cosmology

$$ds^2 = -dt^2 + a^2(t)d\Omega^2,$$

$$g = g_{\mu\nu}dx^\mu \otimes dx^\nu = \eta_{\alpha\beta}e^\alpha \otimes e^\beta = -dt \otimes dt + a^2\sigma^a \otimes \sigma^a,$$

$$e^t = dt, \quad e^a = a\sigma^a, \quad d\sigma^a = \varepsilon_{abc}\sigma^b \wedge \sigma^c,$$

$$de^\alpha = -\frac{1}{2}c^\alpha_{\beta\gamma}e^\beta \wedge e^\gamma = -\omega^\alpha_\beta \wedge e^\beta,$$

$$\omega^\alpha_\beta = \Gamma^\alpha_{\beta\gamma}e^\gamma = \frac{1}{2}(-c^\alpha_{\beta\gamma} + c^\alpha_{\beta\gamma} + c^\alpha_{\gamma\beta})e^\gamma,$$

$$d\omega^\alpha_\beta + \omega^\alpha_\varepsilon \wedge \omega^\varepsilon_\beta = \Omega^\alpha_\beta = \frac{1}{2}R^\alpha_{\beta\gamma\delta}e^\gamma \wedge e^\delta,$$

$$de^t = 0 = \omega^t_a \wedge e^a,$$

$$\begin{aligned} de^a &= \dot{a}dt \wedge \sigma^a + a\varepsilon_{abc}\sigma^b \wedge \sigma^c \\ &= \frac{\dot{a}}{a}e^t \wedge e^a + \frac{1}{a}\varepsilon_{abc}e^b \wedge e^c = -\omega^a_t \wedge e^t - \omega^a_b \wedge e^b, \end{aligned}$$

$$\omega^a_t = \frac{\dot{a}}{a}e^a + v^a e^t = \dot{a}\sigma^a = \Gamma^a_{tt}e^t + \Gamma^a_{tb}e^b,$$

$$\omega^a_b = \frac{1}{a}\varepsilon_{abc}e^c + w^a e^b = \varepsilon_{abc}\sigma^c = \Gamma^a_{bt}e^t + \Gamma^a_{bc}e^c,$$

$$\omega^t_a = \frac{\dot{a}}{a}e^a = \dot{a}\sigma^a = \Gamma^t_{at}e^t + \Gamma^t_{ab}e^b,$$

$$\begin{aligned}\Omega^t_{\ a} &= \mathrm{d}\omega^t_{\ a} + \omega^t_{\ b} \wedge \omega^b_{\ a} \\ &= \ddot{a}\mathrm{d}t \wedge \sigma^a + \dot{a}\varepsilon_{abc}\sigma^b \wedge \sigma^c + \dot{a}\varepsilon_{bac}\sigma^b \wedge e^c \\ &= \frac{\ddot{a}}{a}e^t \wedge e^a = R^t_{\ atb}e^t \wedge e^b + \frac{1}{2}R^t_{\ abc}e^b \wedge e^c,\end{aligned}$$

$$\begin{aligned}\Omega^a_{\ b} &= \mathrm{d}\omega^a_{\ b} + \omega^a_{\ t} \wedge \omega^t_{\ b} + \omega^a_{\ c} \wedge \omega^c_{\ b} \\ &= \varepsilon_{abc}\varepsilon_{cde}\sigma^d \wedge \sigma^e + \dot{a}^2\sigma^a \wedge \sigma^b + \varepsilon_{acd}\varepsilon_{cbe}\sigma^d \wedge \sigma^e \\ &= (\delta_{ad}\delta_{be} - \delta_{ae}\delta_{bd} - \delta_{ab}\delta_{de} + \delta_{ae}\delta_{db})\sigma^d \wedge \sigma^e + \dot{a}^2\sigma^a \wedge \sigma^b \\ &= \frac{1}{a^2}(1+\dot{a}^2)e^a \wedge e^b = R^a_{\ btc}e^t \wedge e^c + \frac{1}{2}R^a_{\ bcd}e^c \wedge e^d.\end{aligned}$$

The nonvanishing components of the curvature tensor in the local orthonormal frame are

$$R^t_{\ atb} = \frac{\ddot{a}}{a}\delta_{ab}, \quad R^a_{\ bcd} = \frac{1}{a^2}(1+\dot{a}^2)(\delta_{ac}\delta_{bd} - \delta_{ad}\delta_{bc}),$$

and those obtained from these by the algebraic identities. The nonvanishing components of the Ricci tensor are

$$R_{tt} = R_{tata} = -3\frac{\ddot{a}}{a},$$

$$\begin{aligned}R_{ab} &= R^t_{\ atb} + R^c_{\ acb} \\ &= \frac{\ddot{a}}{a}\delta_{ab} + \frac{1}{a^2}(1+\dot{a}^2)(\delta_{cc}\delta_{ab} - \delta_{cb}\delta_{ac}) = \left[\frac{\ddot{a}}{a} + \frac{2}{a^2}(1+\dot{a}^2)\right]\delta_{ab}.\end{aligned}$$

The curvature scalar is

$$R = -R_{tt} + R_{aa} = 6\left[\frac{\ddot{a}}{a} + \frac{1}{a^2}(1+\dot{a}^2)\right].$$

The nonvanishing components of the Einstein tensor are

$$G_{tt} = R_{tt} + \frac{1}{2}R = \frac{3}{a^2}(1+\dot{a}^2),$$

$$G_{ab} = R_{ab} - \frac{1}{2}R\delta_{ab} = -\left[2\frac{\ddot{a}}{a} + \frac{1}{a^2}(1+\dot{a}^2)\right]\delta_{ab}.$$

Let us check that the Einstein tensor satisfies the contracted Bianchi identity. The nonvanishing connection components are

$$\Gamma^t_{\ ab} = \frac{\dot{a}}{a}\delta_{ab}, \quad \Gamma^a_{\ tb} = \frac{\dot{a}}{a}\delta_{ab}, \quad \Gamma^a_{\ bc} = \frac{1}{a}\varepsilon_{abc},$$

so

$$\begin{aligned} G^{t\alpha}_{;\alpha} &= e_\alpha G^{t\alpha} + \Gamma^t_{beta\alpha} G^{\beta\alpha} + \Gamma^\alpha_{\beta\alpha} G^{t\beta} \\ &= e_t G^{tt} + \Gamma^t_{ab} G^{ab} + \Gamma^a_{ta} G^{tt} \\ &= \frac{\mathrm{d}}{\mathrm{d}t}\left[\frac{3}{a^2}(1+\dot{a}^2)\right] - 3\frac{\dot{a}}{a}\left[2\frac{\ddot{a}}{a} + \frac{1}{a^2}(1+\dot{a}^2)\right] + 3\frac{\dot{a}}{a}\frac{3}{a^2}(1+\dot{a}^2) = 0, \end{aligned}$$

$$\begin{aligned} G^{a\alpha}_{;\alpha} &= e_\alpha G^{a\alpha} + \Gamma^a_{\beta\alpha} G^{\beta\alpha} + \Gamma^a_{\beta\alpha} G^{\alpha\beta} \\ &= e_b G^{ab} + \Gamma^a_{bc} G^{bc} + \Gamma^c_{bc} G^{ab} = 0, \end{aligned}$$

using the fact that $e_b G^{ab}$ vanishes by homogeneity and $\Gamma^a_{bc} G^{bc} + \Gamma^c_{bc} G^{ab}$ vanishes by symmetry.

E.1: Cosmic Dust

The nonvanishing component of the stress–energy density is $T^{tt} = \rho$, the mass–energy density. The equations of motion for cosmic dust are

$$T^{a\alpha}_{;\alpha} = 0,$$

which is automatically satisfied, and

$$0 = T^{t\alpha}_{;\alpha} = e_t T^{tt} + \Gamma^a_{ta} T^{tt} = \left(\frac{\mathrm{d}}{\mathrm{d}t} + 3\frac{\dot{a}}{a}\right)\rho.$$

This has solution

$$\rho = \frac{a_0^3}{a^3}\rho_0.$$

Einstein's equations for a cosmic dust are

$$G^{tt} = 8\pi G T^{tt}, \quad G^{ab} = 8\pi G T^{ab},$$

$$\frac{3}{a^2}(1+\dot{a}^2) = 8\pi G \frac{a_0^3}{a^3}\rho_0 = \frac{3A}{a^3}, \quad A = \frac{8\pi G}{3} a_0^3 \rho_0,$$

so that A is $8\pi G/3$ times the mass contained in a cube of side a_0 at epoch t_0. We can in fact choose $A = a_0$. Now

$$\dot{a} = \sqrt{\frac{A}{a} - 1}, \quad \ddot{a} = \frac{-\frac{1}{2}\frac{A}{a^2}\dot{a}}{\sqrt{\frac{A}{a} - 1}} = -\frac{1}{2}\frac{A}{a^2},$$

$$2\frac{\ddot{a}}{a} + \frac{1}{a^2}(1+\dot{a}^2) = -\frac{A}{a^3} + \frac{A}{a^3} = 0.$$

Consequently, the equation $G^{ab} = 8\pi G T^{ab} = 0$ is automatically satisfied.

The equation

$$\dot{a} = \sqrt{\frac{A}{a} - 1}$$

has solution

$$t = \frac{1}{2}A(\theta - \sin\theta), \quad a = \frac{1}{2}A(1 - \cos\theta),$$

as can be seen from

$$1 = \frac{1}{2}A(1 - \cos\theta)\dot{\theta} = a\dot{\theta}, \quad \cos\theta = 1 - \frac{2a}{A},$$

$$\sin\theta = \sqrt{1 - \cos^2\theta} = \sqrt{\frac{4a}{A} - \frac{4a^2}{A^2}},$$

$$\dot{a} = \frac{1}{2}A\sin\theta\dot{\theta} = \frac{A}{2a}\sqrt{\frac{4a}{A} - \frac{4a^2}{A^2}} = \sqrt{\frac{A}{a} - 1},$$

as claimed.

E.2: Photon Gas

The stress–energy density for a photon gas has nonvanishing components

$$T^{tt} = \rho, \quad T^{ab} = p\delta_{ab} = \frac{1}{3}\rho\delta_{ab}.$$

Now

$$p = \frac{1}{3}\rho,$$

so that

$$T^{\alpha}_{\alpha} = 0.$$

The equation of motion for a photon gas is

$$0 = T^{t\alpha}_{;\alpha} = e_t T^{tt} + \Gamma^{t}_{ab}T^{ab} + \Gamma^{a}_{ta}G^{tt} = \left(\frac{\mathrm{d}}{\mathrm{d}t} + 4\frac{\dot{a}}{a}\right)\rho,$$

$$0 = T^{a\alpha}_{;\alpha} = e_b T^{ab} + \Gamma^{a}_{bc}T^{bc} + \Gamma^{c}_{bc}T^{ab},$$

the latter being automatically satisfied. This has solution

$$\rho = \frac{a_0^4}{a^4}\rho_0.$$

The Einstein equations for the photon gas are

$$G^{tt} = 8\pi G T^{tt}, \quad G^{ab} = 8\pi G T^{ab},$$

$$\frac{3}{a^2}(1+\dot{a}^2) = 8\pi G \frac{a_0^4}{a^4}\rho_0 = \frac{3A^2}{a^4}, \quad A^2 = \frac{8\pi G}{3} a_0^4 \rho_0,$$

where the latter can be set equal to a_0^2. Hence,

$$\dot{a} = \sqrt{\frac{A^2}{a^2} - 1}, \quad \ddot{a} = \frac{-\frac{A^2}{a^3}\dot{a}}{\sqrt{\frac{A^2}{a^2} - 1}} = -\frac{A^2}{a^3},$$

$$-2\frac{\ddot{a}}{a} - \frac{1}{a^2}(1+\dot{a}^2) - \frac{8\pi G}{3}\rho = \frac{2A^2}{a^4} - \frac{A^2}{a^4} - \frac{8\pi G a_0^4}{3\ a^4}\rho_0 = 0.$$

Therefore the equation $G^{ab} = 8\pi G\ T^{ab} = 0$ is automatically satisfied.
The equation

$$\dot{a} = \sqrt{\frac{A^2}{a^2} - 1}$$

has the solution

$$a = \sqrt{A^2 - t^2}, \quad t = -\sqrt{A^2 - a^2} \quad \text{(expansion phase)},$$

since one then has

$$\dot{a} = -\frac{t}{\sqrt{A^2 - t^2}} = \frac{1}{a}\sqrt{A^2 - a^2} = \sqrt{\frac{A^2}{a^2} - 1}.$$

E.1. Find the connection one-form, the connection components and the curvature tensor for the flat three-space cosmology, for which the metric tensor is given by

$$ds^2 = -dt^2 + a^2(dx^2 + dy^2 + dz^2),$$

where a is a function of t, i.e.,

$$g = g_{\mu\nu} dx^\mu \otimes dx^\nu = \eta_{\alpha\beta} e^\alpha \otimes e^\beta = -dt \otimes dt + a^2 dx^a \otimes dx^a,$$

$$e^t = dt, \quad e^a = a dx^a.$$

Find the Ricci tensor, the curvature scalar and the Einstein tensor, and obtain the differential equations for a in the case in which this universe is filled with cosmic dust.

If the age of this universe (from the Big Bang) is known to be t, what is the current mass density in this universe as a function of t? This is known as the *critical mass density* required to make three-space flat. If the mass density is greater than the critical density, the universe will recollapse, otherwise it will not.

E.1 We have

$$de^t = 0 = -\omega^t_{\ a} \wedge e^a,$$

$$de^a = \dot{a}dt \wedge dx^a = \frac{\dot{a}}{a}e^t \wedge e^a = -\omega^a_{\ t} \wedge e^t - \omega^a_{\ b} \wedge e^b,$$

$$\omega^a_{\ t} = \frac{\dot{a}}{a}e^a + v^a e^t = \dot{a}dx^a = \Gamma^a_{\ tt}e^t + \Gamma^a_{\ tb}e^b,$$

$$\omega^a_{\ b} = \omega^a e^b = 0 = \Gamma^a_{\ bt}e^t + \Gamma^a_{\ bc}e^c,$$

$$\omega^t_{\ a} = \frac{\dot{a}}{a}e^a = \dot{a}dx^a = \Gamma^t_{\ at}e^t + \Gamma^t_{\ ab}e^b,$$

$$\Omega^t_{\ a} = d\omega^t_{\ a} + \omega^t_{\ b} \wedge \omega^b_{\ a} = \ddot{a}dt \wedge dx^a$$
$$= \frac{\ddot{a}}{a}e^t \wedge e^a = R^t_{\ atb}e^t \wedge e^b + \frac{1}{2}R^t_{\ abc}e^b \wedge e^c,$$

$$\Omega^a_{\ b} = d\omega^a_{\ b} + \omega^a_{\ t} \wedge \omega^t_{\ b} + \omega^a_{\ c} \wedge \omega^c_{\ b} = \frac{\dot{a}^2}{a^2}e^a \wedge e^b$$
$$= R^a_{\ btc}e^t \wedge e^c + \frac{1}{2}R^a_{\ bcd}e^c \wedge e^d.$$

The nonvanishing components of the curvature tensor are

$$R^t_{\ atb} = \frac{\ddot{a}}{a}\delta_{ab}, \quad R^a_{\ bcd} = \frac{\dot{a}^2}{a^2}(\delta_{ac}\delta_{bd} - \delta_{ad}\delta_{bc}).$$

The nonvanishing components of the Ricci tensor are

$$R_{tt} = R_{tata} = -3\frac{\ddot{a}}{a}, \quad R_{ab} = R^t_{\ atb} + R^c_{\ acb} = \left(\frac{\ddot{a}}{a} + 2\frac{\dot{a}^2}{a^2}\right)\delta_{ab}.$$

The curvature scalar is

$$R = -R_{tt} + R_{aa} = 6\left(\frac{\ddot{a}}{a} + \frac{\dot{a}^2}{a^2}\right).$$

The nonvanishing components of the Einstein tensor are

$$G_{tt} = R_{tt} + \frac{1}{2}R = 3\frac{\dot{a}^2}{a^2}, \quad G_{ab} = R_{ab} - \frac{1}{2}R\delta_{ab} = -\left(2\frac{\ddot{a}}{a} + \frac{\dot{a}^2}{a^2}\right)\delta_{ab}.$$

The nonvanishing connection components are

$$\Gamma^t_{\ ab} = \frac{\dot{a}}{a}\delta_{ab}, \quad \Gamma^a_{\ tb} = \frac{\dot{a}}{a}\delta_{ab}.$$

The contracted Bianchi identity is easily checked.

The equations of motion for a cosmic dust and a photon gas are the same as in the Friedmann case. The Einstein equations for a cosmic dust are

$$3\frac{\dot{a}^2}{a^2} = 3\frac{A}{a^3}, \quad A = \frac{8\pi G}{3}a_0^3\rho_0 = \frac{8\pi G}{3}a^3\rho,$$

$$\dot{a} = \sqrt{\frac{A}{a}}, \quad a^{1/2}\dot{a} = A^{1/2}.$$

Let $a = 0$ when $t = 0$. Then

$$\frac{2}{3}a^{3/2} = A^{1/2}t = \sqrt{\frac{8\pi G\rho}{3}}a^{3/2}t,$$

or $1 = \sqrt{6\pi G\rho}t$, or

$$\rho = \frac{1}{6\pi Gt^2}.$$

If the age of the universe is known to be t, then $1/6\pi Gt^2$ is the *critical mass density* required to make three-space flat. If the mass density is greater than this value then the universe will recollapse.

Appendix F Dynamical Equations and Diffeomorphisms

$$S_{\mathrm{G}}[\varphi] + S_{\mathrm{M}}[\varphi, \Phi], \quad g \longrightarrow \varphi, \quad \mu, \nu, x^0, x^1, x^2, x^3 \longrightarrow i,$$

$$S_{\mathrm{G},i} + S_{\mathrm{M},i} = 0, \quad S_{\mathrm{M},A} = 0.$$

Under diffeomorphisms,

$$\delta\varphi^i = Q^i_\alpha[\varphi]\delta\xi^\alpha, \quad \delta\Phi^A = Q^A_\alpha[\varphi, \Phi]\delta\xi^\alpha = Q^A_{\alpha B}\Phi^B\delta\xi^\alpha,$$

$$\delta g_{\mu\nu} = \mathcal{L}_{\delta\xi} g_{\mu\nu} = -\delta\xi_{\mu;\nu} - \delta\xi_{\nu;\mu} = \int Q_{\mu\nu\sigma'}\delta\xi^{\sigma'} \mathrm{d}^4x',$$

$$Q_{\mu\nu\sigma'} = -(\delta_{\mu\sigma';\nu} + \delta_{\nu\sigma';\mu}), \quad \delta_{\mu\nu'} = g_{\mu\sigma}\delta^\sigma_{\nu'} = \delta_{\mu\nu}\delta(x, x'),$$

$$0 \equiv \delta S_{\mathrm{G}} = S_{,i}\delta\varphi^i = S_{,i}Q^i_\alpha\delta\xi^\alpha,$$

and therefore

$$S_{,i}Q^i_\alpha \equiv 0.$$

Similarly,

$$\left(S_{\mathrm{M},i}Q^i_\alpha + S_{\mathrm{M},A}Q^A_{\alpha B}\Phi^B\right)\delta\xi^\alpha \equiv 0,$$

whence

$$S_{\mathrm{M},i}Q^i_\alpha = 0, \quad \text{when } S_{\mathrm{M},A} = 0.$$

Defining

$$T_i \equiv 2S_{\mathrm{M},i}, \quad T^{\mu\nu} = 2\frac{\delta S_{\mathrm{M}}}{\delta g_{\mu\nu}},$$

we have

$$T_i Q^i_\alpha = 0, \quad \text{when } S_{\mathrm{M},A} = 0.$$

This can be rewritten

$$0 = \int T^{\nu'\sigma'} Q_{\nu'\sigma'\mu} \mathrm{d}^4x' = -2 \int T^{\nu'\sigma'} \delta_{\nu'\mu;\sigma'} \mathrm{d}^4x'$$
$$= 2 \int T^{\nu'\sigma'}_{;\sigma'} \delta_{\nu'\mu} \mathrm{d}^4x' = 2T^{\sigma}_{\mu;\sigma}.$$

We now write φ^i as a sum

$$\varphi^i = \varphi^i_{\mathrm{B}} + \phi^i,$$

where φ_{B} is an empty space background field with

$$S_{\mathrm{G},i}[\varphi_{\mathrm{B}}] = 0.$$

The dynamical equations are

$$S_{\mathrm{G},i}[\varphi_{\mathrm{B}} + \phi] + S_{\mathrm{M},i}[\varphi_{\mathrm{B}} + \phi, \Phi] = 0, \quad S_{\mathrm{M},A}[\varphi_{\mathrm{B}} + \phi, \Phi] = 0.$$

Let

$$\overline{S}_{\mathrm{G}}[\varphi_{\mathrm{B}}, \phi] \equiv S_{\mathrm{G}}[\varphi_{\mathrm{B}} + \phi] - S_{\mathrm{G}}[\varphi_{\mathrm{B}}] - S_{\mathrm{G},i}[\varphi_{\mathrm{B}}]\phi^i$$
$$= \frac{1}{2} S_{\mathrm{G},ij}[\varphi_{\mathrm{B}}]\phi^i\phi^j + \frac{1}{6} S_{\mathrm{G},ijk}[\varphi_{\mathrm{B}}]\phi^i\phi^j\phi^k + \cdots.$$

where

$$\overline{S}_{\mathrm{G}}[\varphi_{\mathrm{B}}, \phi] \equiv S_{\mathrm{G}}[\varphi_{\mathrm{B}} + \phi] - S_{\mathrm{G}}[\varphi_{\mathrm{B}}] - S_{\mathrm{G},i}[\varphi_{\mathrm{B}}]\phi^i$$
$$= \frac{1}{2} S_{\mathrm{G},ij}[\varphi_{\mathrm{B}}]\phi^i\phi^j + \frac{1}{6} S_{\mathrm{G},ijk}[\varphi_{\mathrm{B}}]\phi^i\phi^j\phi^k + \cdots.$$

Define

$$\mathcal{T}_i \equiv 2\frac{\delta \overline{S}_{\mathrm{G}}}{\delta \varphi^i_{\mathrm{B}}} = 2S_{\mathrm{G},i}[\varphi_{\mathrm{B}} + \phi] - 2S_{\mathrm{G},i}[\varphi_{\mathrm{B}}] - 2S_{\mathrm{G},ij}[\varphi_{\mathrm{B}}]\phi^j$$
$$= 2S_{\mathrm{G},ijk}[\varphi_{\mathrm{B}}]\phi^j\phi^k + \frac{1}{3} S_{\mathrm{G},ijkl}[\varphi_{\mathrm{b}}]\phi^j\phi^k\phi^l + \cdots,$$

$$\frac{\delta \overline{S}}{\delta \phi^i} = \frac{\delta \overline{S}_{\mathrm{G}}}{\delta \phi^i} + \frac{\delta S_{\mathrm{M}}}{\delta \phi^i} = S_{\mathrm{G},i}[\varphi_{\mathrm{B}} + \phi] - S_{\mathrm{G},i}[\varphi_{\mathrm{B}}] + S_{\mathrm{M},i}[\varphi_{\mathrm{B}} + \phi, \Phi]$$
$$= 0$$
$$= S_{\mathrm{G},ij}[\varphi_{\mathrm{B}}]\phi^j + \frac{1}{2}(\mathcal{T}_i + T_i),$$

or

$$S_{\mathrm{G},ij}[\varphi_{\mathrm{B}}]\phi^j = -\frac{1}{2}(\mathcal{T}_i + T_i) = -\frac{\delta \overline{S}[\varphi_{\mathrm{B}}, \phi, \Phi]}{\delta \varphi^i_{\mathrm{B}}},$$

$$T_i = 2S_{\mathrm{M},i}[\varphi_{\mathrm{B}} + \phi, \Phi] = 2\frac{\delta S_{\mathrm{M}}}{\delta \varphi^i_{\mathrm{B}}} = 2\frac{\delta S_{\mathrm{M}}}{\delta \phi^i},$$

$$\frac{\delta \overline{S}}{\delta \Phi^A} = S_{\mathrm{M},A}[\varphi_{\mathrm{B}} + \phi, \Phi] = 0.$$

A possible diffeomorphism (coordinate transformation) law is

$$\delta\varphi^i_{\mathrm{B}} = Q^i_\alpha[\varphi_{\mathrm{B}}]\delta\xi^\alpha, \qquad \delta\varphi^i = Q^i_\alpha[\varphi]\delta\xi^\alpha = Q^i_\alpha[\varphi_{\mathrm{B}} + \phi]\delta\xi^\alpha,$$

$$\delta\phi^i = \delta\varphi^i - \delta\varphi^i_{\mathrm{B}} = Q^i_{\alpha,j}\phi^j\delta\xi^\alpha.$$

When the realization is linear, there are no more terms in the series on the right hand side of the last equation. Now

$$\begin{aligned}\delta\overline{S}_{\mathrm{G}}[\varphi_{\mathrm{B}}, \phi] &\equiv S_{\mathrm{G},i}[\varphi_{\mathrm{B}} + \phi]Q^i_\alpha[\varphi_{\mathrm{B}} + \phi]\delta\xi^\alpha - S_{\mathrm{G},i}[\varphi_{\mathrm{B}}]Q^i_\alpha[\varphi_{\mathrm{B}}]\delta\xi^\alpha \\ &\quad - S_{\mathrm{G},ij}[\varphi_{\mathrm{B}}]\phi^i Q^j_\alpha[\varphi_{\mathrm{B}}]\delta\xi^\alpha - S_{\mathrm{G},j}[\varphi_{\mathrm{B}}]Q^j_{\alpha,i}\phi^i\delta\xi^\alpha \\ &\equiv -\frac{\delta(S_{\mathrm{G},j}[\varphi_{\mathrm{B}}]Q^j_\alpha[\varphi_{\mathrm{B}}])}{\delta\varphi^i_{\mathrm{B}}}\phi^i\delta\xi^\alpha \equiv 0,\end{aligned}$$

$$\delta S_{\mathrm{M}}[\varphi_{\mathrm{B}} + \phi, \Phi] \equiv 0,$$

under the above transformations together with

$$\delta\Phi^A = Q^A_{\alpha B}\Phi^B\delta\xi^\alpha.$$

Hence,

$$(T_i + \mathcal{T}_i)Q^i_\alpha[\varphi_{\mathrm{B}}] = 0, \qquad \text{when} \frac{\delta\overline{S}}{\delta\phi^i} = 0 \text{ and } \frac{\delta\overline{S}}{\delta\Phi^A} = 0.$$

As a consistency check,

$$0 \equiv S_{\mathrm{G},i}[\varphi_{\mathrm{B}}]Q^i_\alpha[\varphi_{\mathrm{B}}],$$

$$0 \equiv S_{\mathrm{G},ij}[\varphi_{\mathrm{B}}]Q^i_\alpha[\varphi_{\mathrm{B}}] + S_{\mathrm{G},i}[\varphi_{\mathrm{B}}]Q^i_{\alpha,j} = S_{\mathrm{G},ij}[\varphi_{\mathrm{B}}]Q^i_\alpha[\varphi_{\mathrm{B}}],$$

whence

$$0 = Q^i_\alpha[\varphi_{\mathrm{B}}]S_{\mathrm{G},ij}[\varphi_{\mathrm{B}}]\phi^j = -\frac{1}{2}Q^i_\alpha[\varphi_{\mathrm{B}}](\mathcal{T}_i + T_i).$$

The transformations laws for T_i and $\mathcal{T}_i$ are

$$\begin{aligned}
\delta T_i &\equiv 2\delta S_{\mathrm{M},i}[\varphi_{\mathrm{B}} + \phi, \Phi] \\
&\equiv 2\delta S_{\mathrm{M},ij}[\varphi_{\mathrm{B}} + \phi, \Phi] Q^j_\alpha[\varphi_{\mathrm{B}} + \phi]\delta\xi^\alpha + S_{\mathrm{M},iA}[\varphi_{\mathrm{B}} + \phi, \Phi] Q^A_{\alpha B}\Phi^B\delta\xi^\alpha \\
&\equiv 2\frac{\delta}{\delta\phi^i}\{S_{\mathrm{M},j}[\varphi_{\mathrm{B}} + \phi, \Phi] Q^j_\alpha[\varphi_{\mathrm{B}} + \phi] + S_{\mathrm{M},A}[\varphi_{\mathrm{B}} + \phi, \Phi] Q^A_{\alpha B}\Phi^B\}\delta\xi^\alpha \\
&\quad - 2S_{\mathrm{M},j}[\varphi_{\mathrm{B}} + \phi, \Phi] Q^j_{\alpha,i}\delta\xi^\alpha \\
&\equiv -T_j Q^j_{\alpha,i}\delta\xi^\alpha,
\end{aligned}$$

$$\begin{aligned}
\delta \mathcal{T}_i &\equiv 2\delta\{S_{\mathrm{G},i}[\varphi_{\mathrm{B}} + \phi] - S_{\mathrm{G},i}[\varphi_{\mathrm{B}}] - S_{\mathrm{G},ij}[\varphi_{\mathrm{B}}]\phi^j\} \\
&\equiv 2\{S_{\mathrm{G},ij}[\varphi_{\mathrm{B}} + \phi] Q^j_\alpha[\varphi_{\mathrm{B}} + \phi] - S_{\mathrm{G},ij}[\varphi_{\mathrm{B}}] Q^j_\alpha[\varphi_{\mathrm{B}}] \\
&\quad - S_{\mathrm{G},ijk}[\varphi_{\mathrm{B}}]\phi^j Q^k_\alpha[\varphi_{\mathrm{B}}] - S_{\mathrm{G},ij}[\varphi_{\mathrm{B}}] Q^j_{\alpha,k}\phi^k\Big\}\delta\xi^\alpha \\
&\equiv 2\Big\{-S_{\mathrm{G},j}[\varphi_{\mathrm{B}} + \phi] Q^j_{\alpha,i} + S_{\mathrm{G},j}[\varphi_{\mathrm{B}}] Q^j_{\alpha,i} \\
&\quad - \frac{\delta}{\delta\varphi^i_{\mathrm{B}}}\Big(S_{\mathrm{G},jk}[\varphi_{\mathrm{B}}] Q^j_\alpha[\varphi_{\mathrm{B}}] + S_{\mathrm{G},j}[\varphi_{\mathrm{B}}] Q^j_{\alpha,k}[\varphi_{\mathrm{B}}]\Big)\phi^k \\
&\quad + S_{\mathrm{G},jk}[\varphi_{\mathrm{B}}] Q^j_{\alpha,i}\phi^k\Big\}\delta\xi^\alpha \\
&\equiv -\mathcal{T}_j Q^j_{\alpha,i}\delta\xi^\alpha.
\end{aligned}$$

These transformations, however, are physically meaningless.
Practical diffeomorphism laws are

$$\delta\varphi^i_{\mathrm{B}} = 0, \qquad \delta\phi^i = \delta\varphi^i = Q^i_\alpha[\varphi_{\mathrm{B}} + \phi]\delta\xi^\alpha,$$
$$\delta\Phi^A = Q^A_{\alpha B}\Phi^B\delta\xi^\alpha.$$

The transformation law for T_i remains the same, but for $\mathcal{T}_i$ it becomes

$$\begin{aligned}
\delta \mathcal{T}_i &\equiv 2\delta\{S_{\mathrm{G},i}[\varphi_{\mathrm{B}} + \phi] - S_{\mathrm{G},i}[\varphi_{\mathrm{B}}] - S_{\mathrm{G},ij}[\varphi_{\mathrm{B}}]\phi^j\} \\
&\equiv 2\Big\{S_{\mathrm{G},ij}[\varphi_{\mathrm{B}} + \phi] Q^j_\alpha[\varphi_{\mathrm{B}} + \phi] - S_{\mathrm{G},ij}[\varphi_{\mathrm{B}}]\Big(Q^j_\alpha[\varphi_{\mathrm{B}}] + Q^j_{\alpha,k}\phi^k\Big)\Big\}\delta\xi^\alpha \\
&\equiv 2\Big\{-S_{\mathrm{G},j}[\varphi_{\mathrm{B}} + \phi] Q^j_{\alpha,i} + S_{\mathrm{G},j}[\varphi_{\mathrm{B}}] Q^j_{\alpha,i} - S_{\mathrm{G},ij}[\varphi_{\mathrm{B}}] Q^j_{\alpha,k}\phi^k\Big\}\delta\xi^\alpha \\
&\equiv -\mathcal{T}_j Q^j_{\alpha,i}\delta\xi^\alpha - 2\Big\{S_{\mathrm{G},jk}[\varphi_{\mathrm{B}}] Q^j_{\alpha,i} + S_{\mathrm{G},ij}[\varphi_{\mathrm{B}}] Q^j_{\alpha,k}\Big\}\phi^k\delta\xi^\alpha \\
&\equiv -\mathcal{T}_j Q^j_{\alpha,i}\delta\xi^\alpha + 2S_{\mathrm{G},ijk}[\varphi_{\mathrm{B}}]\phi^k Q^j_\alpha[\varphi_{\mathrm{B}}]\delta\xi^\alpha,
\end{aligned}$$

in which we have used, in passing to the last line, the identity

$$S_{\mathrm{G},ijk} Q^j_\alpha + S_{\mathrm{G},ij} Q^j_{\alpha,k} + S_{\mathrm{G},jk} Q^j_{\alpha,i} \equiv 0,$$

obtained from $S_{\mathrm{G,j}} Q^j_\alpha \equiv 0$ by functionally differentiating twice. In general, only if

$$Q^j_\alpha[\varphi_{\mathrm{B}}]\delta\xi^\alpha = 0,$$

i.e., if $\delta\xi^\alpha$ is a Killing vector field of the background field, does $\mathcal{T}_i$ transform like a contravariant tensor density. It is often called a *pseudo-tensor density*.

Because $\mathcal{T}^{\mu\nu}$ does not transform like a tensor density, it is impossible to assign a definite location to the energy, momentum and stress in the gravitational field. Nevertheless, the integral

$$\int_\Sigma \xi_\mu (T^{\mu\nu} + \mathcal{T}^{\mu\nu}) \mathrm{d}\Sigma_\nu,$$

with Σ a complete Cauchy hypersurface, is an absolutely conserved quantity for every Killing vector field ξ^μ that the background geometry possesses. Moreover, it is diffeomorphism invariant!

Printed in the United States
By Bookmasters